The science and practice of welding

VOLUME 2

The practice of welding

A.C.DAVIES

B.Sc (London Hons. and Liverpool), C.Eng., M.I.E.E., Fellow of the Welding Institute

EIGHTH EDITION

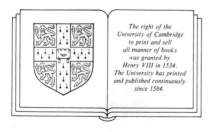

The right of the
University of Cambridge
to print and sell
all manner of books
was granted by
Henry VIII in 1534.
The University has printed
and published continuously
since 1584.

CAMBRIDGE UNIVERSITY PRESS

Cambridge

London   New York   New Rochelle

Melbourne   Sydney

Published by the Press Syndicate of the University of Cambridge
The Pitt Building, Trumpington Street, Cambridge CB2 1RP
32 East 57th Street, New York NY 10022, USA
10 Stamford Road, Oakleigh, Melbourne 3166, Australia

First published 1941
Second edition 1943
Third edition 1945
Reprinted 1947, 1950
Fourth edition 1955
Reprinted 1959
Fifth edition 1963
Reprinted 1966, 1969, 1971
Sixth edition 1972
Reprinted 1975
Seventh edition 1977
Reprinted with revisions 1981
Eighth edition 1984
Reprinted with revisions 1986

Printed in Great Britain at The Bath Press, Avon

Library of Congress catalogue card number: 83-23225

British Library cataloguing in publication data
Davies, A. C.
The science and practice of welding. – 8th ed.
Vol. 2: The practice of welding.
1. Welding
I. Title
671.5′2   TS227

ISBN 0 521 26114 7 hard covers
ISBN 0 521 27840 6 paperback

621.977 D

Contents

Preface

The Practice of Welding, Volume 2 of *The Science and Practice of Welding* has been brought up to date by rewriting many of the sections.

The advent of silicon-controlled rectifiers and integrated circuits has resulted in the continuous control of welding current by one knob, and together with pulse facilities and 'square wave' units has placed very sophisticated equipment at the service of the welding engineer and which should result in even more reliable welding results.

The section on the practice of MMA welding has largely been rewritten and there are new sections on underwater welding and cutting, cold pressure welding and the application of mixed gases to the various welding processes.

The sections on MIG and TIG and their practical applications have been largely rewritten although descriptions of the older units are still retained. Pulse, wave balance and magnetic arc control are discussed and there is a section on robotics with the application of MIG to this growing method of automatic welding.

The latest British Standards are used for the classification of steels, aluminium and magnesium.

It is hoped that at this particular time when great change is taking place in the industry this volume will prove of help to the operator and technician.

My thanks are due to the following firms without whose help and co-operation this book could not have been written. They have assisted in every way with supplying technical information and specialized photographs where indicated.

Air Products Ltd, Crewe: special gases and mixtures, welding of aluminium, stainless and heat resistant steels.

Air Products Ltd, Ruabon: welding of aluminium and its alloys, stainless and 9% nickel steels and plasma cutting.

American Welding Society: welding symbols and classifications.

Alcan Wire Ltd: aluminium welding techniques and applications.

Aluminium Federation: aluminium and its alloys. Welding techniques and applications.

Babcock Wire Equipment Ltd: cold pressure welding with photographs.

Bernard Division, Armco Ltd: MIG welding guns.

British Oxygen Gases Ltd; special gases and mixtures.

Bullfinch Gas Equipment Ltd: brazing techniques and torch photographs.

Deloro-Stellite Ltd and Cabot Corporation: wear technology and laying down of wear resistant surfaces.

Distillers Company (Carbon Dioxide) Ltd; heaters and gauges for CO_2 cylinders with photographs and details.

ESAB Ltd: TIG and MIG and submerged arc processes and accessories including backing strips, robot welding systems and MIG applications, power units for MMA, MIG and TIG welding, positioners, manipulators and add on units, photographs of equipment.

Wescol Ltd: photographs of gauges for gases and flashback arrestor.

Hobart Bros Ltd: MIG unit wire drive with photographs.

Interlas Ltd: Miller synchrowave TIG unit and Pulsar MIG unit techniques and photographs.

Johnson Matthey and Co.: brazing alloys and fluxes.

Messer Grieshiem Ltd: Photograph of cutting machine.

Murex Welding Products Ltd: Schaeffler diagram and MMA electrodes, plant and techniques of welding TIG and MIG units and photographs.

Norman Butters and Co.: MIG welding units.

Filarc Welding Industries BV (successors to Philips Export BV): low hydrogen electrode, downhill pipeline welding.

Oerlikon Bührle Ltd: electrode coatings and their manufacture.

Palco Ltd: details of seam tracking and automatic magnetic arc welding with illustrations.

Wharton Williams Taylor 2W. Underwater welding and cutting techniques.

Union Carbide Co. Ltd: plasma cutting and photographs.

I would also like to express my thanks to the city and Guilds of London Institute for permission to reproduce, with some amendments, examination questions set in recent years and to the following gentlemen, who, by their advice and suggestions, have helped me greatly in the preparation of this volume: Mr B. J. Bennett, Mr D. G. J. Brunt, Mr J. C. Crouch, Mr H. J. Davies, Mr A. Ellis, Mr W. J. Hamlett, Mr D. J. Simpson, Mr W. F. J. Thomas, Mr R. Wilson.

Extracts from British Standards are reproduced by permission of the British Standards Institution. Copies of the latest complete standards can be obtained from: British Standards Institution (Sales Department), Linford Wood, Milton Keynes MK14 6LE.

Oswestry, 1984   *A. C. Davies*

Preface to the seventh edition

The book has been extensively revised with new sections on submerged arc, stud (arc and capacitor), explosive and gravity processes. A new chapter has been added on resistance welding and many sections have been brought up to date including the new processes in iron and steel production, and additional information is included in the chapters on TIG and MIG processes. The new electrode classification to BS 639 1976 has been included together with impact machines and testing.

My thanks are due to each and all of the following firms who have helped me in every way by offering advice and supplying information and photographs as indicated.

A.I. Welders Ltd: flash butt welding technology and photograph.

Air Products Cryogenic Division: the welding of aluminium alloys and stainless steel, with illustrations and details of impact tests.

Avery-Denison Ltd: impact testing machines and photographs.

British Railways Board: details of flash butt and thermit welding of rails and diagrams of 'adjustment switch'.

British Oxygen Co. Ltd: oxy-acetylene welding equipment and photographs, industrial gases and diagrams, manual metal arc welding electrodes and filler wires, manual metal arc, TIG, MIG and plasma welding plant and plasma cutting with diagrams and photographs.

Copper Development Association: the welding of copper and its alloys.

Crompton Parkinson Ltd: stud welding with diagrams.

ESAB Ltd: manual metal arc welding electrodes and filler wires, manual metal arc, TIG, MIG, submerged arc, gravity, and electroslag welding equipment, positioners, and robot welding with illustrations and photographs.

G.K.N. Lincoln Ltd: submerged arc welding equipment, wire electrodes and fluxes.

British Steel Corporation, Library and Information Services of the

Sheffield Laboratories: modern blast furnaces, direct reductions of iron ores, basic oxygen steel, electric arc steel with illustrations.

KSM Stud Welding: stud welding with diagrams.

Pirelli General Cable Co.: welding cables.

The Welding Institute: information on the classification of electrodes.

Copperheat Ltd: pre- and post-heating equipment with photographs.

Sciaky Ltd: spot, seam projection and other types of resistance welding with photographs, laser beam welding with photograph.

Henry Wiggin Ltd: the welding of nickel and nickel alloys.

Union Carbide UK Ltd: TIG and MIG technology, plasma welding, cutting and surfacing technology with diagrams.

Yorkshire Imperial Metals Ltd: explosive welding with diagrams.

Birlec Ltd: induction furnace photograph.

Rockwell Ltd: photographs of CO_2 welding equipment.

Gamma-Rays Ltd: information on non-destructive testing with photographs of radiographic equipment.

I would again like to express my thanks to the City and Guilds of London Institute for permission to reproduce, with some amendments, examination questions set in recent years and to Mr D. G. J. Brunt, T. ENG (C.E.I), F.I.T.E., ASSOC. MEM.I.E.E., M.WELD.I., for help in the reading of the proofs, and to Mr M. S. Wilson, B.SC., M.MET., for help in the revised sections in metallurgy.

Abstracts of British Standards are included by permission of the British Standards Institution, 2 Park Street, London, from whom copies of the latest complete standards may be obtained.

The terms TIG, MIG and CO_2 have been retained for these welding processes, pending revision of BS 499, as they are so widely used. The use of gas mixture of inert and active gases (argon–oxygen, argon–oxygen–CO_2, argon–hydrogen, etc.) as the shielding gas together with the pulsed, modulated feed, modulated arc length and flux cored processes, etc. have made the present terminology rather inadequate.

Oswestry, 1977   *A. C. Davies*

1

*Manual metal arc welding**

The electric arc

An electric arc is formed when an electric current passes between two electrodes separated by a short distance from each other. In arc welding (we will first consider direct-current welding) one electrode is the welding rod or wire, while the other is the metal to be welded (we will call this the plate). The electrode and plate are connected to the supply, one to the + ve pole and one to the − ve pole, and we will discuss later the difference which occurs when the electrode is connected t, − ve or + ve pole. The arc is started by momentarily touching the electrode on to the plate and then withdrawing it to about 3 to 4 mm from the plate. When the electrode touches the plate, a current flows, and as it is withdrawn from the plate the.current continues to flow in the form of a 'spark' across the very small gap first formed. This causes the air gap to become ionized or made conducting, and as a result the current is able to flow across the gap, even when it is quite wide, in the form of an arc. The electrode must always be touched on to the plate before the arc can be started, since the smallest air gap will not conduct a current (at the voltages used in welding) unless the air gap is first ionized or made conducting.

The arc is generated by electrons (small negatively charged particles) flowing from the − ve to the + ve pole and the electrical energy is changed in the arc into heat and light. Approximately two-thirds of the heat is developed near the + ve pole, which burns into the form of a crater, the temperature near the crater being about 6000–7000°C, while the remaining third is developed near to the − ve pole. As a result an electrode connected to the + pole will burn away 50% faster than if connected to the − ve pole. For this reason it is usual to connect medium-coated electrodes and bare

*American designation: shielded metal arc welding (SMAW).

1

rods to the −ve pole, so that they will not burn away too quickly. Heavily coated rods are connected to the +ve pole because, due to the extra heat required to melt the heavy coating, they burn more slowly than the other types of rods when carrying the same current. The thicker the electrode used, the more heat is required to melt it, and thus the more current is required. The welding current may vary from 20 to 600 A in manual metal arc welding.

When alternating current is used, heat is developed equally at plate and rod, since the electrode and plate are changing polarity at the frequency of the supply.

If a bare wire is used as the electrode it is found that the arc is difficult to control, the arc stream wandering hither and thither over the molten pool. The globules are being exposed to the atmosphere in their travel from the rod to the pool and absorption of oxygen and nitrogen takes place even when a short arc is held. The result is that the weld tends to be porous and brittle.

The arc can be rendered easy to control and the absorption of atmospheric gases reduced to a minimum by 'shielding' the arc. This is done by covering the electrode with one of the various types of covering previously discussed, and as a result gases such as hydrogen and carbon dioxide are released from the covering as it melts and form an envelope around the arc and molten pool, excluding the atmosphere with its harmful effects on the weld metal. Under the heat of the arc chemical compounds in the electrode covering also react to form a slag which is liquid and lighter than the molten metal. It rises to the surface, cools and solidifies, forming a protective covering over the hot metal while cooling and protecting it from atmospheric effects, and also slows down the cooling rate of the weld. Some slags are self-removing while others have to be lightly chipped (Fig. 1.1).

The electrode covering usually melts at a higher temperature than the wire core so that it extends a little beyond the core, concentrating and directing the arc stream, making the arc stable and easier to control. The difference in controllability when using lightly covered electrodes and various medium- and heavily covered electrodes will be quickly noticed by the operator at a very early stage in practical manual metal arc welding.

With bare wire electrodes much metal is lost by volatilization, that is turning into a vapour. The use of covered electrodes reduces this loss.

An arc cannot be maintained with a voltage lower than about 14 V and is not very satisfactory above 45 V. With d.c. sources the voltage can be varied by a switch or regulator, but with a.c. supply by transformer the open circuit voltage (OCV) choice is less, being 80 or 100 V on larger units, down to 50 V on small units.

The greater the volts drop across the arc the greater the energy liberated in heat for a given current.

Arc energy is usually expressed in kilojoules per millimetre length of the weld (kJ/mm) and

$$\text{Arc energy (kJ/mm)} = \frac{\text{arc voltage} \times \text{welding current}}{\text{welding speed (mm/s)} \times 1000}.$$

The volts drop can be varied by altering the type of gas shield liberated by the electrode covering, hydrogen giving a higher volts drop than carbon dioxide for example. As the length of the arc increases so does the voltage drop, but since there is an increased resistance in this long arc the current is decreased. Long arcs are difficult to control and maintain and they lower the efficiency of the gas shield because of the greater length. As a result, absorption of oxygen and nitrogen from the atmosphere can take place, resulting in poor mechanical properties of the weld. It is essential that the welder should keep as short an arc as possible to ensure sound welds.

Transference of metal across the arc gap

When an arc is struck between the electrode and plate, the heat generated forms a molten pool in the plate and the electrode begins to melt away, the metal being transferred from the electrode to the plate. The transference takes place whether the electrode is positive or negative and also when it has a changing polarity, as when used on a.c. Similarly it is

Fig. 1.1. The shielded arc. Manual arc weld on steel base plate with a covered electrode.

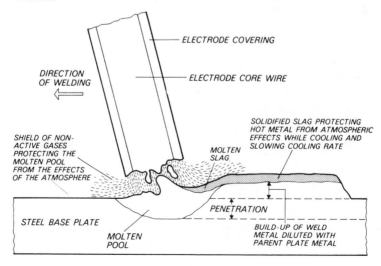

transferred upwards against the action of gravity, as when making an overhead weld. Surface tension plays an important part in overhead welding and a very short arc must be held to weld in the overhead position successfully.

The forces which cause the transfer appear to be due to: (1) its own weight, (2) the electro-magnetic (Lorentz) forces, (3) gas entrainment, (4) magneto-dynamic forces producing movement and (5) surface tension. The globule is finally necked off by the magnetic pinch effect.

If the arc is observed very closely, or better still if photographs are taken of it with a slow-motion cine-camera, it can be seen that the metal is transferred from the electrode to the plate in the form of drops or globules, and these globules vary in size according to the current and type of electrode covering. Larger globules are transferred at longer intervals than smaller globules and the globules form, elongate with a neck connecting them to the electrode, the neck gets reduced in size until it breaks, and the drop is projected into the molten pool, which is agitated by the arc stream, and this helps to ensure a sound bond between weld and parent metal. Drops of water falling from a tap give an excellent idea of the method of transference (see Fig. 1.2). Other methods of transfer known as dip (short circuiting arc) and spray (free flight transfer) are discussed in the section on MIG welding process.

Arc blow

We have seen that whenever a current flows in a conductor a magnetic field is formed around the conductor. Since the arc stream is also a flow of current, it would be expected that a magnetic field would exist around it, and that this is so can be shown by bringing a magnet near the arc. It is seen that the arc is blown to one side by the magnet, due to the interaction of its field with that of the magnet (just as two wires carrying a current will attract each other if the current flows in the same direction in each, or repel if the currents are in opposite directions), and the arc may even be extinguished if the field due to the magnet is strong enough. When welding, particularly with d.c., it is sometimes found that the arc tends to wander and becomes rather uncontrollable, as though it was being blown to and fro. This is known as arc blow and is experienced most when using

Fig. 1.2. Detachment of molten globule in the metal arc process.

currents above 200 or below 40 A, though it may be quite troublesome, especially when welding in corners, in between this range. It is due to the interaction of the magnetic field of the arc stream with the magnetic fields set up by the currents in the metal of the work or supply cables. The best methods of correction are:

(1) Weld away from the earth connexion.

(2) Change the position of the earth wire on the work.

(3) Wrap the welding cable a few turns around the work, if possible, on such work as girders, etc.

(4) Change the position of the work on the table if working on a bench.

In most cases the blow can be corrected by experimenting on the above lines, but occasionally it can be very troublesome and difficult to eliminate. Alternating-current welding has the advantage that since the magnetic field due to the arc stream is constantly alternating in direction at the frequency of the supply, there is much less trouble with arc blow, and consequently this is very advantageous when heavy currents are being used. Arc blow can be troublesome in the TIG and MIG processes, particularly when welding with d.c.

Spatter

At the conclusion of a weld small particles or globules of metal may sometimes be observed scattered around the vicinity of the weld along its length. This is known as 'spatter' and may occur through:

(1) Arc blow making the arc uncontrollable.

(2) The use of too long an arc or too high an arc voltage.

(3) The use of an excessive current.

The latter is the most frequent cause.

Spatter may also be caused by bubbles of gas becoming entrapped in the molten globules of metal, expanding with great violence and projecting the small drops of metal outside the arc stream, or by the magnetic pinch effect, by the magnetic fields set up, and thus the globules of metal getting projected outside the arc stream.

Spatter can be reduced by controlling the arc correctly, by varying current and voltage, and by preventing arc blow in the manner previously explained. Spatter release sprays ensure easy removal.

Eccentricity of the core wire in an MMA welding electrode

If the core wire of a flux-coated electrode is displaced excessively from the centre of the flux coating because of errors in manufacture, the arc may not function satisfactorily. The arc tends to be directed towards one side as if influenced by 'arc blow' and accurate placing of the deposited metal is prevented (Fig. 1.3*a*). A workshop test to establish whether the core wire is displaced outside the manufacturer's tolerance is to clean off the flux covering on one side at varying points down the length of the electrode and measure the distance *L* (Fig. 1.3*b*). The difference between the maximum and minimum reading is an approximate indication of the eccentricity.

Electrode efficiency (metal recovery and deposition coefficient). The efficiency of an electrode is the weight of metal actually deposited compared with the weight of that portion of the electrode consumed. It can be expressed as a percentage thus:

$$\text{Efficiency \%} = \frac{\text{weight of metal deposited}}{\text{weight of metal of the electrode consumed}} \times 100.$$

With ordinary electrodes the efficiency varies from 75 to 95%, but with iron powder electrodes the efficiency can be up to 200%. Efficiencies of 110% and above are indicated in BS 639 (1976) and ISO 2560 by a three-digit figure, giving the efficiency percentage rounded off to the nearest 10 at the beginning of the optional part of the classification.

The efficiency of a particular type of electrode can be obtained by stripping five electrodes of their coverings and weighing them. Take a piece of clean steel plate about $300 \times 75 \times 12$ mm and weigh it and deposit five similar electrodes to the ones stripped, retain the five stub ends and weigh them. Weigh the plate with deposit and subtract the original weight of the plate to give the weight of actual weld metal deposited, say *W* g. Subtract

Fig. 1.3. (*a*)

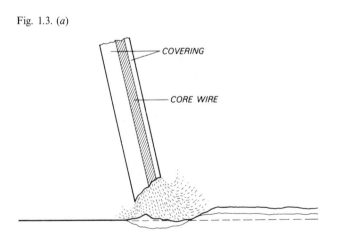

the weight of the stub ends from the weight of the electrodes stripped and this gives the weight of the electrode used up,

say w g. Then the percentage efficiency $= \dfrac{W}{w} \times 100$.

Iron powder electrodes

The deposition rate of a given electrode is dependent upon the welding current used, and for maximum deposition rate, maximum current should be used. This maximum current depends upon the diameter of the core wire, and for any given diameter of wire there is a maximum current beyond which increasing current will eventually get the wire red hot and cause overheating and hence deterioration of the covering.

To enable higher currents to be used an electrode of larger diameter core wire must be used, but if sufficient iron powder is added to the covering of the electrode, this covering becomes conducting, and a higher welding current can now be used on an electrode of given core wire diameter. The deposition rate is now increased and in addition the iron powder content is added to the weld metal, giving greater efficiency, that is enabling more than the core wire weight of metal to be deposited because of the extra iron powder. Efficiencies of up to 200% are possible, this meaning that twice the core wire weight of weld metal is being deposited. These electrodes can have coverings of rutile or basic type or a mixture of these. The iron powder ionizes easily, giving a smoother arc with little spatter, and the cup which forms as the core wire burns somewhat more quickly than the covering gives the arc directional properties and reduces loss due to metal volatilization. See also Electrode efficiency (metal recovery and deposition coefficient).

Fig. 1.3. (*b*) Eccentric core wire.

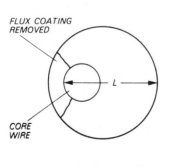
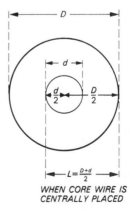

FLUX COATING
REMOVED

CORE
WIRE

WHEN CORE WIRE IS
CENTRALLY PLACED

Hydrogen-controlled electrodes (basic covered)*

If oxygen is present with molten iron or steel a chemical reaction occurs and the iron combines chemically with the oxygen to form a chemical compound, iron oxide. Similarly with nitrogen, iron nitride being formed if the temperature is high enough as in metal arc welding. When hydrogen is present however there is no chemical reaction and the hydrogen simply goes into solution in the steel, its presence being described as x millilitres of hydrogen in y grams of weld metal.

This hydrogen can diffuse out of the iron lattice when in the solid state resulting in a lowering of the mechanical properties of the weld and increasing the tendency to cracking. By the use of basic hydrogen-controlled electrodes, and by keeping the electrodes very dry, the absorption of hydrogen by the weld metal is reduced to a minimum and welds can be produced that have great resistance to cracking even under conditions of very severe restraint.

The coverings of these electrodes are of calcium or other basic carbonates and fluorspar bonded with sodium or potassium silicate. When the basic carbonate is heated carbon dioxide is given off and provides the shield of protective gas thus:

calcium carbonate (limestone) heated → calcium oxide (quicklime) + carbon dioxide.

There is no hydrogen in the chemicals of the covering, so that if they are kept absolutely dry, the deposited weld metal will have a low hydrogen content. Electrodes which will give deposited metal having a maximum of 15 millilitres of hydrogen per 100 grams of deposited metal (15 ml/100 g) are indicated by the letter H in BS 639 (1976) classification. The absence of diffusible hydrogen enables free cutting steels to be welded with absence of porosity and cracking and the electrodes are particularly suitable for welding in all conditions of very severe restraint. They can be used on a.c. or d.c. supply according to the makers' instructions and are available also in iron powder form and for welding in all positions. Low and medium alloy steels which normally would require considerable pre-heat if welded with rutile-coated electrodes can be welded with very much less pre-heating, the welds resisting cracking under severe restraint conditions and also being very suitable for welding in sub-zero temperature conditions. By correct storage and drying of these electrodes the hydrogen content can be reduced to 5 ml/100 g of weld metal for special applications. Details of these drying methods are given in the section on storage and drying of electrodes (q.v.).

Experiment to illustrate the diffusible hydrogen content in weld metal. Make a run of weld metal about 80 mm long with the metal arc on a small square

* Typical AWS classification of these electrodes may be E 7015 or E 7018, for example.

of steel plate using an ordinary steel welding rod with a cellulose, rutile or iron oxide coating. Deslag, cool out quickly and dry off with a cloth and place the steel plate in a beaker or glass jar of paraffin. It will be noted that minute bubbles of gas stream out of the weld metal and continue to do so even after some considerable time. If this gas is collected as shown in Fig. 1.4 it is found to be hydrogen which has come from the flux covering and the moisture it contains. A steel weld may contain hydrogen dissolved in the weld metal and also in the molecular form in any small voids which may be present. Hydrogen in steel produces embrittlement and a reduction in fatigue strength. If a run of one of these hydrogen-controlled electrodes is made on a test plate and the previous experiment repeated it will be noted that no hydrogen diffuses out of the weld.

Appendix C to BS 639 gives the recommendation method of determining the quantity of diffusible hydrogen present together with a drawing of a diffusible hydrogen meter.

Deep penetration electrodes

A deep penetration electrode is defined in BS 499, Part 1 (*Welding terms and symbols*) as 'A covered electrode in which the covering aids the production of a penetrating arc to give a deeper than normal fusion in the root of a joint'. For butt joints with a gap not exceeding 0.25 mm the penetration should be not less than half the plate thickness, the plate being twice the electrode core thickness. For fillet welds the gap at the joint should not exceed 0.25 mm and penetration beyond the root should be 4 mm minimum when using a 4 mm diameter electrode.

Fig. 1.4. Collecting diffusible hydrogen from a mild steel weld.

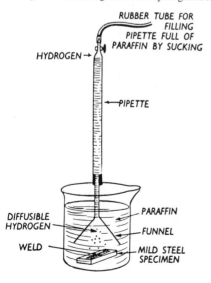

Welding position

Weld slope is the angle between line of the root of the weld and the horizontal (Fig. 1.5).

Weld rotation. Draw a line from the foot of the weld at right angles to the line welding to bisect the weld profile. The angle that this line makes with the vertical is the angle of weld rotation.

The table indicates the five welding positions used for electrode classification. Any intermediate position not specified may be referred to as 'inclined'. (See also Figs. 1.6, 1.7, 1.8.)

Position	Slope	Rotation	Symbol	Fig.
Flat	0–5°	0–10°	F	1.5
Horizontal–vertical	0–5°	30–90°	H	1.6
Vertical–up	80–90°	0–180°	V	1.7a
Vertical–down	80–90°	0–180°	D	1.7b
Overhead	0–15°	115–180°	O	1.8

Storage of electrodes

The flux coverings on modern electrodes are somewhat porous and absorb moisture to a certain extent. The moisture content (or humidity)

Fig. 1.5. Flat position.

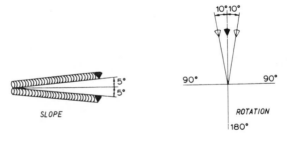

Fig. 1.6. Horizontal–vertical position.

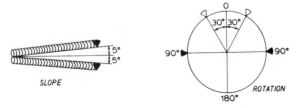

of the atmosphere is continually varying and hence the moisture content of the covering will be varying. Moisture could be excluded by providing a non-porous covering, but any moisture entrapped would be liable to cause rupture of the coating when the moisture was turned to steam by the heating effect of the passage of the current through the electrode. Cellulosic electrodes absorb quite an appreciable amount of moisture, and it does not affect their properties since they function quite well with a moisture content. They should not be over-dried or the organic compounds of which they are composed tend to char, affecting the voltage and arc properties. The extruded electrodes with rutile, iron oxide and silicate coatings do not pick up so much moisture from the atmosphere and function quite well with a small absorbed content. If they get damp they can be satisfactorily dried out, but it should be noted that if they get excessively wet, rusting of the core wire may occur and the coating may break away. In this case the electrodes should be discarded.

Storage temperatures should be about 12°C above that of external air temperature with 0–60% humidity. Cellulose covered electrodes are not so critical: but they should be protected against condensation and stored in a humidity of 0–90%.

Drying of electrodes. The best drying conditions are when the electrodes are removed from their package and well spaced out in a drying oven which has

Fig. 1.7. (*a*) Vertical–up position. (*b*) Vertical–down position.

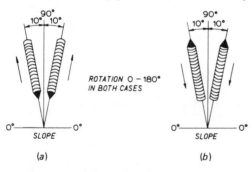

Fig. 1.8. Overhead position.

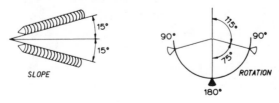

a good circulation of air. Longer drying times are required if the electrodes are not spaced out. The following table gives an indication of temperatures and times required, but see also the special conditions for drying basic electrodes (Fig. 1.9).

Drying of electrodes: approximate times and temperatures with electrodes spaced apart. Times will vary with air circulation, electrode spacing and oven loading

Electrode type	Diameter mm	Temperature °C	Time in mins air circulation	
			good	poor
Rutile mild steel	1.6–2.5	110	10–30	20–30
	3.2–5.0	110	20–45	30–60
	6.0–10.0	110	30–60	45–120
Cellulose	2.5–6.0	110	10–15	15–20

Hydrogen-controlled (basic) electrodes

The coatings of these electrodes contain no hydrogen-forming compounds, but if moisture is absorbed by the coating it becomes a source of hydrogen and cannot be tolerated. They must therefore be stored in a dry, heated and well-ventilated store on racks above floor level and unused electrodes should be returned to the store rather than left in the colder and moister conditions of the workshop where they could absorb moisture. A temperature of about 12°C above that of the external air temperature is

Fig. 1.9

suitable. Before use they should be removed from their package and spread out in the drying oven, the drying time and temperature depending upon the permissible volume of hydrogen in the weld deposit. Suggested figures are given in the following table.

Hydrogen content in millilitres of hydrogen per 100 grams of weld metal	Temperature °C	Time minutes	Use
10–15 ml H_2/100 g	150	60	To give resistance to HAZ cracking in thick sections of mild steel, high restraint.
5–10 ml H_2/100 g	200	60	High quality welds in pressure vessel and structural applications.
below 5 ml H_2/100 g	450	60	Thick sections to avoid lamellar tearing and critical applications.

In order to obtain high radiographic standards of deposited weld metal the drying periods given above may be extended. The following periods are given as an indication of prolonged drying times such that the electrode coating will not suffer a decrease in coating strength.

Drying temperature	Maximum time
150°C	72 hours
250°C	12 hours
450°C	2 hours

The makers' instructions for drying should be strictly adhered to.

Many electrodes if stored in damp situations get a white fur on their coverings. This is sodium carbonate produced by the action of the carbon dioxide (carbonic acid) of the atmosphere on the sodium silicate of the binder in the flux covering. The fur appears to have little detrimental effect on the weld but shows that the electrodes are being stored in too damp a situation.

Electrode classification

Classification of covered electrodes for the manual metal arc welding of carbon and carbon–manganese steels. BS 639

There is a compulsory and an optional part of the classification. In the compulsory part a covered electrode for manual metal arc welding is indicated by the letter E. This is followed by a two-digit figure, which gives the tensile strength and the yield stress of the weld metal thus:

Table 1

Electrode designation	Tensile strength N/mm² or MPa	Minimum yield stress N/mm² or MPa
E 43	430–550	330
E 51	510–650	360

The following two digits each indicate elongation and impact strength thus:

Table 2. *First digit for elongation and impact strength*

First digit	Minimum elongation % E 43 E 51	Temperature for impact value of 28 J.°C
0	not specified	not specified
1	20 18	+ 20
2	22 18	0
3	24 20	− 20
4	24 20	− 30
5	24 20	− 40

Table 3. *Second digit for elongation and impact strength*

Second digit	Minimum elongation %		Impact properties Impact value J		Temperature
	E43	E 51	E 43	E 51	°C
0	not specified		not specified		
1	22	22	47	47	+ 20
2	22	22	47	47	0
3	22	22	47	47	− 20
4	not relevant	18	not relevant	41	− 30
5	There are no electrodes designated by this digit				− 40
6	not relevant	18	not relevant	47	− 50

Following these digits is a letter (or letters) which indicates the type of covering.

A	iron (iron oxide)	O	oxidizing
AR	acid (rutile)	R	rutile (medium coated)
B	basic	RR	rutile (heavy coated)
C	cellulose	S	other types

This completes the compulsory portion of the classification, which therefore indicates tensile strength, yield strength, elongation and impact values and type of electrode covering.

The optional part of the classification begins with a three-digit number for the nominal electrode efficiency, included only if this is equal to or greater than 110. It is given to the nearest multiple of 10, with values ending in five being rounded off to the next higher ten.

Following this is a digit indicating the welding positions as shown and a digit for the current and voltage conditions as in Table 4.

Welding positions

1	all positions.
2	all positions except vertically down.
3	flat and, for fillet welds, horizontal–vertical.
4	flat.
5	flat, vertical down, and, for fillet welds, horizontal–vertical.
6	any position or combination of positions not classified above.

Table 4. *Welding current and voltage conditions*

Code	Direct current: recommended electrode polarity	Alternating current: minimum open circuit voltage
0	Polarity as recommended by the manufacturer	Not suitable for use on A.C.
1	+ or –	50
2	–	50
3	+	50
4	+ or –	70
5	–	70
6	+	70
7	+ or –	90
8	–	90
9	+	90

Hydrogen-controlled electrodes. The letter H shall be included in the classification for those electrodes which deposit not more than 15 ml of diffusible hydrogen per 100 g of deposited weld metal when stored according to the manufacturer's instructions. Recommended drying conditions shall be shown on the packet for hydrogen levels in the following

Types of electrode flux coverings

Class	Composition of covering	Characteristics	Uses
Cellulose (C)	Organic material containing cellulose and with some titanium oxide. Hydrogen releasing.	Thin, easily removable slag. Rather high spatter loss. Considerable envelope of shielding gas. Coarse ripple on weld surface, deeply penetrating arc with rapid burn-off rate.	All classes of mild steel welding in all positions; a.c. or d.c. electrode positive.
Acid (A)	Oxides and carbonates of iron and manganese with deoxidizers such as ferro-manganese.	Generally a thick coating which produces a fluid slag of large volume and solidifies in a 'puffed up' manner, is full of holes and easily detached. Smooth weld finish with small ripples. Good penetration. Weld liable to solidification cracking if plate weldability is not good.	Usually in the flat position only but can be used in other positions; a.c. or d.c.
Acid rutile (AR)	Generally a thick coating containing up to 35% rutile. Ilmenite (iron oxide and titanium oxide) is also used.	A fluid slag with other characteristics similar to the acid type of covering.	Similar to class (A).
Rutile (medium coating)	Mixture of titanium oxide and up to 15% organic (cellulose) matter with additions to produce a fluid slag. Coating thickness less than 50% of the core wire diameter.	Heavier coating than the AR type. Smooth weld finish, medium penetration, little spatter; fast freezing, easily detachable slag even from deep grooves.	Widely used for steel welding of all types. All positions; a.c. or d.c. Especially suitable for vertical and overhead positions.
Rutile (RR) (heavy coating)	Coverings of titanium oxide with up to 5% cellulosic matter with calcium fluoride. Coating thickness at least 50% greater than the core wire diameter. R and RR type coatings contain up to 50% titanium oxide (rutile and/or ilmenite).	Viscous slag, easy to remove except in deep Smooth weld contour.	Mainly in flat position but suitable for all positions; a.c. or d.c.

Oxidizing (O)	Iron oxide with or without manganese oxide and silicates.	Oxidizing slag so that the weld metal has a low carbon and manganese content referred to as 'dead soft'. Reduction of area and impact values are lower than for other types of electrodes. Core wire melts up inside coating forming a cup so that the electrode can be used for 'touch-welding'. Low penetration; solid slag often self-deslagging, with weld of neat appearance.	d.c. or a.c. supply with OCV as low as 45 V.
Basic (B)	Calcium or other basic carbonates and fluorspar bonded with sodium or potassium carbonates. Medium coating. Coating compounds contain no hydrogen. CO_2 releasing.	Brown slag easy to remove. Medium ripple on weld metal, medium penetration. Fillet profile flat or convex. Deposited metal has high resistance to hot and cold cracking because there is a low hydrogen content in the weld. Electrodes must be stored under warm dry conditions and dried before use.	Suitable for d.c. (electrode positive) or a.c. with OCV of 70 V. Used for mild, low alloy, high tensile and structural steels. Particularly for conditions of high restraint. For flat, vertical and over-head positions, the latter having a flat deposit.
Any other type (S)	This category is for any electrode coverings not included in the foregoing list. Iron powder electrodes do not come into this category but should be indicated by their efficiency with a three-digit figure.		

ranges: 10–15 ml, 5–10 ml, 0–5 ml per 100 g of deposited weld metal respectively.

Examples of the use of BS 639 classification.

Example (1)

E .51 .33 .RR .130 .3 .1 .H

E Flux-covered electrode for manual metal arc welding.

51 (Table 1) The all weld metal (AWM) tensile strength lies between 510 and 650 N/mm² and the yield stress is not less than 360 N/mm².

3 (Table 2) The minimum elongation would be 20% with an impact strength of at least 28 J at − 20°C.

3 (Table 3) A second value of elongation, 22% minimum with a minimum impact strength of 47 J at − 20°C.

RR The predominant coating ingredient is rutile and the flux coating diameter to core wire diameter ratio shall be greater than 1.5:1.

130 The electrode is of a type containing iron powder in the flux covering and will deposit weld metal in excess of the core wire weight. The deposit in this case would be from 125 to 134% of the core wire weight.

3 The electrode can be used in the flat position and for horizontal vertical fillet welds.

1 (Table 4) The electrode can be used on a d.c. power supply, with the electrode connected to either positive or negative pole, or on an a.c. supply having at least 50 OCV.

H The weld metal shall contain not more than 15 ml hydrogen in every 100 g weld metal when the electrode has been stored and used in accordance with the manufacturers instructions.

Note that this is a hydrogen-controlled *rutile* electrode.

Example (2)

E .43 .21 .C .1.9

E Flux-covered electrode for manual metal arc welding.

43 AWM tensile strength of 430–550 N/mm², yield stress 330 N/mm² minimum.

2 AWM elongation of 22% with an impact strength of 28 J minimum at 0°C.

1 AWM elongation of 22% with an impact strength of 47 J at 20°C.

C A cellulosic covered electrode.

1 Can be used in all positions.

9 Can be used on a d.c. supply with electrode positive or on an a.c. supply with an OCV of 90 V.

Note on ISO 2560 – 1973 (E). (Covered electrodes for manual arc welding of mild steel and low alloy steel – Code of symbols for identification.) BS

639 (1976) follows closely the ISO 2560 specification, the main differences being that in the ISO standard the tensile strengths are 430–510 N/mm² and 510–610 N/mm², but in each case in view of possible variations in welding and testing, the upper limits of 510 and 610 N/mm² are allowed to be exceeded by 40 N/mm² giving the BS 639 values of 550 and 650 N/mm². The other main difference is that of impact strength and elongation, where the ISO standard has one digit only as from Table 2 whereas BS 639 has two digits, one each from Tables 2 and 3, the second digit covering the amalgamation of the rules of the International Association of Classification Societies and the BS 639 (1972) requirements.

The extra Charpy testing at varying temperatures gives the designer a better guide to the selection of electrodes to give resistance to brittle fracture.

American Welding Society (AWS) electrode classification A5. 1–69

Mild steel electrodes. A four-digit number is used, preceded by the letter E indicating electrode. The first two digits indicate the tensile strength of the weld metal in thousands of pounds force per square inch (1000 psi) in the 'as-welded' condition. The third digit indicates the welding position and the fourth digit the current to be used and the type of flux coating. An example of the first and second digits is: E 60xx, which indicates that it is a metal arc welding electrode with an 'as-welded' deposit having a UTS (ultimate tensile strength) of 60 000 lbf/in² minimum or 412 N/mm².

The third and fourth digits are:

E xx10 High cellulose coating, bonded with sodium silicate. Deeply penetrating, forceful, spray type arc, thin friable slag. All positional, d.c. electrode positive only.

E xx11 Similar to E xx10 but bonded with potassium silicate to allow it to be used on a.c. or d.c. positive.

E xx12 High rutile coating, bonded with sodium silicate. Quiet arc, medium penetration, all positional, a.c. or d.c. negative.

E xx13 Coating similar to E xx12 but with the addition of easily ionized materials and bonded with potassium silicate to give a steady arc on a low voltage supply. Slag is fluid and easily removed. All positional; a.c. or d.c. negative.

E xx14 Coating similar to E xx12 and E xx13 types with the addition of a medium quantity of iron powder. All positional; a.c. or d.c.

E xx15 Lime-fluoride coating (basic, low hydrogen) type bonded with sodium silicate. All positional. For welding high tensile steels; d.c., positive only.

E xx16 Similar coating to E xx15 but bonded with potassium silicate; a.c. or d.c. positive.

E xx18 Coating similar to E xx15 and E xx16 but with the addition of iron powder. All positional; a.c. or d.c.

E xx20 High iron oxide coating bonded with sodium silicate. For welding in the flat or HV (horizontal–vertical) positions. Good X-ray quality; a.c. or d.c.

E xx24 Heavily coated electrode having flux ingredients similar to E xx12 and E xx13 with the addition of a high percentage of iron powder for fast deposition rates. Flat and horizontal positions only; a.c. or d.c.

E xx27 Very heavily coated electrode having flux ingredients similar to E xx20 type, with the addition of a high percentage of iron powder. Flat or horizontal positions. High X-ray quality; a.c. or d.c.

E xx28 Similar to E xx18 but heavier coating and suitable for use in flat or HV positions only; a.c. or d.c.

E xx30 High iron oxide type coating but produces less fluid slag than E xx20. For use in flat position only (primarily narrow groove butt welds). Good radiographic quality; a.c. or d.c.

Example

E 6013. Welding electrode having weld metal of UTS 60000 lbf/in^2 or 412 N/mm^2, with high rutile coating bonded with potassium silicate. All positional; a.c. or d.c. negative.

Example

E 7018. Welding electrode with weld metal of UTS 70000 lbf/in^2 or 480 N/mm^2, with basic coating hydrogen controlled, with the addition of iron powder. All positional; a.c. or d.c.

See Appendix 1 for further lists of American symbols and classifications.

Welders' accessories (Fig. 1.10*a*)

Electrode holder

This is an arrangement which enables the welder to hold the electrode when welding. It has an insulated body and head which reduces the danger of electric shock when working in damp situations and also reduces the amount of heat conducted to the hand whilst welding. The electrode is clamped between copper jaws which are usually spring loaded, and a simple movement of a side lever enables the electrode to be changed easily and quickly. In another type the electrode end fits into a socket in the holder head and is held there by a twist on the handle. The welding flexible cable is attached to the holder by clamping pieces or it may be sweated into a terminal lug. The holder should be of light yet robust construction and well insulated, and electrode changing must be a simple operation. Fig. 1.10*b* shows typical holders.

Fig. 1.10. (*a*)

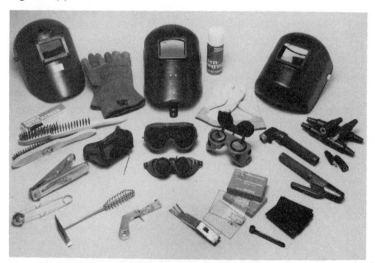

Fig. 1.10. (*b*)

Head shields, lenses and general protective gear. The rays from the metallic arc are rich in infra-red and ultra-violet radiation, and it is essential that the eyes and face of the welder should be protected from these rays and from the intense brightness of the arc. The welding shield can either fit on to the head (Fig. 1.11*a*), leaving both hands free, or may be carried in one hand. It should extend so as to cover the sides of the face, especially when welding is done in the vicinity of other welders, so as to prevent stray flashes reaching the eyes. The shield must be light and, because of this, preferably made of fibre.

Helmets and hand shields are now available, weighing little more than ordinary shields, that give maximum comfort when welding in restricted conditions. They are similar in appearance to the standard shield but have double-wall construction, the inner face being perforated with small holes, through which pure air is supplied to the interior of the helmet. Head and face are kept cool and the operator's eyes and lungs are protected from dust and fumes as well as radiation and spatter as with ordinary shields. The

Fig. 1.11. (*a*)

air is supplied from any standard compressor or air line, the supply tube being fitted with a pressure-reducing valve to give a pressure of 1.7 bar (Fig. 1.11*b*).

The arc emits infra-red and ultra-violet radiation in addition to light in the visible spectrum. The filter or lens is designed to protect the eyes from the ultra-violet and infra-red radiation which would injure the eyes and also to reduce the amount of visible light so that there is no discomfort. The filters are graded by numbers, followed by the letters EW (electrical welding) denoting the process, according to BS 679, and increase in shade depth with increasing number of increasing currents. The filters recommended are: up to 100 A, 8 or 9/EW; 100–300 A, 10 or 11/EW; over 300 A, 12, 13, 14/EW. The choice of the correct filter is the safeguard against eye damage. Occasional accidental exposure to direct or reflected rays may result in the condition known as arc flash. This causes no permanent damage but is painful, with a feeling of sand in the eyes accompanied by watering. Bathing the eyes with eye lotion and wearing dark glasses reduces the discomfort and the condition usually passes with no adverse effects in from 12 to 24 hours. If it persists a doctor should be consulted.

A lens of this type is expensive and is protected from metallic spatter on *both* sides by plain glass, which should be renewed from time to time, as it becomes opaque and uncleanable, due to spatter and fumes.

The welding area must be adequately screened so that other personnel are not affected by rays from the arc.

The eyes must also be protected against foreign bodies such as particles

Fig. 1.11. (*b*)

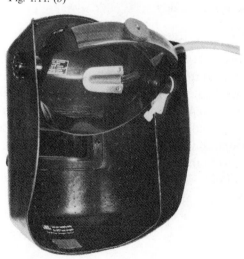

of dust and metal, which are all potential dangers in the welding shop.

Spectacles with clear or tinted lenses are available and also fit over normal spectacles if required. For use particularly on grinding operations there are clear goggles and there are also face shields with plain or tinted visors which fit on the head like a welding head mask but have the shield hinged so that it is easily raised when required. For conditions where there is much noise there are hearing protectors which fit snugly over the ears in the same way as radio headphones. There are also ear plugs which are made in a soft foam and which are easily inserted, and are very effective.

Protection against inhalation of polluted atmosphere is given by single cartridge dust respirators and various types of moulded plastic masks with adjustable headbands in which the filter is replaceable. A protective cap is also available with a hard shell, foam lining and elastic grip giving good fitting to the head.

Leather or skin aprons are excellent protection for the clothes against sparks and molten metal. Trouser clips are worn to prevent molten metal lodging in the turn-ups, and great care should be taken that no metal can drop inside the shoe, as very bad burns can result before the metal can be removed. Leather spats are worn to prevent this. Gauntlet gloves are worn for the same reason, especially in vertical and overhead welding. In welding in confined spaces, the welder should be fully protected, so that his clothes cannot take fire due to molten metal falling on him, otherwise he may be badly burnt before he can be extracted from the confined space.

The welding bench in the welding shop should have a flat top of steel plate, about 1.5 m × 0.75 m being a handy size. On one end a vice should be fitted, while a small insulated hook on which to hang the electrode holder when not in use is very handy.

Jigs and fixtures

These are a great aid to the rapid setting up of parts and holding them in position for welding. In the case of repetition work they are essential equipment for economical working. Any device used in this way comes under this heading, and jigs and fixtures of all types can be built easily, quickly and economically by arc welding. They are of convenience to the welder, reduce the cost of the operation, and standardize and increase the accuracy of fabrication.

Jigs may be regarded as specialized devices which enable the parts being welded to be easily and rapidly set up, held and positioned. They should be rigid and strong since they have to stand contractional stresses without deforming unless this is required; simple to operate, yet they must be accurate. Their design must be such that it is not possible to put the work in them the wrong way, and any parts of them which have to stand continual wear should be faced with wear-resistant material. In some cases, as in inert gas welding (q.v.), the jig is used as a means of directing the inert gas on to the underside of the weld (backpurge) and jigs may also incorporate a backing strip.

Fixtures are of a more general character and not so specialized as jigs. They may include rollers, clamps, wedges, etc., used for convenience in manipulation of the work. Universal fixtures are now available, and these greatly reduce the amount of time of handling of the parts to be welded and can be adapted to suit most types of work.

Manipulators, positioners, columns and booms

Positioners are appliances which enable work to be moved easily, quickly and accurately into any desired position for welding – generally in the downhand position since this speeds up production by making welding easier, and is safer than crane handling. Universal positioners are mounted on trunions or rockers and can vary in size from quite small bench types to very large ones with a capacity of several tons. Manually operated types are generally operated through a worm gearing with safety locks to prevent undesired movement after positioning. The larger types are motor-driven, and on the types fitted with a table, for example, the work can be swung through any angle, rotated, and moved up and down so that if required it can be positioned under a welding head.

As automatic welding has become more and more important so has the design of positioners and rollers improved. Welding-columns and booms may be fixed or wheel mounted and have the automatic welding head mounted on a horizontal boom which can slide up and down a vertical column. The column can be swivel mounted to rotate through 360° and can be locked in any position. In the positioning ram-type boom there is horizontal and vertical movement of the boom carrying the welding head and they are used for positioning the head over work which moves beneath them at welding speed. In the manipulating ram-type boom, the boom is provided with a variable speed drive of range of about 150–1500 mm per min enabling the boom to move the welding head over the stationary work. Both types of boom in the larger sizes can be equipped with a platform to carry the operator, who can control all movements of head and boom from his position in Fig. 1.12. Various types are shown.

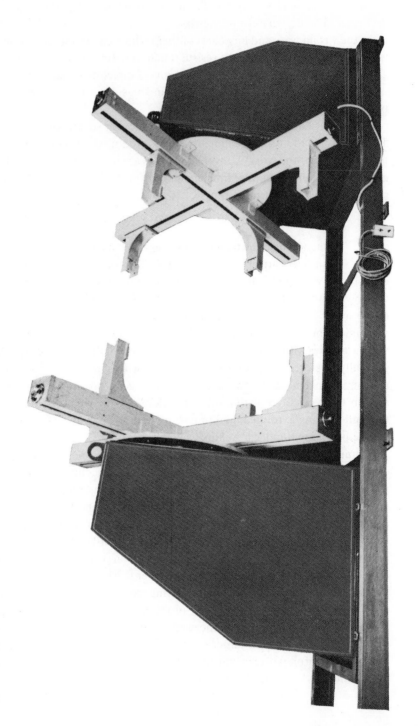

Fig. 1.12. (a) Head and tailstock manipulator.

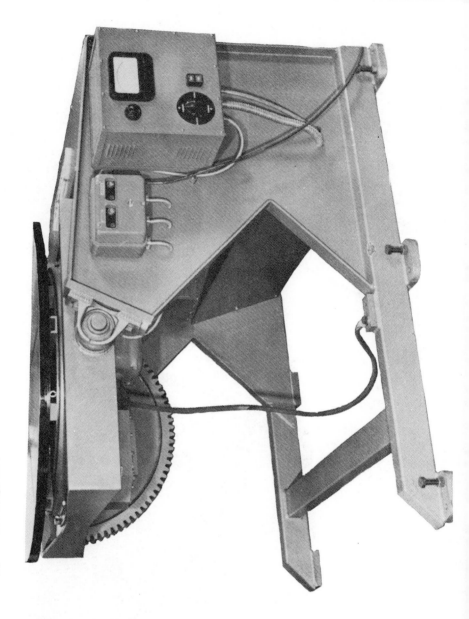

Fig. 1.12. (b) Power rotation, power tilt through 135°.

Fig. 1.12.   Roller bed, travelling type, motorized.

Fig. 1.12. (*d*) Power rotation, manual tilt.

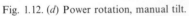

The practice of manual metal arc welding*

Electrode lengths. The actual length must be within ± 2 mm of the nominal value

Diameter, mm	Length
2	200
	250
	300
	350
2.5	250
	300
	350
3.15 or 3.25	350
	450
4 to 10	350
	450
	500
	500
	600
	700
	900

Note: Because the 3.25 mm diameter is widely used this size is being retained until such time as it can be discarded through its lesser usage.

Electrode diameter		Nearest fractional equivalent, in.
mm	inch	
1.6	0.06	—
2.0	0.08	$\frac{5}{64}$
2.5	0.10	$\frac{3}{32}$
3.0	0.12	—
3.25	0.13	$\frac{1}{8}$
4.0	0.16	$\frac{5}{33}$
5.0	0.19	$\frac{3}{16}$
6.0	0.23	—
6.3	0.25	$\frac{1}{4}$
8.0	0.32	$\frac{5}{16}$
10.0	0.37	$\frac{3}{8}$

Safety precautions

Protection of the skin and eyes. Welding gloves should be worn and no part of the body should be exposed to the rays from the arc otherwise burning will result. Filter glasses (p. 23) should be chosen according to BS

* American designation: shielded metal arc welding (SMAW).

recommendations. Where there are alternatives the lower shade numbers should be used in bright light or out of doors and the higher shade numbers for use in dark surroundings.

Do not weld in positions where other personnel may receive direct or reflected radiation. Use screens. If possible do not weld in buildings with light coloured walls (white) as this increases the reflected light and introduces greater eye strain.

Do not chip or deslag metal unless glasses are worn.

Do not weld while standing on a damp floor. Use boards.

Switch off apparatus when not in use.

Make sure that welding return leads make good contact, thus improving welding conditions and reduced fire risk.

Avoid having inflammable materials in the welding shop.

Degreasing, using chemical compounds such as trichlorethylene, perchlorethylene, carbon tetrachloride or methyl chloride, should be carried out away from welding operations, and the chemical allowed to evaporate completely from the surface of the component before beginning welding. The first-named compound gives off a poisonous gas, phosgene, when heated or subjected to ultra-violet radiation, and should be used with care. Special precautions should be taken before entering or commencing any welding operations on tanks which have contained inflammable or poisonous liquids, because of the risk of explosion or suffocation. The student is advised to study the following publications of the Department of Employment (SHW–Safety, health and welfare):

SHW 386. *Explosions in drums and tanks following welding and cutting.* HMSO.

SHW Booklet 32. *Repair of drums and tanks. Explosions and fire risks.* HMSO.

SHW Booklet 38. *Electric arc welding.* HMSO.

Memo No 814. *Explosions – Stills and tanks.* HMSO.

Technical data notes No. 18. *Repair and demolition of large storage tanks.* No. 2. *Threshold limit values.* HMSO.

Health and safety in welding and allied processes. The Welding Institute.

Filters for use during welding. BS 679. British Standards Institution.

Protection for eyes, face and neck. BS 1542. British Standards Institution.

Metal arc welding

This section on practical welding techniques applies equally well to all the following steels:

BS 970, Pt 1: *Carbon and carbon–manganese steels*;

Pt 2: *Direct hardening and alloy steels*;

Pt 3: *Steels for case hardening*;

Pt 4: *Stainless, heat resisting and valve steels.*

BS 1501, Pt 1: *Fired and unfired pressure vessels*;

Pt 2: *Alloy steels for fired and unfired pressure vessels.*

BS 4360: *Structural steels and branded steels.*

In each and every case the electrode recommended by the manufacturer for the particular steel being welded should be used and the procedure advised for the electrode (such as pre- or post-heat) strictly adhered to.

Sources of supply. The source of welding supply can be a.c. or d.c. The relatively cheap a.c. sets with an output maximum of about 150 A are extremely useful for repair and maintenance in small workshops, but are used mainly for steel. Larger a.c. sets are available for fabrication shops including shipyards.

The d.c. output units are used for both steel and non-ferrous metals and can be obtained in current ranges of 300 A, 400 A, as required. The latest sets have thyristor controlled rectifiers with a one knob stepless control of current, and a remote control can also be supplied.

In order to assist the operator, tables are given indicating the approximate current values with various types and sizes of electrodes. These tables are approximate only and the actual value of the current employed will depend to a great extent upon the work. In general the higher the current in the range given for one electrode size, the deeper the penetration and the faster the rate of deposit. Too much current leads to undercutting and spatter. Too small a current will result in insufficient penetration and too small a deposit of metal. As a general rule, slightly increase the arc voltage as the electrode size increases. The angle of the electrode can be varied between 60° and 90° to the line of the weld. As the angle between electrode and plate is reduced, the gas shield becomes less effective, the possibility of adverse effects of the atmosphere on the weld increases, and penetration is reduced.

Full details of currents suitable for the particular electrodes being used are usually found on the electrode packet and should be adhered to; in the following pages it is assumed that, if a.c. is being used, no notice should be taken of the polarity rules, and covered electrodes only should be used.

The technique of welding both mild steel and wrought iron is similar.

Diameter of rod (mm)	Current (amperes)		
	Min.	Max.	Average
1.6	25	45	40
2.5	50	90	90
3.2	60	130	115
4.0	100	180	140
5.0	150	250	200
6.0	200	310	280
6.3	215	350	300
8.0	250	500	350

Note. American designation: straight polarity, electrode −ve, DCEN; reverse polarity, electrode +ve, DCEP. These apply when welding with d.c.

Striking and maintaining the arc. With an electrode of 4 mm diameter and a current of 130–150 A and a mild steel plate 6–8 mm thick, the first exercise is to strike and maintain the arc. The operator should be comfortably placed with the cable to the electrode exerting no pull on the holder (e.g. cable over the operator's shoulder). The electrode tip is scratched on to the plate at the same time as the hand shield with filter glass is drawn across the face. Never expose eyes, face or any part of bare skin to arc flashes, as the effects can be most painful. The electrode is lifted to about 3 mm from the plate, drawing the arc, and the molten pool can be clearly seen (Fig. 1.13). The lighter, brighter colour is the slag while the more viscous, darker colour is the molten metal.

Straight runs on the plate can now be made once the arc can be struck and maintained. As the electrode melts it deposits the metal on the plate and this deposit should be in a straight line and continuous with no gaps or entrapped slag. Too high a current for a given size of electrode will result in

Fig. 1.13. Striking and maintaining the arc.

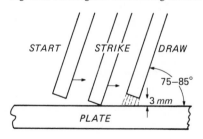

the electrode becoming overheated and eventually red hot. A 4 mm electrode with about 160 A will deposit about 200–235 mm (about 9 in.) of weld metal, if the speed is correct. Slower travel will result in a wider run and it will not be so long, while faster travel results in a longer and narrower bead with the danger of gaps and slag inclusions. The fault of most beginners is that the run is made too quickly. The depth of the crater indicates the amount of peretration (Fig. 1.14*a*).

If, when striking the arc or welding, the electrode touches the plate and 'freezes' it should be freed with a sharp twist of the holder. If this fails, switch off the electric supply at the set and remember that the electrode will still be very hot.

When welding with a d.c. machine, considerable difference can be seen according to whether the electrode is made positive or negative polarity. Since greater heat is generated at the positive pole, making the electrode negative reduces the burn-off rate and deepens the creater and thus the penetration. With the electrode positive the penetration is reduced but the burn-off rate is increased, most heat being generated near the electrode. For this reason (together with the type of electrode coating) most special steels (stainless, heat resistant, etc.), vertical and overhead runs in steel, cast iron and aluminium and bronze are nearly always welded with electrode positive. The instructions on the electrode package issued by the makers should be strictly adhered to.

It will be noticed that when welding on a cold plate the deposited bead rises well above the level of the metal when it is cold. As welding progresses, however, and the parent metal gets heated up, the penetration becomes deeper and the head has a lower contour. For this reason the current setting can be higher when welding is commenced and then should be reduced slightly as the metal heats up, resulting in a bead having an even contour throughout its length.

This exercise should be continued until a straight, even, uniform bead about 250 mm long can be run with good penetration.

If a bead is to be continued after stopping, as for example to change an

Fig. 1.14. (*a*)

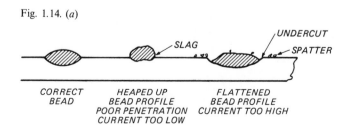

CORRECT BEAD

HEAPED UP BEAD PROFILE POOR PENETRATION CURRENT TOO LOW

SLAG

FLATTENED BEAD PROFILE CURRENT TOO HIGH

UNDERCUT

SPATTER

electrode, the end of the bead and the crater must be deslagged and brushed clean and the arc struck at the forward end of the crater. The electrode is then quickly moved to the rear end and the bead continued, so that no interruption can be detected (Fig. 1.14b). This should be practised until no discontinuity in the finished bead can be observed. Since in welding long runs, the welder may have to change rods many times, the importance of this exercise will be appreciated, otherwise a weakness or irregularity would exist whenever the welding operation was stopped.

Beads can now be laid welding away from the operator and also welding from left to right or right to left. A figure-of-eight about 120 mm long provides good practice in this and in changing direction of the bead when welding. The bead should be laid continuously around the figure.

A transformer unit usually has two voltage settings, one 80 open circuit volts and the other 100 OCV, the latter being used for thinner sheet when the current setting is low. Other sets may have lower OCV's, the figure required generally being given on the electrode packet. With an a.c. volts setting of 80 V the current setting for a given electrode can be varied from a low value, giving poor penetration and a shallow crater, with the metal heaping up on the plate producing overlap, to a high value, giving an excessively deep crater, too much penetration with a fierce hissing and not well controllable arc, much spatter, and a flat bead with undercut and an overheated electrode. Intermediate between these extreme values is a correct value of current, say 115–125 A for a 3.2 mm electrode. There is good penetration, the metal flows well, the arc is controllable with no spatter or undercut, and the arc sound is a steady crackle.

These results are tabulated below. If a d.c. set is being used with a voltage control this also can be varied for the operator to find out the effect of this control.

Fig. 1.14 (b) Arc struck in front of crater and moved back to continue the run preserves weld profile and helps eliminate any start porosity.

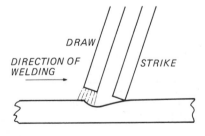

DRAW

DIRECTION OF
WELDING

STRIKE

Welding circuit	Effect
Too low current	Poor penetration; shallow crater; metal heaps up on plate with overlap; arc has unsteady spluttering sound.
Too high current	Deep crater; too deep penetration; flat bead; fierce arc with loud crackle; electrode becomes red hot; much spatter.
Too high voltage	Noisy hissing arc; fierce and wandering arc; bead tends to be porous and flat; spatter.
Correct voltage and current and welding speed	Steady crackle; medium crater giving good penetration; easily controlled stable arc; smooth even bead.

Weaving (see pp. 126–8 for automatic weaving)

This may be attempted before or after the preceding exercise according to the inclination of the operator. Weaving is a side to side motion of the electrode, as it progresses down the weld, which helps to give better fusion on the sides of the weld, and also enables the metal to be built up or reinforced along any desired line, according to the type of weave used. It increases the dilution of weld metal with parent metal and should be reduced to a minimum when welding alloy steel.

There are many different methods of weaving, and the method adopted depends on the welder and the work being done. The simplest type is shown in Fig. 1.15*a*, and is a simple regular side to side motion, the circular portion helping to pile the metal in the bead in ripples. Fig. 1.15*b* is a circular motion favoured by many welders and has the same effect as Fig. 1.15*a*. Care should be taken with this method when using heavy coated rods that the slag is not entrapped in the weld. Fig. 1.15*c* is a figure-of-eight method and gives increased penetration on the lines of fusion, but care must again be taken that slag is not entrapped in the overlap of the weave on the edges. It is useful when reinforcing and building up deposits of wear-

Fig. 1.15

(a)

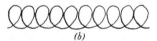

(b)

NORMAL WEAVES FOR HORIZONTAL BEAD

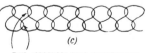

(c)

(d)

THIS PRODUCES DEEP PENETRATION
AND REINFORCEMENT ON THESE LINES

USEFUL FOR HORIZONTAL BEAD ON VERTICAL
PLATE. HESITATING ACTION CONTROLS METAL
ON EACH SIDE OF BEAD

resisting steels. Fig. 1.15*d* is a weave that is useful when running horizontal beads on a vertical plane, since by the hesitating movement at the side of the bead, the metal may be heaped up as required. The longer the period of hesitation at any point in a weave, the more metal will be deposited at this point.

Weaving should be practised until the bead laid down has an even surface with evenly spaced ripples. The width of the weave can be varied, resulting in a narrow or wide bead as required.

Padding and building up shafts

This exercise provides a good test of continued accuracy in laying a weld. A plate of 8–9.5 mm mild steel about 150 mm square is chosen and a series of parallel beads are laid side by side across the surface of the plate and so as to almost half overlap each other. If the beads are laid side by side with no overlap, slag becomes entrapped in the line where the beads meet, being difficult to remove and causing blowholes (Fig. 1.16). Each bead is

Fig. 1.16

SLAG TENDS TO GET
ENTRAPPED HERE

BY OVERLAPPING THE RUNS THEY ARE
EASIER TO DESLAG. NO SLAG IS
ENTRAPPED THUS NO BLOWHOLES ARE
FORMED

Fig. 1.17

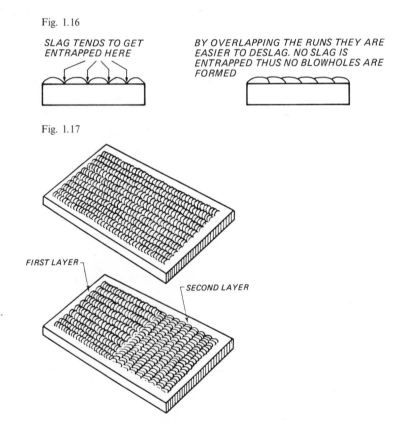

FIRST LAYER

SECOND LAYER

deslagged before the next is laid. The result is a build-up layer of weld metal.

After thoroughly cleaning and brushing all slag and impurities from this layer, another layer is deposited on top of this with the heads at right angles to those of the first layer (Fig. 1.17), or they may be laid in the same direction as those of the first layer. This can be continued for several layers, and the finished pad can then be sawn through and the section etched, when defects such as entrapped slag and blowholes can at once be seen.

Odd lengths of steel pipe, about 6 mm or more thick, may be used for the next exercise, which again consists in building up layers as before. The beads should be welded on opposite diameters to prevent distortion (Fig. 1.18a). After building up two or more layers, the pipe can be turned down on the lathe and the deposit examined for closeness of texture and absence of slag and blowholes. Let each bead overlap the one next to it as this greatly reduces the liability of pin-holes in the finished deposit.

The same method is adopted in building up worn shafts and a weld may be run around the ends as shown to finish the deposit. Another method sometimes used consists of mounting the shaft on V blocks and welding spirally (Fig. 1.18a and b).

Tack welding

Tack welds are essential in welding and fabrication to ensure that there is correct gap and line-up of the components to be welded. The tack welds should be made with a higher current than normal to ensure good fusion and penetration and should be strong enough and of sufficient length and frequency of spacing to ensure rigidity against the distortional

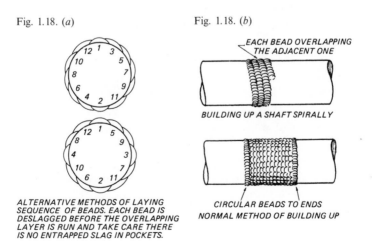

Fig. 1.18. (a)   Fig. 1.18. (b)

EACH BEAD OVERLAPPING THE ADJACENT ONE

BUILDING UP A SHAFT SPIRALLY

ALTERNATIVE METHODS OF LAYING SEQUENCE OF BEADS. EACH BEAD IS DESLAGGED BEFORE THE OVERLAPPING LAYER IS RUN AND TAKE CARE THERE IS NO ENTRAPPED SLAG IN POCKETS.

CIRCULAR BEADS TO ENDS NORMAL METHOD OF BUILDING UP

stresses set up in the welding process. The tack welds should be deslagged and examined for cracks, which may occur because of the rapid cooling of the tack weld if there is no pre-heat. If any cracks are found they should be gouged or ground out and the tack weld remade. Cracks left in may cause cracking in the finished weld.

Thicker plate may need stronger tack welds. These should be made in two runs, the second tack weld being made after the first has been deslagged and examined. Remember when making the first run that the tack weld must be fused into the run so that there is no evidence of the tack weld. This may mean welding at a slightly increased rate at these points. Tack welds are often made by the TIG process in fabrication of plate and pipe work for lining up and then welded by the MIG process.

Plates set up for fillet welding should have the tack welds made at about twice the frequency of pitch used for butt welds.

The operator should now be able to proceed to the making of welded joints, and it will be well to consider first of all the method of preparation of plates of various thicknesses for butt welding. Although a U preparation may prove to be more expensive than a V preparation, it requires less weld metal and distortion will be lessened. For thicker sections the split-weave or stringer bead technique is often used (Fig. 1.21) since the stresses due to contraction are less with this method than with a wide weave.

When a weld is made on a plate inclined at not more than 5° to the horizontal, this is termed the flat position, and wherever possible welding should be done in this position, since it is the easiest from the welder's point of view.

Fig. 1.19 shows some simple welding joints.

Butt welding in the flat position

Joints are prepared with an included angle of 60° with root face and root gap as required for the plate thickness. (An angle of 60° is chosen because it combines accessibility to the bottom of the joint with the least amount of weld metal (Fig. 1.21).

The plates can be tack welded in position to prevent movement due to expansion and contraction during welding and the tack welds must be strong enough to prevent movement of the plates. Using say a 4 mm electrode with about 125–135 A for a 3 mm gap and 160–170 A for a 1.5 mm gap the welds are kept to about 3–3.5 mm thickness. The length of the run depends upon the size of the electrode and the width of the weld, which gets greater as the joint is filled up. To avoid having the top deposits in thick plate too wide, the stringer or split-weave method is used. If the plates are to be welded from one side only, the penetration *must* be right through to the bottom of the V and the underbead must be continuous along the length of the joint (Fig. 1.20). If a sealing run is to be made the underside of the joint should be back chipped (by electrically or pneumatically operated chipping hammer) so that any defects such as inclusions, porosity, etc., are removed and a sealing run is then made, using a rod about 5 mm diameter.

In many cases the joint to be welded is accessible from one side only, so that the penetration bead cannot be examined for defects and it is not possible to employ a sealing run. In this case a backing bar can be used. The plates are prepared to a sharp edge and the backing bar, usually about 25 mm wide and 5 mm thick, is tack welded to one plate and the plates set apart to the required gap. The first run is made with a higher current than previously used as there is no possibility of burn through. This ensures that the roots of the parent plate and the backing bar are well fused together, and the other runs are then made, as previously, with lower current.

Double-V butt joint preparation is often used with thicker plate because

Fig. 1.19. Simple welding joints.

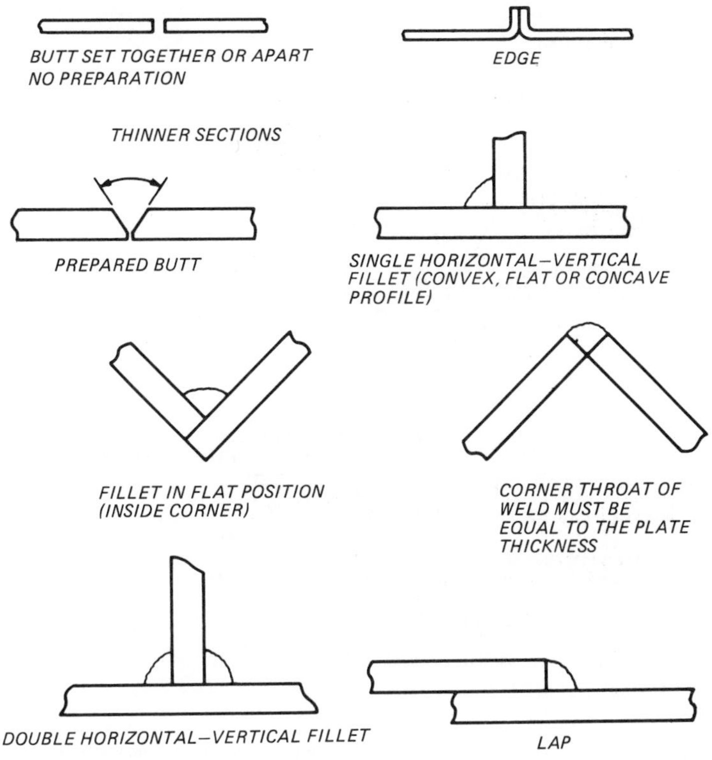

BUTT SET TOGETHER OR APART
NO PREPARATION

EDGE

THINNER SECTIONS

PREPARED BUTT

SINGLE HORIZONTAL–VERTICAL
FILLET (CONVEX, FLAT OR CONCAVE
PROFILE)

FILLET IN FLAT POSITION
(INSIDE CORNER)

CORNER THROAT OF
WELD MUST BE
EQUAL TO THE PLATE
THICKNESS

DOUBLE HORIZONTAL–VERTICAL FILLET

LAP

of the saving in weld metal, which in the case of a 19 mm ($\frac{3}{4}''$) plate is about half of that required for single-V preparation. The plates are prepared to a sharp edge (Fig. 1.21) and the first run made with little or no weave and absolutely continuous. The plates are turned over, the joint root deslagged, after which the next run is made on to the root of the first run, ensuring that there is no entrapped slag or porosity in the centre of the joint. The remaining runs can be made on alternate sides of the joint where possible, reducing the distortional stresses.

Butt welds on pipes provide good exercise in arc manipulation. The pipes are prepared in the same way as for plates by V'ing. They are then lined up in a clamp, or are lined up and tack welded in position and then placed across two V blocks.

Right angle outside corner joints

Tack weld two 12.5 mm ($\frac{1}{2}$ in.) steel plates together with a 2 mm gap between them and an angle of 90° between the plates (Fig. 1.22), making a right angled corner joint. With an electrode of 4 mm diameter and

Fig. 1.20

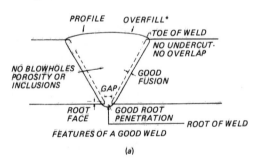

FEATURES OF A GOOD WELD

(a)

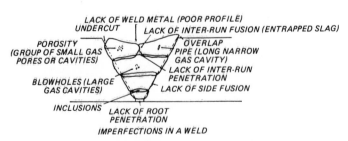

IMPERFECTIONS IN A WELD

(b)

* The BS term is 'excess weld metal' but the term 'overfill' is now used b
circumstances, such as under fatigue loading, excess weld metal (or ·
detrimental to the weld.

about 160 A current make a run about 180 mm long keeping the electrode at an angle of about 60° and ensuring that penetration is adequate but not excessive. The speed of welding should be increased where the tack welds occur to ensure the uniform appearance of the upper side of the weld. Deslag upper and underside of the joint and observe that correct penetration has been achieved. Then make the second (and subsequent) runs with an electrode angle of about 70–75° using a slight weave and avoiding any undercut. Deslagging is followed by the final run to obtain the correct throat thickness and profile as shown. A sealing run can be made if required on the underside of the joint using a somewhat higher current.

Fig. 1.21. Flat butt welds, preparation.

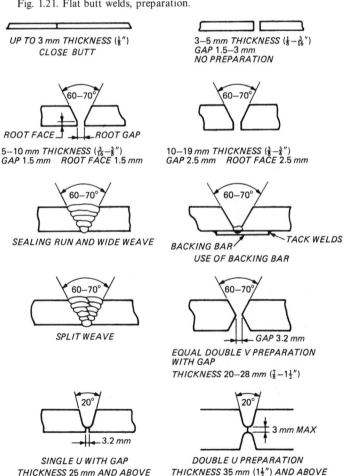

UP TO 3 mm THICKNESS ($\frac{1}{8}$")
CLOSE BUTT

3–5 mm THICKNESS ($\frac{1}{8}$–$\frac{3}{16}$")
GAP 1.5–3 mm
NO PREPARATION

ROOT FACE   ROOT GAP

5–10 mm THICKNESS ($\frac{3}{16}$–$\frac{3}{8}$")
GAP 1.5 mm   ROOT FACE 1.5 mm

10–19 mm THICKNESS ($\frac{3}{8}$–$\frac{3}{4}$")
GAP 2.5 mm   ROOT FACE 2.5 mm

SEALING RUN AND WIDE WEAVE

BACKING BAR   TACK WELDS
USE OF BACKING BAR

SPLIT WEAVE

GAP 3.2 mm
EQUAL DOUBLE V PREPARATION
WITH GAP
THICKNESS 20–28 mm ($\frac{7}{8}$–1$\frac{1}{2}$")

3.2 mm
SINGLE U WITH GAP
THICKNESS 25 mm AND ABOVE

3 mm MAX
DOUBLE U PREPARATION
THICKNESS 35 mm (1$\frac{1}{2}$") AND ABOVE

Fig. 1.22. Right angle corner welds.

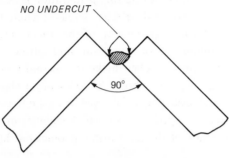

NO UNDERCUT

90°

PLATES TACK WELDED IN POSITION. AFTER DESLAGGING,
FIRST RUN GIVES UNIFORM PENETRATION ON UNDERSIDE
OF JOINT, TACK WELD MUST BE INCORPORATED INTO THE
MAIN WELD, WITHOUT TRACE

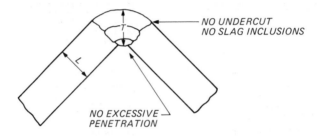

NO UNDERCUT
NO SLAG INCLUSIONS

NO EXCESSIVE
PENETRATION

THROAT THICKNESS EQUALS PLATE THICKNESS

Fig. 1.23. Fillet weld flat position.

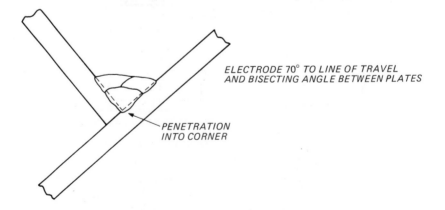

ELECTRODE 70° TO LINE OF TRAVEL
AND BISECTING ANGLE BETWEEN PLATES

PENETRATION
INTO CORNER

Fillet welds in the flat position

Two plates about 9 mm thick are tack welded to make a tee joint and set at an angle of 45° to the horizontal as shown in Fig. 1.23. The weld is made with the electrode at an angle of about 70° to the line of travel and bisecting the angle between the plates. The weld metal must penetrate and fuse into the corner of the joint and there must be no undercutting with equal leg length. A 5 mm electrode at about 200 A will give a 6 mm fillet (throat) about 280 mm long. Make sure that the tack welds, which have been previously deslagged and examined for cracks, etc., are well incorporated into the weld by slightly increasing the welding speed at these points.

Fillet welds in the horizontal–vertical position

With the one plate in the vertical position there will be more tendency for the weld metal to fall downwards to the horizontal plate. To counteract this the electrode is directed at about 40° to the horizontal plate and more vertically to the line of travel – say about 60–80° increasing with the rod diameter. Although the slag, which is more fluid, tends to fall to the horizontal plate, make sure that the weld has equal leg length and avoid the most common error of undercutting in the vertical plate (Fig. 1.24).

Electrode diam. (mm)	Fillet leg length (mm)	Length of run (mm)
3.25 ($\frac{1}{8}$ in.)	4	300 (12 in.)
4.0	5	300
6.0 ($\frac{1}{4}$ in.)	6.0	300

Fig. 1.24. Horizontal–vertical fillet weld.

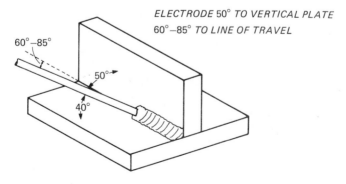

Too long a length of run gives lack of root fusion and too small leg length. Too short a length of weld causes excessive undercut and uncontrolled slag trouble. Root penetration (Figs. 1.26*d* and 1.27) is essential, and avoid the other faults shown.

A lap joint is a particular case of a fillet weld. Take care not to cause breaking away of the edges as shown (Fig. 1.25). Fig. 1.26*a, b* and *c* shows various types of fillet welds while Fig. 1.26*e* shows profiles and definitions of leg and throat.

Fig. 1.25. Lap joint.

Fig. 1.26. Horizontal–vertical fillet welds. (*a*) Double fillet, no preparation. (*b*) Prepared fillet, welded one side only. (*c*) Multi-run fillet, equal leg length. (*d*) Faults in horizontal–vertical fillet welds. (*e*) Weld profiles.

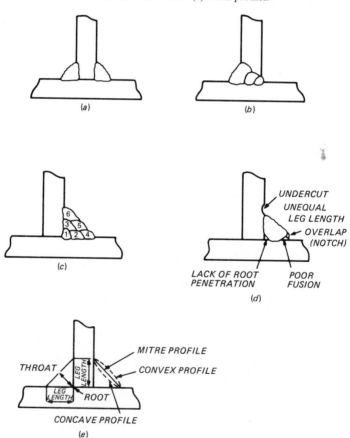

Special electrodes for fillet welding help greatly in producing welds having uniform surface, and good penetration with no undercut. Control of the slag often presents difficulties. Keeping the rod inclined at about 70° as before stated helps to prevent the slag running ahead of the molten pool. Too fast a rate of travel will result in the slag appearing on the surface in uneven thicknesses, while too slow a rate of travel will cause it to pile up and flow off the bead. Observation of the slag layer will enable the welder to tell whether his speed is correct or not.

The plain fillet or tee joint (Fig. 1.19) is suitable for all normal purposes and has considerable strength. The prepared fillet (Fig. 1.26*b*) is suitable for heavier loads, and is welded from one side only.

In thick sections a thinner electrode is used for the first bead, followed by final runs with thicker rods, each run being well deslagged before the next is laid.

Vertical welding

All welds inclined at a greater angle than 45° to the horizontal can be classed as vertical welds.

Vertical welding may be performed either upwards or downwards, and in both methods a short arc should be held to enable surface tension to pull the drop across into the molten pool. Electrodes 4 mm diameter or smaller

Fig. 1.27. (*a*) A horizontal–vertical fillet weld showing good penetration. (*b*) Showing poor penetration at corner of fillet.

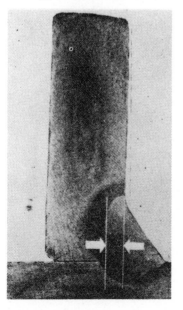 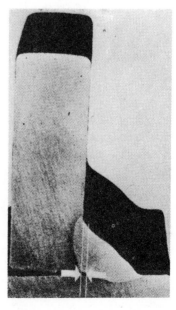

are generally used, and special rods having light coatings to reduce difficulties with slag are available. Vertical beads should first be run on mild steel plate, the electrode being held at 75–90° to the plate (Fig. 1.28).

Downward welding produces a concave bead, and is generally used for lighter runs, since a heavy deposit cannot be laid. If it is, the metal will not freeze immediately it is deposited on the plate, and will drop and run down the plate. This method is, therefore, usually only used as a finishing or washing run over an upward weld because of its neat appearance, or for thin sections.

Fig. 1.28

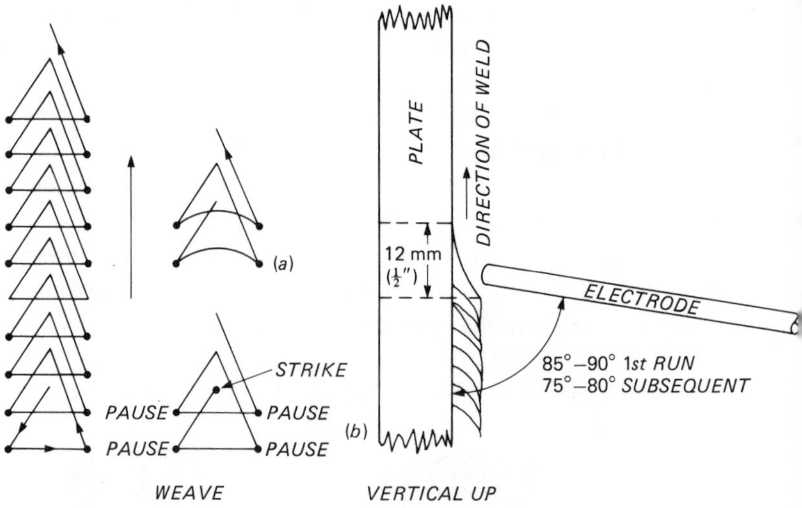

Fig. 1.29

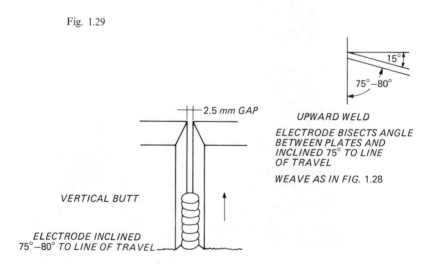

Upward welding (Fig. 1.29) produces a convex bead, and is used on sections of about 6 mm thickness. The metal just deposited is used as a step on which to continue the deposit, and the slag flows away from the pool and does not hinder penetration as it does in the downward method.

See Fig. 1.28 for angle of electrode and method of weaving. Hesitate at each side and travel from left to right by moving upwards as shown. This keeps the upper side of the pool shallow and enables the slag to run away easily with no danger of entrapping. Fig. 1.30 shows a vertical fillet weld.

Horizontal butt welding

This type of weld is very useful in fabrication, repair and reinforcement. A weaving motion is used for the first runs having prepared the plates as shown in Fig. 1.31 with a medium current and a short arc. Subsequent runs can be made with very little weave and again a short arc – avoid undercutting the upper plate. (Note that there is little if any preparation of the lower plate.)

In reinforcement of vertical surfaces, the horizontal layer may be first laid along the bottom of the surface and succeeding beads built up above this, using the lower bead in each case as a step. The next layer is then deposited with the beads at right angles to those in the first layer as in padding, this second layer consisting of normal vertical beads. An alternative method is to lay all the beads vertically, though this is scarcely so satisfactory.

Fig. 1.30. Vertical fillet weld.

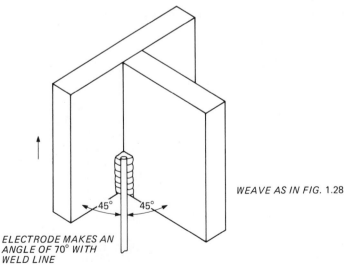

WEAVE AS IN FIG. 1.28

ELECTRODE MAKES AN ANGLE OF 70° WITH WELD LINE

45°   45°

Overhead welding

Overhead welding takes a great deal of practice before the operator can deposit an even weld with no blemishes. The weld should be made with electrodes of 3.2 or 4 mm diameter specially designed for all positional welding. The slag from these rods is fluid and does not give much trouble when welding. Because the electrodes are relatively small in diameter, overhead welding is a relatively slow process and should only be used when it is not possible to perform the welding in any other position such as downhand. An example is the overhead position when welding on the underside of a pipe joint in pipeline welding.

The operator must wear suitable protective clothes, tight wrist protection, gauntlet gloves, protective apron over shoulders and neck, no turn-ups on trouser bottoms and protective leather apron. The welding mask must completely cover face, head and neck. The welding position must be comfortable with the weight of the welding cable taken over the shoulder and the position should be such that molten slag (or even metal) does not fall on the operator.

Overhead butt weld (Fig. 1.32a). Using about 110 A for a 3.2 mm diameter rod make the first or root run with no weave and the rod at about 80–85° to the weld line. The arc must be as short as possible to enable surface tension and the magnetic forces to pull the metal up into the pool against the pull of gravity. After cleaning the first run, the subsequent runs are made with a little weave using a 4 mm diameter electrode and a current of about 160 A.

Overhead fillet welds are made with the same electrode and current settings as the overhead butt joint. Keep the arc very short and use no weave. If,

Fig. 1.31. Horizontal butt welds.

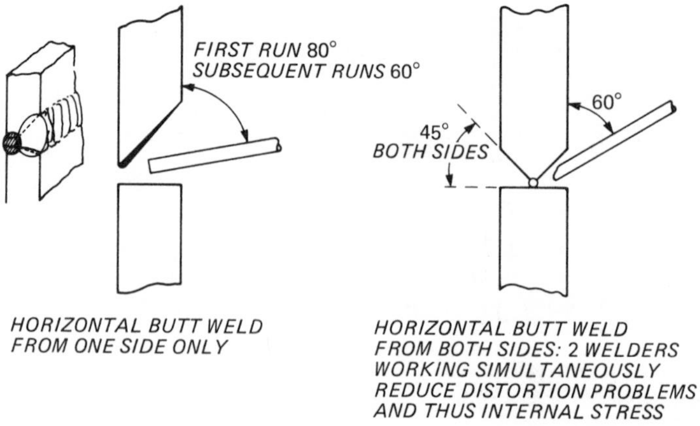

FIRST RUN 80°
SUBSEQUENT RUNS 60°

60°

45°
BOTH SIDES

HORIZONTAL BUTT WELD
FROM ONE SIDE ONLY

HORIZONTAL BUTT WELD
FROM BOTH SIDES: 2 WELDERS
WORKING SIMULTANEOUSLY
REDUCE DISTORTION PROBLEMS
AND THUS INTERNAL STRESS

because of the short arc, the electrode persists in sticking, raise the current a little, and avoid undercutting of the upper plate (Fig. 1.32b).

Aluminium and aluminium alloys

Aluminium can be arc welded using electrodes coated with fluxes consisting of mixtures of fluorides and chlorides. The flux dissolves the layer of aluminium oxide (alumina) on the surface of the metal, and also prevents oxidation during welding. The melting point is about 660 °C but aluminium has a high thermal conductivity and as a result, since there is little if any change of colour before melting point is reached, as with steel, the rapid melting which occurs presents the operator with some difficulty when first beginning welding aluminium. Except for the thinnest sheets, pre-heating to the range 100–300 °C should be carried out and great pains taken that the area of welding is free from all oxides, paints, and other contamination and the joint should be cleaned by wire brushing or machining immediately before beginning to weld.

Fig. 1.32. (a) Overhead butt weld with backing bar.

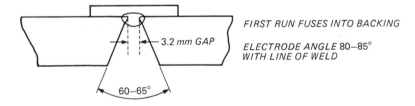

FIRST RUN FUSES INTO BACKING

ELECTRODE ANGLE 80–85°
WITH LINE OF WELD

Fig. 1.32. (b) Overhead fillet weld.

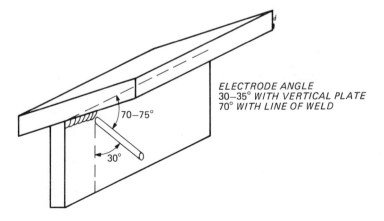

ELECTRODE ANGLE
30–35° WITH VERTICAL PLATE
70° WITH LINE OF WELD

Preparation. Fig. 1.33 shows the preparation for various sheet thicknesses. Tack welds, jigs, or other clamps should be used to position the work and tack welding should be done at very frequent intervals, then deslagged and smoothed before welding commences. The thermal expansion is about twice that of mild steel so allowance must be made accordingly. Molten aluminium absorbs hydrogen and this results in porosity so that pre-heating the work and making sure that the electrodes are kept dry and heating to 130–160°C prevents this.

The table shows the more generally welded alloys. The electrodes used are of the 10–12% silicon type, so that if heat treatable alloys are welded the heat treatment is lost and the alloy needs heat treatment again. The Al–Si electrode does not have the same mechanical properties as the parent plate in every case, and it should be remembered that the higher Al–Mg alloys are not satisfactorily arc welded. Aluminium and the work hardening Al–Mg alloys are now fabricated by the TIG and MIG processes (q.v.).

Electrodes	Material examples
Aluminium–10% Si	Pure aluminium, 5251 (N4), 6063 (H9), 6061 (H20), 6082 (H30),
	Castings LM18, LM6, LM8, LM9, LM20.

Technique. The rod is connected to the positive pole of a d.c. supply, and the arc struck by scratching action, as explained for mild steel. It will be found that, as a layer of flux generally forms over the end of the rod, it has to be struck very hard to start the arc. The rod is held at right angles to the work and a *short* arc must be held (Fig. 1.34), keeping the end pushed down into the molten pool. This short arc, together with the shielding action of the coating of the rod, reduces oxidation to a minimum. A long arc will result

Fig. 1.33. Preparation of aluminium.

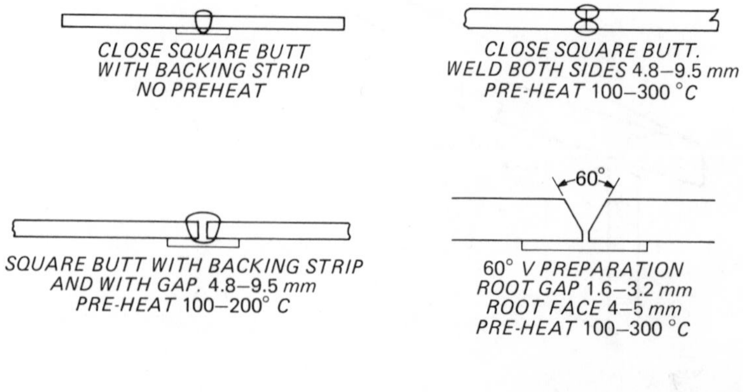

CLOSE SQUARE BUTT
WITH BACKING STRIP
NO PREHEAT

CLOSE SQUARE BUTT.
WELD BOTH SIDES 4.8–9.5 mm
PRE-HEAT 100–300 °C

SQUARE BUTT WITH BACKING STRIP
AND WITH GAP. 4.8–9.5 mm
PRE-HEAT 100–200° C

60° V PREPARATION
ROOT GAP 1.6–3.2 mm
ROOT FACE 4–5 mm
PRE-HEAT 100–300 °C

in a weak, brittle weld. No weaving need be performed, and the rate of welding must be uniform, As the metal warms up, the speed of welding must be increased.

Castings are welded in the same way after preparation, but owing to their larger mass, care must be taken to get good fusion right down into the parent metal, since if the arc is held for too short a time on a given portion of the weld the deposited aluminium is merely 'stuck' on the surface as a bead with no fusion. This is a very common fault. Castings should be pre-heated to 200°C to reduce the cracking tendency and to make welding easier.

Lap joints and fillet joints should be avoided since they tend to trap corrosive flux, where it cannot be removed by cleaning. Fillet welds are performed with no weave and with the rod bisecting the angle between the plates.

After treatment. The flux is very corrosive and the weld must be thoroughly washed and brushed in hot water after it has cooled out. Immersion in a 5% solution of nitric acid in water is an even better method of removing the flux, this being followed by brushing and washing in hot water.

Cast iron
The following types of electrode are in general use for the welding of cast iron.
(1) Mild steel, low carbon, basic coated (low hydrogen), electrode d.c. +ve, a.c. 40 OCV.
(2) Nickel cored, electrode d.c. +ve, a.c. 40 OCV.
(3) Monel (nickel–copper) cored electrode, d.c. +ve, a.c. 40 OCV.
(4) Nickel iron cored electrode, d.c. −ve, a.c. 40 OCV.

(1) When steel-based weld metal is deposited on cold cast iron, quick cooling results, due to the large mass of cold metal near the weld. This quick

Fig. 1.34. Aluminium welding.

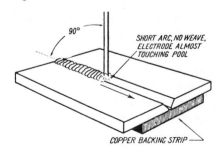

90°

SHORT ARC, NO WEAVE, ELECTRODE ALMOST TOUCHING POOL

COPPER BACKING STRIP

cooling results in much of the carbon in an area adjacent to the weld being retained in the combined form (cementite) and thus a hardened zone exists near the weld. In addition, the steel weld metal absorbs carbon and the quick cooling causes this to harden also. As a result welds made with this type of rod have hard zones and cannot always be machined.

If, however, there can be pre-heat to about 500–600 °C followed by slow cooling the deposit may produce machinable welds. In many cases, however, machining is not necessary, thus this drawback is no disadvantage. The weld is strong and electrodes of about 3.2 mm diameter are generally used with a low current which introduces a minimum of heat into the work.

(2) The nickel cored electrode may be used on cast iron without any pre-heating and it gives a deposit that is easily machinable with easy deslagging. It can be used in all positions and is used for buttering layers.

(3) The monel cored electrode is easy to deposit on cold cast iron and is again easily machinable. Nickel and monel electrodes have reduced carbon pick-up and thus a reduced hardening effect, but pre-heating should be done with castings of complicated shape, followed by slow cooling though, with care, even a complicated casting may be welded satisfactorily without pre-heat if the welding is performed slowly. The lower the heat input the less the hardening effect in the HAZ.

(4) Electrodes which deposit nickel–iron alloy are generally used where high strength is required, as for example with the SG irons.

In the repair of cast iron using high nickel electrodes the weld metal is strong and ductile. The electrode coating has a high carbon content and gives up to $1\frac{1}{2}\%$ carbon, as graphite, in the weld. The carbon has a low solubility in the nickel and excess carbon appears in the weld, increasing weld volume and reducing shrinkage. Pre-heating may be done whenever the casting is of complicated shape and liable to fracture easily, though, with care, even a complicated casting may be welded satisfactorily without pre-heating if the welding is done slowly.

Nickel alloy electrodes are also used for the welding of SG cast irons, but the heat input will affect the pearlitic and ferritic structures in the heat-affected zone, precipitating eutectic carbide and martensite in a narrow zone at the weld interface even with slow cooling. For increased strength, annealing or normalizing should be carried out after welding. The lower the heat input the less the hardening effect in the HAZ.

Preparation. Cracks in thin castings should be V'd or, better still, U'd, as for example with a bull-nosed chisel. Thicker castings should be prepared with a single V below 9.5 mm thick and a double U above this. Studding (q.v.)

can be recommended for thicker sections and the surrounding metal should be thoroughly cleaned. The polarity of the electrode depends on the electrode being used and the maker's instructions should be followed, though with d.c. it is generally +ve. With a.c. an OCV of 80 V is required.

Since the heat in the work must be kept to a minimum, a small-gauge electrode, with the lowest current setting that will give sufficient penetration, should be used. A 3.2 mm rod with 70 to 90 A is very suitable for many classes of work. Thick rods with correspondingly heavier currents may be used, but are only advisable in cases where there is no danger of cracking. Full considerations of the effect of expansion and contraction must be given to each particular job.

Technique. The rod is held as for mild steel, and a slight weave can be used as required. Short beads of about 50 to 60 mm should be run. If longer beads are deposited, cracking will occur unless the casting is of the simplest shape. In the case of a long weld the welding can be done by the skip method, since this will reduce the period of waiting for the section welded to cool. It may be found that with steel base rods, welding fairly thin sections, fine cracks often appear down the centre of the weld on cooling. This can often be prevented, and the weld greatly improved, by peening the weld immediately after depositing a run with quick light blows with a ball-paned hammer. If cracks do appear a further light 'stitching' run will seal them. Remember that the cooler the casting is kept, the less will be the risk of cracking, and the better the result. Therefore take time and let each bead cool before laying another. The weld should be cool enough for the hand to be held on it before proceeding with the next bead. In welding a deep V, lay a deposit on the sides of the V first and follow up by filling in the centre of the V. This reduces risk of cracking. If the weld has been prepared by studding (q.v.), take care that the studs are fused well into the parent metal.

In depositing non-ferrous rods, the welding is performed in the same way, holding a short arc and welding *slowly*. Too fast a welding speed results in porosity. In many cases a nickel–copper rod may be deposited first on the cast iron and then a steel base rod used to complete the weld. The nickel–copper rod deposit prevents the absorption of carbon into the weld metal and makes the resulting weld softer. Where a soft deposit is required on the surface of the weld for machining purposes, the weld may be made in the ordinary way with a steel base rod and the final top runs with a non-ferrous rod. The steel base rod often gives a weld which has hard spots in it that can only be ground down, hence this weld can never be completely guaranteed machinable.

Studding

We have seen that whenever either steel base or non-ferrous rods are used for cast iron welding, there is a brittle zone near the line of fusion, and since contraction stresses are set up, a weakness exists along this line. This weakened area can be greatly strengthened in thick section castings by studding. Welds made by this method have proved to be exceptionally strong and durable. Studding consists of preparing the casting for welding with the usual single or double V, and then drilling and tapping holes along the V and screwing steel studs into the holes to a depth slightly deeper than the diameter of the studs. Studs of 4.8 mm diameter and larger are generally used, depending on the thickness of the casting, and they must project about 6 mm above the surface. The number of studs can be such that their area is about $\frac{1}{5}$ to $\frac{1}{4}$ of the area of the weld, though a lesser number can be used in many cases (Fig. 1.35). Welding is performed around the area near each stud, using steel or steel base electrodes, so as to ensure good fusion between the stud and the parent casting. These areas are then welded together with intermittent beads, as before explained, always doing a little at a time and keeping the casting as cool as possible. This method should always be adopted for the repair, by arc welding, of large castings subjected

Fig. 1.35. Methods of studding.

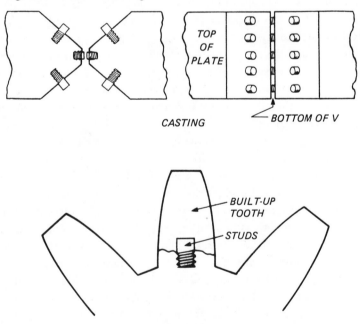

GEAR TOOTH

to severe stress. An alternative method is to weld steel bars across the projecting studs as additional reinforcement.

Copper, bronze and brass welding

If the weld must have the same characteristics as the parent metal, as for example for electrical conductivity, great difficulty is encountered with the welding of copper by the MMA process because the weld is very porous and unsatisfactory and the TIG and MIG processes should be chosen. Similarly for brass but with the use of an alloy electrode of bronze (about 80% copper, 20% tin or similar) welding can be satisfactorily carried out on most of the copper alloys with certain exceptions. Welding is performed with d.c. and the electrode positive.

Copper, copper–tin (bronze) and copper–zinc (brass) have a much greater coefficient of thermal conductivity than mild steel and the thermal expansion is also greater so that more heat is required in the welding process, so that it is almost imperative to pre-heat except with the smallest sections, and since tack welds lose their strength at elevated temperatures, line up of the parts to be welded should be with clamps, jigs or fixtures, and due allowance made for expansion and contraction during and after welding. In general, great care must be taken to avoid porosity due to the presence of gas which is expelled on solidification. Stress relief (post weld) can be carried out on the work hardening alloys to about 250–300 °C.

Copper. Tough pitch copper should not be welded as the weld is always porous. Phosphorus deoxidized copper is satisfactorily welded but the weld, made with a rod of copper–tin (bronze) will not have the same properties as the parent plate. Pre-heating is up to 250 °C.

Bronze. The tin bronzes up to about 9% Sn are weldable, but those with higher tin content are not welded due to solidification cracking and any alloy with more than 0.5% Pb is usually not welded.

Brasses. These are alloyed mostly with zinc and those with single-phase composition, as for example cartridge brass (70% Cu, 30% Zn) and those with up to about 37% Zn, are weldable, but yellow or Muntz metal (60% Cu, 40% Zn) and manganese bronze (58% Cu, 38% Zn, rem, Mn, Fe, Ni or Sn) really need pre-heat to about 450 °C.

Preparation. Plates below 3 mm thickness need no preparation, but above this they are prepared as for mild steel with a 60° V and the surfaces thoroughly cleaned. The work must be supported during welding. Sections above about 4 mm must be pre-heated to 250°C because of the high thermal conductivity of the metal.

Technique. The rod should be connected to the +ve pole of a d.c. supply except where otherwise stated. The current value is generally the same as or slightly less than that for the same gauge mild-steel electrode, with an arc voltage drop of 20–25 V. The actual values will depend on the particular job. The electrode should be held steeply to the line of weld, and a short arc held keeping the electrode well down into the molten pool. The weld must be made *slowly*, since a quick rate of welding will produce a porous deposit even when bronze rods are used. As welding proceeds and the part heats up, it is usually advisable to reduce the current, or there is a liability to melt through the weld, especially in thinner sections. Light hammering, while still hot, greatly improves the structure and strength of the weld. The extent to which this should be done depends on the thickness of the metal.

Hard surfacing

Surface resistance to abrasion, impact, corrosion and heat. The advantages of hard surfacing are that the surfaces can be deposited on relatively much cheaper base metal to give the wear-resistant or other qualities exactly where required, with a great saving in cost. In addition built-up parts save time and replacement costs. The chief causes of wear in machine parts are abrasion, impact, corrosion and heat. In order to resist impact the surface must be sufficiently hard to resist deformation yet not hard enough to allow cracks to develop. On the other hand, to resist abrasion a surface must be very hard and if subject to severe impact conditions cracking may occur, so that in general the higher the abrasion resistance the less the impact resistance, and evidently it is not possible to obtain a surface which has the highest values of both impact and abrasion resistance. In choosing an electrode for building up surfaces consideration must be given as to whether high abrasion or high impact resistance is required. High-impact electrodes will give moderate abrasion resistance and vice versa, so that the final choice must be made as to the degree of (1) hardness, (2) toughness, (3) corrosion resistance, (4) temperature of working, (5) type of base metal and whether pre- and post-heating is possible. The main types of wear- and abrasion-resistant surfacing electrodes are:

(1) Fused granules of tungsten carbide in an austenitic matrix. The deposit has highest resistance to wear but is brittle and has medium impact strength. The electrodes are tubular and therefore moisture-resistant with high deposition rates.

(2) Chromium carbide. The basic-coated electrodes deposit a dense network of chromium carbide in an austenitic matrix and have high resistance to wear with good impact resistance.

(3) Cobalt–chromium–tungsten non-ferrous alloys. These have a high carbon content, are corrosion-resistant, have a low coefficient of friction and retain their hardness at red heat.

(4) Nickel-base alloys containing chromium, molybdenum, iron and tungsten. These have good abrasion resistance and metal-to-metal impact resistance. They work-harden and have resistance to high-temperature wear and corrosive conditions.

(5) Air-hardening martensitic steels. These have a high hardness value due to their martensitic structure and there is a variety of electrodes available in this group. Dilution plays an important part in the hardness of the deposit, as with all surfacing applications. A single run on mild steel may be only a little harder than the parent plate, but if deposited on carbon or alloy steel, carbon and alloy pick-up greatly increases the hardness of the deposit.

(6) Austenitic steels: (1) 12/14% manganese deposits develop their hardness by cold work-hardening so that the deposit has the strength of its austenitic core with the hard surface. With approximately 3% Ni there is reduced tendency to cracking and brittleness due to heat compared with plain 12/14% Mn weld metal. (2) Chromium–nickel, chromium molybdenum nickel and chromium–manganese deposits work-harden and give resistance to heavy impact.

Thus, in general, low-alloy deposits give medium abrasion with high impact resistance, medium-alloy deposits give high abrasion and medium impact properties.

Austenitic, including 13% manganese, deposits give high impact resistance and work-harden as a result of this impact and work to give abrasion-resistant qualities. Of the carbides, chromium deposits give the best abrasion and impact properties while tungsten has the hardest surface and thus the highest resistance to abrasion with medium impact properties.

The table overleaf gives a selection of electrodes available.

Preparation and technique

The surface should be ground all over and loose or frittered metal removed. Because of the danger of cracking, large areas should be divided up and the welding done in skipped sections so that the heat is distributed as evenly as possible over the whole area. Sharp corners should be rounded and thick deposits should be avoided as they tend to splinter or spall. The

Type (R – rutile) (B – basic) (T – tubular)		Hardening	HV	Use	Abrasion (H – high) (M – medium)	Impact
pearlitic P	Cr–Mn	air	250	Used also for butter layers.	M	M
martensitic R	Cr–Mn	air	350	For and with buffer layers.	M	H
martensitic R	Cr–Mn–Mo	air	650	After buffer layers.	H	M
martensitic B	Cr–Mo–V	air	700	After buffer layers for heavy reinforcement.	H	M
martensitic B	Cr–Mo with borides	air	800	Surfacing generally restricted to two layers.	H	M
austenitic B	Cr–Mo–Ni	work	250 500	For joining and depositing on 13% Mn steel and joining this steel to carbon steel.	M	M
austenitic B	Cr–Ni–W–B	work (slight)	450 500	Hard at elevated temperatures, corrosion-resistant.	M	M
austenitic B	13% Mn	work	170 500		H	M
austenitic T	Cr–Mn	work	300 480	For reinforcing 13% Mn steel castings and as buffer layer for harder surfaces.	M	H
non-ferrous B	Co–Cr–W		630	Red hard and corrosion-resistant, various grades.	H	M
austenitic T matrix	Chromium–carbide–Mn		560 matrix 1400 carbides	Heat resistant to about 1100 °C.	H	H
non-ferrous T matrix	tungsten–carbide		600 matrix 1800 carbides	Used at all temperatures.	H	M

first runs on any surface are subject to considerable dilution, so it is advisable to lay down a buffer layer using a nickel-based or austenitic stainless steel electrode, especially when welding on high carbon or high alloy steels. The buffer layer should be chosen so as to be of intermediate hardness between the parent metal and the deposit, or two layers should be used with increasing hardness if the parent metal is very soft compared with the deposit.

Mild pre-heating to 150–200°C is advantageous if the base metal has sufficient carbon or alloying elements to make it hardenable but is not usual on steel below 0.3% C or stainless steel. If the base metal is very hard and brittle, slow pre-heating to 400–600°C with slow cooling after welding may be necessary to prevent the formation of brittle areas in the HAZ. Below a hardness value of 350 HV (hardness Vickers), surfaces are machinable generally with carbide tools but above this grinding is usually necessary. In depositing a surface on manganese steel there should be no pre-heating or stress relieving and the electrode should be connected to the +ve pole. Use only sufficient current to ensure fusion, keep a short arc, hold the electrode as in welding mild steel and introduce as little heat as possible into the casting by staggering welding so that the temperature does not rise much above 200°C. Austenitic stainless steel electrodes should be used for joining broken sections as these electrodes work-harden in the same way as the parent metal and 13% manganese, or 13% manganese, 3.5% nickel electrodes used for building up.

Dilution is about 25–35% in the first layer, and to obtain a dilution of about 5% at least three layers must be laid down with a thickness of 6–10 mm.

Note. The manual metal arc method of surfacing is rather slow compared with semi-automatic and automatic processes such as MIG and submerged arc so that it is best suited for smaller areas and complex shapes.

Corrosion-resistant surfacing

Nickel and its alloys can be laid as surfaces on low carbon and other other steel to give corrosion- amd heat-resistant surfaces. Electrode diameter should be as large as possible, with minimum current compatible with good fusion to reduce dilution effects. The first bead is laid down at slow speed and subsequent runs overlapped with minimum weaving, reducing dilution to 15–20% compared with 25–35% for the first bead. Subsequent layers should be put down with interpass temperature below 180°C to minimize dilution and avoid micro-fissures in the deposit. Dilution is reduced to about 5% after three layers with 6–10 mm thickness. Suitable electrodes are in the nickel, Monel, Inconel, Incoloy and Incoweld range, for which the student should refer to manufacturers' instructions.

Tipping tool steel

Cutting tools for lathes, milling machines and high-speed cutting tools of all types can be made by depositing a layer of high-speed steel on to a shank of lower carbon steel. Special electrodes are made for this purpose, and give very good results.

The surface to be tipped is ground so as to receive the deposit. The electrode is connected to the + ve pole when d.c. is used, and held vertically to the line of weld, a narrow deposit being laid as a general rule. More than one layer is generally advisable, since the first layer tends to become alloyed with the parent metal. For this reason the current setting should be as low as possible, giving small penetration. Use a very narrow weave so as to prevent porosity, which is very usual unless great care is taken. Each bead should be allowed to cool out and then be deslagged before depositing the next. The deposited metal when allowed to cool out slowly usually has a Brinell hardness of 500–700, depending upon the rod used. The hardness can be increased by heat treatment in the same way as for high-speed tool steel, and the deposit retains its hardness at fairly high temperatures.

These types of rods are very suitable for depositing cutting edges on drills, chisels, shearing blades, dies and tappets, etc. Fig. 1.36 shows the method of depositing the surface on a lathe tool.

Stellite (Cobalt base and nickel base alloys for hard surfacing)

Stellite is an alloy of cobalt, chromium, tungsten and carbon, and when deposited on steel, steel alloy or cast iron it gives a surface having excellent wear-resisting qualities and one that will stand up well to corrosive action. It preserves its hardness of surface even when red-hot (650 °C), and is thus suitable for use in places where heat is likely to be generated.

It may be deposited very satisfactorily with the arc, though the deposit is not as smooth as one deposited with the oxy-acetylene flame. The arc method of application, however, is specially recommended where it is essential not to introduce undue heat into the part, due to danger of warping or cracking. In many cases, especially where the part has large mass, stelliting by the arc saves time and money compared with the flame.

Fig. 1.36

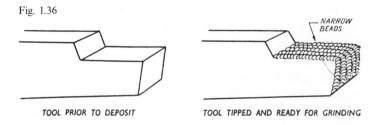

TOOL PRIOR TO DEPOSIT   TOOL TIPPED AND READY FOR GRINDING

NARROW BEADS

Bare rods of stellite may be used when welding with d.c. and are connected to the + ve pole. If a.c. is employed, covered rods must be used, and better results are also obtained with covered rods when using d.c., since the deposit is closer grained and the arc more stable.

Covered rods can be obtained or bare rods may be fluxed with a covering of equal parts of calcium carbonate (chalk), silica flour and either borax or sodium carbonate (baking soda), mixed with shellac as a binder. Coated electrodes give a gas shield which protects the molten metal from oxidation in the welding process and provides a protective slag over the metal when cooling. The deposit resists heat, abrasion, impact, corrosion, galling, oxidation, thermal shock and cavitation erosion to varying degrees depending upon the alloy. Types of Stellite alloys available are:

No. 1. 2.5% C, resistant to abrasion and solid particle erosion, slightly reduces toughness. Used for screw components and pump sleeves. 46 C (Rockwell).

No. 6. Most generally useful cobalt alloy, 1% C, 28% Cr, 4% W, rem. Co. Excellent resistance to mechanical and chemical action over a wide temperature range. Widely used as valve seat material as it has high temperature hardness and high resistance to cavity erosion.

No. 12. Very similar to alloy No. 6 but with higher hardness. Used for cutting edge in carpet, plastics, paper and chemical industries. 40 C.

No. 21. Low carbon, Co–Cr alloy with molybdenum. Has high temperature strength, resistant to cavitation erosion, corrosion and galling. Used for hot die material, fluid valve seat facing.

No. 238. Co–Fe–Cr–Mo alloy, low carbon. Has excellent resistance to mechanical and thermal shock with good hot hardness.

No. 2006. Low cobalt alternative to No. 6 to which it is similar in galling and cavitation erosion properties, but has better abrasive wear and metal-to-metal sliding conditions. Used for hot working dies and shears.

No. 2012. Low cobalt alternative to No. 12. Has improved resistance to abrasion but is not as tough. Can be used for cutting edges.

It should be noted that in some cases the nickel-based alloys such as Mastelloy C (C 0.1%, Cr 17%, Mo 17%, Fe. 6%, Rem. Ni) available in covered electrode form may be superior to the cobalt-base alloys.

See also the section on hard surfacing in Chapter 6.

Preparation. The surface to be stellited must be thoroughly cleaned of all rust and scale and all sharp corners removed. A portable grinder is extremely useful for this purpose. In some cases, where the shape is complicated, pre-heating to prevent cracking is definitely an advantage.

Technique. Approximate currents in amperes are within the following range:

Electrode diameter (mm)	d.c. electrode +ve	a.c.
3.2	85–100	90–120
4.0	120–150	135–160
4.8	150–175	160–180
6.4	200–250	220–270

A slightly longer arc than usual is held, with the rod nearly perpendicular to the surface, as this helps to spread the stellite more evenly. Care must be taken not to get the penetration too deep, otherwise the stellite will become alloyed with the base metal and a poor deposit will result. Since stellite has no ductility, cooling must be at an even rate throughout to avoid danger of cracking. The surface is finally ground to shape. Lathe centres, valve seats, rock drills and tool tips, cams, bucket lips, dies, punches, shear knives, valve tappet surfaces, thrust washers, stillson teeth, etc., are a few examples of the many applications of hard surfacing by this method.

Stainless steels

Note on BS 2926. This standard includes the chromium–nickel austenitic steels and chromium steels and uses a code by which weld metal content and coating can be identified.

The first figure is the % chromium content, the second figure the % nickel content and the third figure the % molybdenum content. The letter L indicates the low carbon version, Nb indicates stabilization with niobium, W indicates that there is tungsten present. R indicates a rutile coating, usually either d.c. or a.c., and B a basic coating, usually d.c. electrode +ve only. A suffix MP indicates a mild steel core.

For example: 19.12.3.Nb.R is a niobium-stabilized 19% Cr, 12% Ni, 3% Mo, rutile-coated electrode.

Stainless steel welding

Martensitic (chromium) stainless steels. These steels contain 12–16% chromium and harden when welded. The carbon content is usually up to 0.3% but may be more. They are used for cutlery and sharp-edged tools and for circumstances in which anti-corrosion properties are important. They harden when welded so that they should be pre-heated to 200–300°C and then allowed to cool slowly. After welding they should then be post-heated to about 700°C to remove brittleness. By using an austenitic electrode of the 25.12.Cr.Ni type with 3% tungsten (see table) the weld is more ductile and freer from cracking. If, however, the weld must match the parent plate as nearly as possible, electrodes of the chromium type must be chosen but are not very satisfactory. Use the austenitic rod where possible.

Ferritic stainless steel. Ferritic steels contain 16–30% Cr, do not harden when welded but suffer from grain growth when heated in the 950–1100°C range, so that they are brittle at ordinary temperatures but may be tougher at red heat. Pre-heating to 200°C should be carried out and the weld completed, followed by post-heating to 750°C. For mildly corrosive conditions, electrodes of matching or higher chromium content should be used, e.g. 25.12.3.W, which give a tough deposit. If joint metal properties need not match the parent plate, electrodes of 26.20 can be used. It should be noted that the weld deposit of these electrodes contain nickel which is attacked by sulphurous atmosphere, so they are only suitable for mildly corrosive conditions.

Austenitic stainless steels. These largely non-magnetic steels form the largest and most important group from the fabrication point of view. They do not harden when welded because of their austenitic structure, the largest group being the 18% Cr 8% Ni type with other groups such as 25% Cr 20% Ni and 18% Cr 12% Ni. Other elements such as molybdenum are added to make them more acid-resistant, and they are available with rutile coatings suitable for d.c. electrode +ve or a.c. minimum OCV 55–80 V depending upon the electrode; or with basic coatings usually for d.c. only, electrode +ve, these being especially suitable for vertical and overhead positions. The table gives a selection of electrodes available.

Technique. These steels have a greater coefficient of expansion than mild steel so that the tendency to distort is greater. Tack welds should be placed at about twice the frequency as for mild steel but clamps and jigs are very suitable. The technique is similar to mild steel but the arc should be struck on a striking plate and avoid any stray arcing. Welding should be done with a short arc to avoid loss of alloying metals in the transference over the arc gap. Keep the electrode at about 80° to the line of travel using as low a current as possible to ensure penetration, since the lower conductivity of stainless steel reduces the width of the HAZ and excess current will cause burn through. After welding, clean with a brush having stainless steel wires – grinding is also satisfactory. Weaving should be about twice rod width and watch for any crack down the centre line of the weld indicating hot cracking due to restraint (see table on p. 66).

Examples of electrodes for stainless and heat resistant steels

Electrode a.c. d.c. OCV	British Standard 2926 Cr Ni Mo				A.W.S. equivalent	Coating	UTS (N/mm²)	Applications
80 +	19	9	–	Nb	E347–16	R	650	For the welding of plain or stabilized stainless steels.
60 +	19	12	3	L	E316L–16	R	610	For welding Mo bearing stainless steels of 316 or 316L type.
60 +	19	9	–	L	E308–16	R	610	For 18/8 Cr–Ni steels with low carbon deposit (0.03).
100 +	25	20	–		E310–15	B	640	For heat- and corrosion-resistant stainless steels of similar composition. Also low alloy and mild steel to stainless steel with low restraint – non-magnetic.
85 +	19	9	–		E308–16L	R	585	For unstabilized stainless steel of this composition. Also for welding and surfacing 14% Mn steels.
65 +	19	9	–	Nb	E347–16	R	650	Austenitic electrode for Nb stabilized stainless steels of the 19/9 class. Can be used all positional.
70 +	19	12	3	Nb	E316–16	R	640	Austenitic electrode for unstabilized stainless steels and similar types.

55	+	19	12	3	Nb	E318–16	R	640	For all plain or Ti or Nb stabilized stainless steels.
—	+	19	12	3	Nb	E318–15	B		Austenitic electrode for Ti or Nb stabilized stainless steels. Suitable for vertical and overhead welding, e.g. fixed pipe work, d.c. only.
60	+	19	13	4	L	E317	R	630	For AISI 317 steels. Carbon below 0.08%.
85	+	17	8	2		E16–8–2	B	630	Ferrite controlled with 2% Mn. Suitable for all similar types especially for long service at elevated temperatures, e.g. in steam pipes and gas turbines.
—	+	23	12	–		E309–15	B	587	Deposits of 22/12 Cr–Ni for welding clad steels of Cr–Ni of similar composition and for dissimilar metals.
55	+	23	12	2		E309–Mo–16	R	569	For clad steels with a deposit of AISI 309Mo type, also for dissimilar metals and stainless steel to mild steel.
70	+	29	10	–		E312–16	R	800	For dissimilar steels, nickel steels, hardenable steels or steel of the same composition. Ferrite about 35%.

Stainless steel butt weld preparation

	Plate thickness (mm)	Preparation	Suggested electrode diameter (mm)
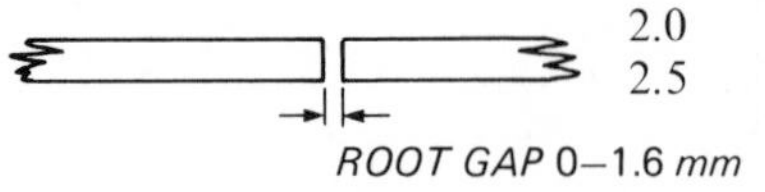 ROOT GAP 0–1.6 mm	1.2–1.6	square edge	1.6
	2.0	square edge	2.0
	2.5	square edge	2.5
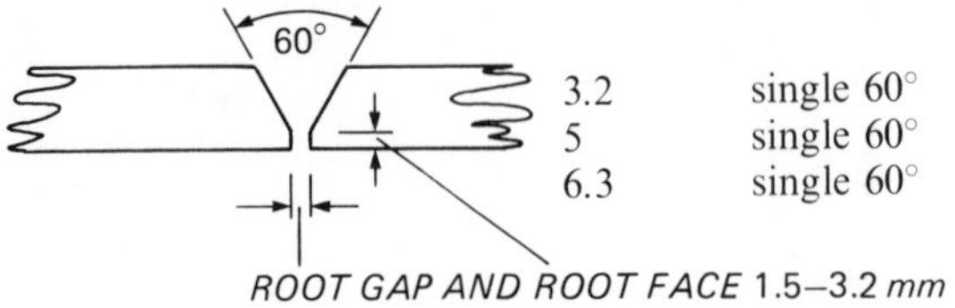 60° ROOT GAP AND ROOT FACE 1.5–3.2 mm	3.2	single 60°	2.5
	5	single 60°	3.2
	6.3	single 60°	3.2 and 4

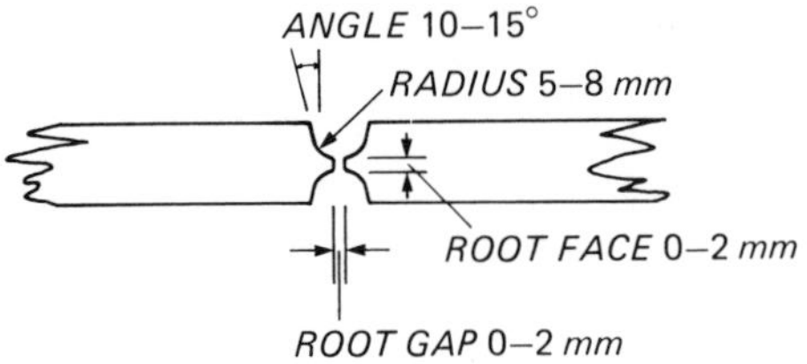

ANGLE 10–15°

RADIUS 5–8 mm

ROOT FACE 0–2 mm

ROOT GAP 0–2 mm

DOUBLE U

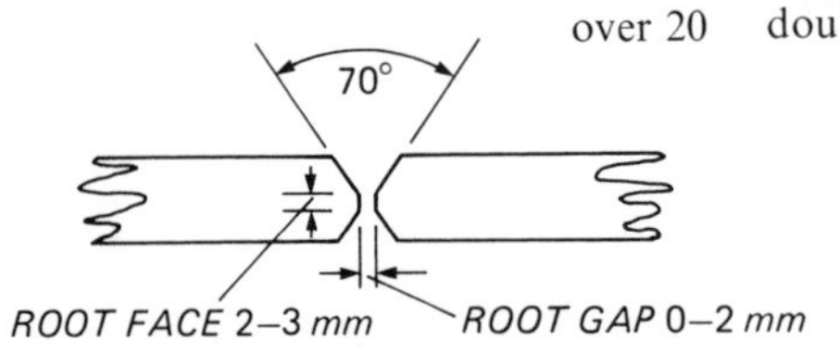

	10–20	single 60° or double 70°	4 and 5
	over 20	double 70° or double	4 and 5

70°

ROOT FACE 2–3 mm   ROOT GAP 0–2 mm

DOUBLE V

Shaeffler constitutional diagram for stainless steel weld metal (Fig. 1.37)

The various alloying elements in a stainless steel are expressed in terms of (1) nickel, which tends to form austenite, and (2) chromium, which tends to form ferrite. Thus

Nickel equivalent $= \%\ Ni + (30 \times \%\ C) + (0.5 \times \%\ Mn)$
and
Chromium equivalent $= \%\ Cr + (\%\ Mo) + (1.5 \times \%\ Si) + (0.5 \times \%$ Nb or Cb).

From the lines on the diagram we see that the low-alloy steels are hardenable because they contain the martensitic phase as welded. With increasing alloying elements, austenite and ferrite phases become more stable and the steel is no longer quench hardenable. Thus steels with high nickel, manganese and carbon become austenitic (shown in diagram) while those with high chromium and molybdenum are more ferritic (also shown). In the area *A + F* the weld will contain both austenite and ferrite (duplex phase) and it can be seen how this leads to the designation of stainless steels as austenitic, martensitic and ferritic.

The shaded areas on the Shaeffler's diagram indicate regions in which defects may under certain circumstances appear in stainless steel welds.

Example

Stainless steel BS 2926. 19.9.L.R. AWS equivalent E.308 L. Carbon 0.03%, Manganese 0.7%, Silicon 0.8%, Chromium 19.0%, Nickel 10.0%.

Nickel equivalent = % Ni + (30 × % C) + (0.5 × % Mn).
$$= 10 + (30 \times 0.03) + (0.5 \times 0.7)$$
$$= 10 + 0.9 + 0.35$$
$$= 11.25.$$

Chromium equivalent = % Cr + % Mo + (1.5 × % Si) + (0.5 × % Nb)
$$= 19 + 0 + (1.5 \times 0.8) + (0.5 \times 0)$$
$$= 19 + 0 + 1.2 + 0$$
$$= 20.2$$

Fig. 1.37. Constitution diagram for stainless steel welds (after Schaeffler) with approximate regions of defects and phase balance, depending on composition.

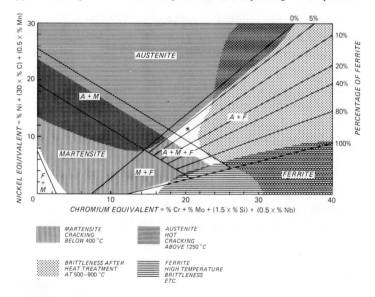

By plotting the two values on the diagram we get the point marked * in the clear part of the diagram, giving the main phases present and giving some indication of how it will behave during welding.

The welding of stainless steel to low alloy steel and mild steel

The technique for welding is similar to that for stainless steel but dilution is the greatest problem with this work. If the electrode used is a mild or low alloy steel, the weld will have a low nickel and chromium enrichment and will be subject to cracking problems. Because of this the joint between the steels should be welded with an austenitic stainless steel electrode of much higher chromium–nickel content, say 25–12 or 23–12–2 or 29–10, which will result in weld metal which will have up to 18% Cr and 8% Ni after allowing for dilution. This now has a high resistance to hot cracking but the 25–20 type electrode must not be subject to high weld restraint.

Nickel and nickel alloys

The welding of nickel and its alloys is widely practised using similar techniques to those used for ferrous metals. The electrodes should be dried before use by heating to 120°C or, if they are really damp, by heating for 1–2 hours at about 260°C. Direct current from generator or rectifier with electrode positive gives the best results. A short arc should be held with the electrode making an angle of 20–30° to the vertical and when the arc is broken it should first be reduced in length as much as possible and held almost stationary, or the arc can be moved backwards over the weld already laid and gradually lengthened to break it. This reduces the crater effect and reduces the tendency to oxidation. The arc should be restruck by striking at the end of a run and moving quickly back over the crater, afterwards moving forward with a slight weave over the crater area, thus eliminating starting porosity.

Fig. 3.28 shows the most satisfactory joint preparation and it should be remembered that the molten metal of the nickel alloys is not as fluid as that of steel so that a wider V preparation is required with a smaller root face to obtain satisfactory penetration.

Each run of a multi-run weld should be deslagged by chipping and wire-brushed. Grinding should not be undertaken as it may lead to particles of slag being driven into the weld surface with consequent loss of corrosion resistance. Stray arcing should be avoided and minimum weaving performed because it results in poorer quality of weld deposit due to increased dilution.

Although it is preferable to make welds in the flat position, vertical welds can be made either upwards or downwards holding the electrode at right angles to the line of weld, using reduced current compared with similar flat conditions.

For fillet welds the electrode angle should roughly bisect the angle between the plates and be at right angles to the line of weld. If the plates are of unequal thickness the arc should be held more on to the thicker plate to obtain better fusion, and tilted fillets give equal leg length more easily.

The table gives a list of the alloys suitable for welding by this process and the type of electrode recommended.

Steels containing 9% nickel used for low temperature ($-196°C$) are welded with basic-coated electrodes using d.c. such as 80/20 Inconel or Nyloid 2.

Alloy	Electrodes recommended	
	Alloy to itself	Alloy to steel
Nickel 200	Nickel 141	Nickel 141
Nickel 201		
Monel 400	Monel 190	Monel 190
Monel K500	Monel 134	Monel 190
Inconel 600	Inconel 132	Inconel 182
	Inconel 182	Incoweld A
Incoloy DS	Incoweld A	Incoweld A
Incoloy 800	Incoweld A	Incoweld A
Incoloy 825	Incoloy 135	Inconel 182
		Incoloy A
Nimonic 75	Inconel 132	Inconel 182
	Inconel 182	Incoweld A
Nilo alloys	Incoweld A	Incoweld A
	Nickel 141	Nickel 141

Clad steels

Stainless clad steel is a mild or low alloy steel backing faced with stainless steel such as 18% Cr 8% Ni or 18% Cr 10% Ni with or without Mo, Ti and Nb, or a martensitic 13% Cr steel, the thickness of the cladding being 10–20% of the total plate thickness.

Preparation and technique. The backing should be welded first and the mild steel root run should not come into contact with the cladding, so that preparation should be either with V preparation close butted with a deep root face, or the cladding should be cut away from the joint at the root (see

Fig. 1.38). The clad side is then back grooved and the stainless side welded with an electrode of similar composition. Generally an austenitic stainless steel electrode of the 25% Cr 20% Ni type should be used for the root run on the clad side because of dilution effects, and at least two layers or more if possible should be laid on the clad side to prevent dilution effects affecting the corrosion-resistant properties. First runs should be made with low current values to reduce dilution effects. For martensitic 13% Cr cladding, pre-heating to 240°C is advisable followed by post-heating and using an austenitic stainless steel electrode such as 22% Cr 12% Ni 3% Mo with about 15% ferrite, which gives weld metal of approximately 18% Cr 8% Ni with about 6% ferrite. Welding beads not adjacent to the backing plate can be made with 18% Cr 12% Ni 3% Mo electrodes. If the heat input is kept as low as possible welding may be carried out without heat treatment, the HAZ being tempered by the heat from successive runs.

Nickel clad steel

Mild and low alloy steel can be clad with nickel, monel or inconel for corrosion resistance at lower cost compared to solid nickel base material and with an increase in thermal conductivity and greater strength, the thickness of cladding usually being not greater than 6 mm. When welding clad steels it is essential to ensure the continuity of the cladding, and because of this butt joints are favoured. Dilution of the weld metal with iron occurs when welding the clad side, but the electrode alloy can accept this with the exception of monel 60: in this case a buffer layer of monel 61 should be laid down first. A minimum of two runs should be used on the clad side and first runs should be made with low current values to reduce the dilution.

Fig. 1.38. Alternative methods of welding clad steel.

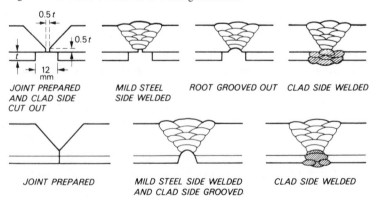

JOINT PREPARED AND CLAD SIDE CUT OUT   MILD STEEL SIDE WELDED   ROOT GROOVED OUT   CLAD SIDE WELDED

JOINT PREPARED   MILD STEEL SIDE WELDED AND CLAD SIDE GROOVED   CLAD SIDE WELDED

Recommended electrodes are: for nickel 200 use a nickel 141 electrode: for monel 400 use a monel 190 electrode and for inconel 600, an inconel 182 electrode. Preparation of joints is similar to that for stainless clad steel welding.

Welding of pipelines

The following brief account will indicate to the welder the chief methods used in the welding of pipelines for gas, oil, water, etc. The lengths of pipes to be welded are placed on rollers, so that they can easily be rotated. The lengths are then lined up and held by clamps and tack welded in four places around the circumference, as many lengths as can be handled conveniently, depending on the nature of the country, being tacked together to make a section. The tack welder is followed by the main squad of welders, and the pipes are rolled on the rollers by assistants using chain wrenches, so that the welding of the joint is entirely done in the flat position; hence the name *roll welding*.

After careful inspection, each welded section is lifted off the rollers by tractor-driven derricks and rested on timber baulks, either over or near the trench in which it is to be laid. The sections are then *bell hole* welded together. The operator welds right down the pipe, the top portion being done downhand, the sides vertical, and the underside as an overhead weld. Electrodes of 4 and 5 mm diameter are used in this type of weld.

Stove pipe welding (see also Appendix 6)
This type of welding has a different technique from conventional positional welding methods and has enabled steel pipelines to be laid across long distances at high rates. The vertically downward technique is used, welding from 12 o'clock to 6 o'clock in multiple runs.

Cellulose or cellulose–iron powder coated electrodes are used. These give a high burn-off rate, forceful arc and a light, fast-freezing slag and are very suitable for the vertical downward technique. The coating provides a gas shield which is less affected by wind than other electrodes but, generally, welding should not be carried out where the quality of the completed weld would be impaired by prevailing weather conditions, which may include rain, blowing sand and high winds. Weather protection equipment can be used wherever practicable.

Preparation is usually with 60° between weld faces, increased sometimes to 70° with a 1.5 mm root face and 1.5 mm root gap; internal alignment clamps are used. The stringer bead is forced with no weave into the root, then the hot pass with increased current fuses the sides and fills up any burn-through which may have occurred. Filler runs, stripper runs and capping run complete the welding (Fig. 1.39*a*). A diesel driven welding generator with tractor unit or mounted with the pipe laying plant is used with the electrode − ve (reverse polarity).

The welding of pipelines is usually performed by a team of welders; the larger the diameter of pipe, the greater the number of welders. In most cases the welder performs the same type of weld on each successive joint. The first team deposits the stringer bead and then moves on to the next joint. The second team carries out the hot pass, and they are followed by the third with the hot fill and the three teams follow along the pipeline from one joint to another. On a 42″ pipe there could be twelve welders spread over three adjacent joints all welding simultaneously. The first group is followed by further groups of fillers and cappers. It is important that the first three passes are deposited without allowing the previous one to cool.

Note. The wire brushes used are of strong stainless steel wire and power driven. As the wire is so strong there must be adequate protection of eyes and face by using a transparent thick face mask.

For clarity Fig. 1.39*a* and *b* shows the various passes and the clock face, while Fig. 1.40 shows pipe preparation including the use of backing rings and Fig. 1.41 shows preparation for a cut and shut bend and a gusset bend.

The CO_2 semi-automatic process is also used for pipeline welding. The supply unit is usually an engine-driven generator set and the technique used is similar to the stove pipe method, but it is important to obtain good line-up to avoid defects in the penetration and correct manipulation to avoid cold shuts. Fully automatic orbital pipe welders are also available using the TIG process and are used on pipes from 25 to 65 mm outside diameter.

For full details of the various methods of preparation and welding standards for pipes, the student should refer to BS 2633, class 1: *Steel pipework*, and BS 2971, class 2: *Steel pipelines*; BS 2910: *Radiographic examination of pipe joints*, and BS 938: *Metal arc welding of structural steel tubes.*

Fig. 1.39. (*a*)

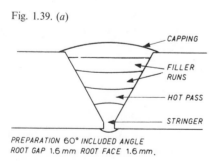

PREPARATION 60° INCLUDED ANGLE
ROOT GAP 1.6 mm ROOT FACE 1.6 mm.

For river crossings the pipe thickness is increased by 50–100% and the lengths are welded to the length required for the crossing. Each joint is further reinforced by a sleeve, and large clamps bolted to the line serve as anchor points. The line is then laid in a trench in the river bed.

For water lines from 1 to $1\frac{1}{2}$ m diameter, bell and spigot joints are often used. On larger diameter pipes the usual joint is the reinforced butt. The

Fig. 1.39. (*b*) Clock face for reference.

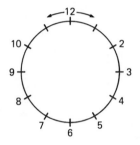

Fig. 1.39. (*c*) Electrode angles (hot fill).

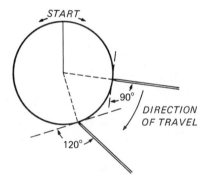

Fig. 1.40

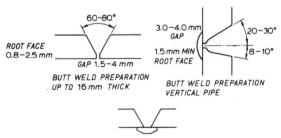

Stove pipe welding – example for pipes of 300–400 mm diameter

Pass	Electrode diam. (mm)	Current (A)	Technique
Stringer bead	3.25	100–130	Preheat to 50–100 °C. Strike electrode at 12 o'clock and keep at approximately 90° to line of travel. To control fine penetration, vary electrode angle to 6 o'clock for more to 12 o'clock for less. Use straight touch technique. If necessary someone can control current from generator according to welder's instructions. Arc should be visible on inside of pipe. Do not stop for small 'windows' or burn throughs, which will be eliminated by the hot pass. Grind bead when finished taking slightly more from sides until weld tracks are just visible.
Hot pass	4	180–220	This is deposited immediately after the stringer bead and melts out any inclusions, fills in 'windows' and *must* be deposited on a hot stringer bead which it anneals. It is a light drag weld with a backward and forward 'flicking' movement of the electrode held generally at 60°–70° to the line of travel but this can vary to 90° at 5 o'clock to 6 o'clock and 7 o'clock to 6 o'clock. Rotary wire brushing with a heavy duty power brush completes the pass.
Hot fill	4	160–190	This is deposited immediately after the hot pass, while the pipe is still warm. It cleans the weld profile and adds to the strength of the joint. The electrode is held at 60° to 70° to the line of travel at 12 o'clock, increasing to 120° at 6 o'clock (see Fig. 1.39c).
Fill and stripper passes	4	150–190	If the joint has cooled, preheat to 50 to 100 °C. All runs are single passes with no weaving and each run is well wire brushed. The electrode is held at 60°–70° to line of travel at 12 o'clock, changing gradually to 90°–100° at 6 o'clock. If the areas between 2 o'clock and 5 o'clock and between 10 o'clock and 7 o'clock are low, stripper passes are made until the weld is of a uniform level.
Capping run	4	130–190	The electrode is held at 60°–80° at 12 o'clock, to 120° to 6 o'clock. A fast side to side weave is permitted, covering not more than 2½ times the diameter of the rod. Lengthening the arc slightly between 5 o'clock and 7 o'clock avoids undercut. Manipulation between 4 o'clock and 8 o'clock can be by a slight 'flicking' technique. After completion the joint is power wire brushed and wrapped in a dry asbestos or similar cloth with waterproof backing and the joint is allowed to cool uniformly.

pipe is V'd on the inside and welded from the inside. A steel reinforcing band is then slipped over the joint and fillet welded in position.

The types of joint are illustrated in Fig. 1.42.

Fig. 1.41. Steel pipe preparation.

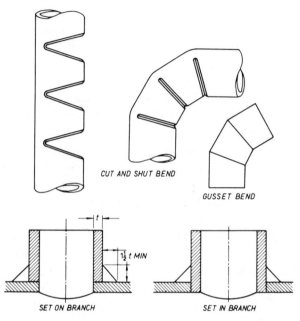

CUT AND SHUT BEND

GUSSET BEND

SET ON BRANCH

SET IN BRANCH

Fig. 1.42

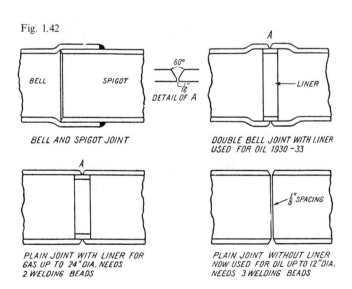

BELL AND SPIGOT JOINT

DETAIL OF A

DOUBLE BELL JOINT WITH LINER USED FOR OIL 1930-33

PLAIN JOINT WITH LINER FOR GAS UP TO 24"DIA. NEEDS 2 WELDING BEADS

PLAIN JOINT WITHOUT LINER NOW USED FOR OIL UP TO 12"DIA. NEEDS 3 WELDING BEADS

Welding of carbon and carbon–manganese and low-alloy steels*

Steels with a carbon content up to about 0.25% carbon (mild steels) are easy to weld and fabricate because, due to their low carbon content, they do not harden by heat treatment so that the weld and HAZ does not have hardened zones even though there is quick cooling. As the carbon content of a steel increases it becomes more difficult to weld, because as the weld cools quickly owing to the quenching action of the adjacent cold mass of metal hardened zones are formed in the HAZ, resulting in brittleness and possible cracking if the joint is under restraint. Pre-heat can be used to overcome this tendency to cracking.

One or more alloying elements such as Ni, Cr, Mo, Mn, Si, V and Cu are added to steel to increase tensile, impact and shear strength, resistance to corrosion and heat and in some cases to give these improved properties either at high temperature or low (cryogenic) temperatures. The alloying elements do not impair the weldability as would an increase in carbon content, but the steels are susceptible to cracking, and it is this problem which has to be considered in more detail.

Cracking may appear as: (1) delayed cold cracking caused by the presence of H_2; (2) high-temperature liquation cracking (hot cracking); (3) solidification cracking; (4) lamellar tearing; hot cracking and cold cracking have been already considered in Vol. 1. Hot cracks usually appear down the centre of a weld, which is the last part to solidify, while cold cracks occur in the HAZ (Fig. 1.43). These latter may not occur until the weld is subject to stress in service, and since they are often below the metal surface they cannot be seen, so that the first indication of their presence is failure of the joint.

The factors which lead to cold cracking are:
(1) The composition of steel being welded.
(2) The presence of hydrogen.
(3) The rate of cooling of the welded joint.
(4) The degree of restraint (stress) on the joint.

(1) Composition of the steel and carbon equivalent (CE). The tendency to crack increases as the carbon and alloying element content increases, and since there is a great variety in the types of steel to be welded it is convenient

---

* See also BS 4360, *Weldable structural steels* and BS 970 Pt 1: Carbon and carbon–manganese steels, Pt 2: Direct hardening steels including nitrogen hardening steels, Pt 3: Steels for case hardening, Pt 4: Stainless, heat resisting and valve steels.

to convert the varying amounts of alloying elements present in a given steel into terms of a simple equivalent carbon steel, thus giving an indication of the tendency to crack. This 'carbon equivalent' can be calculated from a formula such as the following:

$$CE = C\% + \frac{Mn\%}{6} + \frac{Cr\% + Mo\% + V\%}{5} + \frac{Ni\% + Cu\%}{15}$$

(which is a variation of the original formula of Dearden and O'Neill) so that to take a simple example, if steel has the following composition: C 0.13%, Mn 0.3%, Ni 0.5%, Cr 1.0%, application of the formula gives a CE of 0.41%.

(2) The presence of hydrogen in the welded zone greatly increases the tendency to crack, the amount of hydrogen present depending upon the type of electrode used and the moisture content of its coating. A rutile coating may have a high moisture content giving up to 30 ml of hydrogen in 100 g of weld metal. The hydrogen diffuses into the HAZ, and on cooling quickly a hard martensitic zone exists with a liability of cracks occurring. Even a small amount of hydrogen present can result in cracking in severely restrained joints. Basic (hydrogen controlled) electrodes correctly dried before use (p. 8) result in a very low hydrogen content. Austenitic stainless steel electrodes deposit weld metal in which the hydrogen is retained and does not diffuse into the HAZ. The weld has a relatively low yield strength and, when stressed, yields and reduces the restraint on the joint, so they are used, for example, to weld steels such as armour plate which may crack when welded with basic-coated mild steel electrodes. Gas-shielded processes using CO_2 or argon–CO_2 mixtures give welds of a very low hydrogen content.

Fig. 1.43

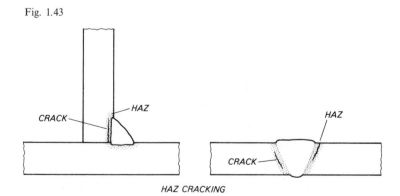

HAZ CRACKING

(3) Rate of cooling of the welded zone. The rate of cooling depends upon (1) the heat energy put into the joint and (2) the combined thickness of the metal forming the joint. Arc energy is measured in kilojoules per mm length of weld and can be found from the formula

$$\text{Arc energy (kJ/mm)} = \frac{\text{arc voltage} \times \text{welding current}}{\text{welding speed (mm/s)} \times 1000}.$$

The greater the heat input into the joint the slower the rate of cooling so that the use of a large-diameter electrode with high current reduces the quenching effect and thus the cracking tendency. Similarly, smaller-diameter electrodes with lower currents reduce the heat input and give a quicker cooling rate, increasing the tendency to crack due to the formation of hardened zones. Subsequent runs made immediately afterwards are not quenched as is the first run, but if the first or subsequent runs are allowed to cool, conditions then return to those of the first run. For this reason interpass temperature is often stipulated so as to ensure that the weld is not allowed to cool too much before the next run or pass is made. The use of large electrodes with high currents, however, does not necessarily give good impact properties at low temperatures. For cryogenic work it is essential to obtain the greatest possible refining of each layer of weld metal by using smaller-diameter electrodes with stringer or split-weave technique. As the 'combined thickness', that is the total thickness of the sections at the joint, increases, so the cooling rate increases, since there is increased section through which the heat can be conducted away from the joint. The cooling rate of a fillet joint is greater than that for a butt weld of the same section plate since the combined thickness is greater (Fig. 1.44).

(4) Restraint. When a joint is being welded the heat causes expansion, which is followed by rapid cooling. If the joint is part of a very

Fig. 1.44

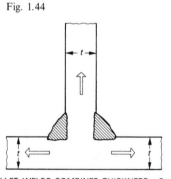
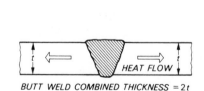
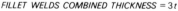

FILLET WELDS COMBINED THICKNESS = 3*t*   BUTT WELD COMBINED THICKNESS = 2*t*

rigid structure the welded zone has to accommodate the stresses due to these effects and if the weld is not ductile enough, cracking may occur. The degree of restraint is a variable factor and is important when estimating the tendency to crack.

Controlled Thermal Severity (CTS) tests in which degrees of restraint are placed upon the joint to be welded and on which pre- and post-heat can be applied are used to establish the liability to crack. (see Reeve test, Volume 1.)

Hydrogen cracking can be avoided by (1) using basic hydrogen-controlled electrodes, correctly dried and (2) pre-heating.

The temperature of pre-heat depends upon (1) the CE of the steel, (2) the process used and in the MMA process, the type of electrode (rutile or basic), (3) the type of weld, whether butt or fillet and the run out length (x mm electrode giving y mm weld), (4) the combined thickness of the joint and (5) arc energy.

Reference tables are given in BS 5135 (1974) from which the pre-heat temperature can be ascertained from the above variables. Pre-heat temperatures may vary from 0 to 150 °C for carbon and carbon–manganese steels and be up to 300 °C for higher-carbon low-alloy steels containing chromium and molybdenum. The pre-heating temperature is specified as the temperature of the plate immediately before welding begins and measured for a distance of at least 75 mm on each side of the joint, preferably on the opposite face from that which was heated. The combined thickness is the sum of the plate thickness up to a distance of 75 mm from the joint. If the thickness increases greatly near the 75 mm zone higher combined thickness values should be used and it should be noted that if the whole unit being welded (or up to twice the distance given above) can be pre-heated, pre-heat temperatures can be reduced by about 50 °C. Austenitic electrodes can generally be used without pre-heat.

Steels used for cryogenic (low temperature) applications can be carbon–manganese types which have good impact properties down to -30 °C and should be welded with electrodes containing nickel; 3% nickel steels are used for temperatures down to -100 °C and are welded with matching electrodes whilst 9% nickel steels, used down to -196 °C (liquid nitrogen), are welded with nickel–Chromium–iron electrodes since the 3% nickel electrodes are subject to solidification cracking.

Creep-resistant steels usually contain chromium and molybdenum and occasionally vanadium and are welded with basic-coated low-alloy electrodes with similar chromium and molybdenum content. Two types of cracking are encountered: (1) transverse cracks in the weld metal, (2) HAZ cracking in the parent plate. Pre-heating and interpass temperatures of 200–300 °C with post-heat stress relief to about 700 °C is usually advisable.

Lamellar tearing

In large, highly stressed structures cracks may occur in the material of the parent plate or the HAZ of a joint, the cracks usually running parallel to the plate surface (Fig. 1.45). This is known as lamellar tearing, and it is the result of very severe restraint on the joint and poor ductility, due to the presence of non-metallic inclusions running parallel to the plate surface which are difficult to detect by the usual non-destructive tests. Certain types of joint such as T, cruciform and corner are more susceptible than others. Should lamellar tearing occur the joint design should be modified and tests made on the parent plate to indicate its sensitivity to tearing, while buttering of the surface may also help.

For structural steels in general, basic-coated electrodes are used and current electrode lists of the electrode makers should be consulted for the most up-to-date information.

When welding these steels the following points should be observed:

(*a*) When tack welds are used to position the work, as is the general practice, they should be well fused into the weld because, due to the rapid cooling and consequent hardening of the area around the tack weld, cracks may develop.

(*b*) Since the chilling effect is most marked on the first run, careful watch should be kept for any cracks which may develop. Any sealing run applied to the back of the joint should preferably be made either while the joint is still hot, or with a large electrode. The least possible number of runs should be used to fill up the V to minimize distortion.

(*c*) When austenitic rods are used to obtain a weld free from cracks, all the runs should be made with this type of rod and ordinary steel rods not used for subsequent runs. Plates thicker than 15 mm are preferably pre-

Fig. 1.45. Lamellar tearing.

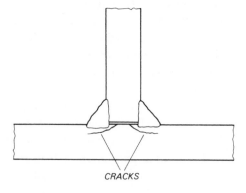

CRACKS

heated to 100–200 °C to avoid cracking. Cold work and interrupted welds should generally be pre-heated.

(*d*) The electrode or holder should not be struck momentarily or 'flashed' by design or accident on to the plate prior to welding, since the rapid cooling of the small crater produced leads to areas of intense hardness that may result in fatigue cracks developing.

(*e*) Hard spots in the parent plate may be softened by post-heat applied locally but this may reduce the endurance value of the joint.

Welding of steel castings

The use of steel castings in welded assemblies generally falls under two methods.

(1) Welding steel castings together to form a complex casting that would be difficult to cast as a whole.

(2) Castings and wrought material welded together to form a 'composite' fabrication.

The same rules regarding weldability apply as when welding wrought material. Castings may be in low-carbon-content mild steel, low-alloy steels, or high-alloy steels (such as austenitic manganese and chrome nickel stainless types) and for the alloy steels the same precautions must be taken against cracking as for the wrought form (q.v.). In general the stringer bead technique is used for thick butt welds since it reduces the cracking tendency and horizontal vertical fillets of greater than 8 mm leg length may be multi-run to avoid the tendency towards undercutting of the vertical plate.

Table giving some of the chief types of metal arc welding electrodes available for welding alloy steels

Electrode	Description and uses
Mild steel heavy duty	Rutile or basic coated for medium and heavy duty fabrications. The latter suitable for carbon and alloy steels and mild steel under restraint and for thick sections and root runs in thick plate.
High-tensile alloy steels	Austenitic rutile or basic coated for high-tensile steels including armour plate and joints between low-alloy and stainless steel and in conditions where pre-heat is not possible to avoid cracking.
Structural steels	Basic coated for high-strength structural steels 300–425 N/mm^2 tensile strength and for copper bearing weathering quality steels.
Notch ductile steels for low-temperature service	Basic coated for steels containing 2–5% Ni, 3% Ni and carbon–manganese steels. Also austenitic high nickel electrodes for 9% Ni steel for service to $-196°C$ and for dissimilar metal welding and for high Ni–Cr alloys for use at elevated temperatures.
Creep-resisting steels	Ferritic, basic coating for (1) 1.25% Cr, 0.5% Mo, (2) 2.5% Cr, 1.0% Mo, (3) 4–6% Cr, 0.5% Mo, steels with pre- and post-heat. Austenitic ferrite controlled for creep-resistant steels and for thick stainless steel sections requiring prolonged heat treatment after welding.
Heat- and corrosion-resisting steels	(1) Rutile or basic coating 19% Cr 9% Ni for extra low carbon stainless steels. (2) Basic coating Nb stabilized for plain or Ti or Nb stabilized 18/8 stainless steels and for a wide range of corrosive- and heat-resisting applications. Variations of these electrodes are for positional welding, smooth finish, and for high deposition rates. (3) Rutile or basic coating Mo bearing for 18/10 Mo steels. (4) Rutile or basic coating, low-carbon austenitic electtodes for low-carbon Mo bearing stainless steels and for welding mild to stainless steel. (5) Basic coating austenitic for heat-resisting 25% Cr, 12% Ni steels, for welding mild and low-alloy steels to stainless steel and for joints in stainless-clad mild steel. (6) Rutile coating austenitic for 23% Cr, 11% Ni heat-resistant steels containing tungsten. (7) Basic coating 25% Cr, 20% Ni (non-magnetic) for welding austenitic 25/20 steels and for mild and low-alloy steels to stainless steel under mild restraint.

Electrode	Description and uses
	(8) Basic coating 60% Ni, 15% Cr for high-nickel alloys of similar composition (incoloy DS, cronite, etc.) and for welding these alloys and stainless steel to mild steel in low restraint conditions.
	(9) A range of electrodes for high-nickel, monel, inconel and incoloy welding. These electrodes are also suitable for a wide range of dissimilar metal welding in these alloys.
Hard-surfacing and abrasion-resisting alloys	(1) High impact moderate abrasion-resistance rutile coating for rebuilding carbon steel rails, shafts, axles and machine parts subject to abrasive wear. 250 HV.
	(2) Medium impact medium abrasion-resistance rutile or basic coating for rebuilding tractor links and rollers, roller shafts, blades, punching die sets, reasonably machinable. The basic coated electrode gives maximum resistance to underbead cracking and is suitable for resurfacing low-alloy and hardenable steels. 360 HV.
	(3) High abrasion medium impact. (*a*) Rutile or basic coating for hard-surfacing bull-dozer blades, excavator teeth, bucket lips, etc. The basic coated electrode has greater resistance to underbead cracking and eliminates the need for a buffer layer on high-carbon steels. 650 HV. Unmachinable. (*b*) Tubular electrodes depositing chromium carbide. For worn carbon steel such as dredger buckets, excavator shovels etc. Matrix 560 HV. Carbide 1400 HV.
	(4) Severe abrasion moderate impact. (*a*) Tubular electrode depositing fused tungsten carbide giving highest resistance to abrasion with moderate impact resistance. Matrix 600 HV. Carbide 1800 HV. (*b*) High-alloy weld deposit for severe abrasion, suitable for use on sand and gravel excavators. Resistant to oxidation. Matrix 700 HV. Carbides 1400 HV.
12–14% Manganese steels	(1) Basic coating, 14% manganese steel (work-hardening) for 12–14% Mn steel parts – steel excavators and mining equipment. 240 HV.
	(2) Austenitic stainless steel weld deposit suitable for joining 12–14% Mn steels and for reinforcing and for butter layer for 12–14% Mn electrodes. 250 HV, work-hardened to 500 HV.
	(3) Tubular-type electrodes for hardfacing 12–14 Mn steel parts with high resistance to abrasion and heavy impact. Matrix 640 HV. Carbides 1400 HV.

2

Gas shielded metal arc welding

Metal inert gas (MIG), metal active gas (MAG) including CO_2 and mixed gas processes*

The MIG semi-automatic and automatic processes are increasing in use and are displacing some of the more traditional oxy-acetylene and MMA uses.

For repair work on thin sheet as in the motor trade, semi-automatic MIG using argon–CO_2 mixtures has displaced the traditional oxy-acetylene methods because of the reduced heat input and narrower HAZ, thus reducing distortion. For larger fabrication work, mechanical handling equipment with automatic MIG welding heads has revolutionized the fabrication industry, while the advent or robots, which are program controlled and use a fully automated MIG welding head with self-contained wire feed, make less demands on the skilled welder.

Argon could not be used alone as a shielding gas for mild, low-alloy and stainless steel because of arc instability but now sophisticated gas mixtures of argon, helium, CO_2 and oxygen have greatly increased the use of the process. Much research is proceeding regarding the welding of stainless and 9% nickel steels by this method, using magnetic arc oscillation and various gas combinations to obtain positional welds of great reliability and freedom from defects.

The process has very many applications and should be studied by the student as one of the major processes of the future.

It is convenient to consider, under this heading, those applications which involve shielding the arc with argon, helium and carbon dioxide (CO_2) and mixtures of argon with oxygen and/or CO_2 and helium, since the power source and equipment are essentially similar except for the gas supply. These processes fall within the heading MIG/MAG.

* American designation: gas metal-arc welding (GMAW).

With the tungsten inert gas shielded are welding process, inclusions of tungsten become troublesome with currents above 300 A. The MIG process does not suffer from these disadvantages and larger welding currents giving greater deposition rates can be achieved. The process is suitable for welding aluminium, magnesium alloys, plain and low-alloy steels, stainless and heat-resistant steels, copper and bronze, the variation being filler wire and type of gas shielding the arc.

The consumable electrode of bare wire is carried on a spool and is fed to a manually operated or fully automatic gun through an outer flexible cable by motor-driven rollers of an adjustable speed, and rate of burn-off of the electrode wire must be balanced by the rate of wire feed. Wire feed rate determines the current used.

In addition, a shielding gas or gas mixture is fed to the gun together with welding current supply, cooling water flow and return (if the gun is water cooled) and a control cable from gun switch to control contractors. A d.c. power supply is required with the wire electrode connected to the positive pole (Fig. 2.1).

Fig. 2.1. Components of gas shielded metal arc welding process.

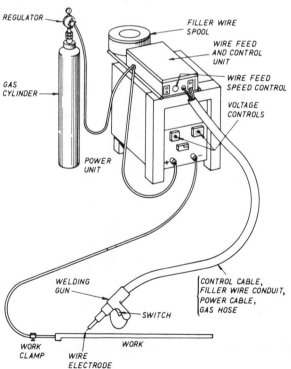

Spray transfer

In manual metal arc welding, metal is transferred in globules or droplets from electrode to work. If the current is increased to the continuously fed, gas-shielded wire, the rate at which the droplets are projected across the arc increases and they become smaller in volume, the transfer occurring in the form of a fine spray.

The type of gas being used as a shield greatly affects the values of current at which spray transfer occurs. Much greater current densities are required with CO_2 than with argon to obtain the same droplet rate. The arc is not extinguished during the operation period so that arc energy output is high, rate of deposition of metal is high, penetration is deep and there is considerable dilution. If currents become excessively high, oxide may be entrapped in the weld metal, producing oxide enfoldment or puckering (in Al). For spray transfer therefore there is a high voltage drop across the arc (30–45 V) and a high current density in the wire electrode, making the process suitable for thicker sections, mostly in the flat position.

The high currents used produce strong magnetic fields and a very directional arc. With argon shielding the forces on the droplets are well balanced during transfer so that they move smoothly from wire to work with little spatter. With CO_2 shielding the forces on the droplet are less balanced so that the arc is less smooth and spatter tendency is greater (Fig. 2.2). The power source required for this type of transfer is of the

Fig. 2.2. Types of arc transfer. (*a*) Spray transfer: arc volts 27–45 V. Shielding gases: argon, argon–1 or 2% oxygen, argon–20% CO_2, argon–2% oxygen–5% CO_2. High current and deposition rate, used for flat welding of thicker sections. (*b*) Short-circuit or dip transfer: arc volts 15–22 V. Shielding gases as for spray transfer. Lower heat output and lower deposition rate than spray transfer. Minimizes distortion, low dilution. Used for thinner sections and positional welding of thicker sections.

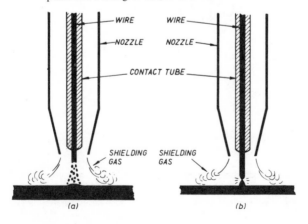

WIRE   WIRE

NOZZLE   NOZZLE

CONTACT TUBE

SHIELDING GAS   SHIELDING GAS

(a)   (b)

constant voltage type described later. Spray transfer is also termed free flight transfer.

Short circuit or dip transfer

With lower arc volts and currents transfer takes place in globular form but with intermittent short-circuiting of the arc. The wire feed rate must just exceed the burn-off rate so that the intermittent short-circuiting will occur. When the wire touches the pool and short-circuits the arc there is a momentary rise of current, which must be sufficient to make the wire tip molten, a neck is then formed in it due to magnetic pinch effect and it melts off in the form of a droplet being sucked into the molten pool aided by surface tension. The arc is then re-established, gradually reducing in length as the wire feed rate gains on the burn-off until short-circuiting again occurs (Fig. 2.2). The power source must supply sufficient current on short-circuit to ensure melt-off or otherwise the wire will stick into the pool, and it must also be able to provide sufficient voltage immediately after short-circuit to establish the arc. The short-circuit frequency depends upon arc voltage and current, type of shielding gas, diameter of wire, and the power source characteristic. The heat output of this type of arc is much less than that of the spray transfer type and makes the process very suitable for the welding of thinner sections and for positional welding, in addition to multi-run thicker sections, and it gives much greater welding speed than manual arc on light gauge steel, for example. Dip transfer has the lowest weld metal dilution value of all the arc processes.

Semi-short-circuiting arc

In between the spray transfer and dip transfer ranges is an intermediate range in which the frequency of droplet transfer is approaching that of spray yet at the same time short-circuiting is taking place, but is of very short duration. This semi-short-circuiting arc has certain applications, as for example the automatic welding of medium-thickness steel plate with CO_2 as the shielding gas.

d.c. power supply and arc control

There are two methods of automatic arc control:

(1) Constant voltage or potential, known as the self-adjusting arc.

(2) Drooping characteristic or controlled arc (constant current).

The former is more usually employed both on MIG and CO_2 welding plant though the latter may be used very occasionally with larger diameter wires and higher currents and with the flux cored welding process.

Constant voltage d.c. supply

Power can be supplied from a welding generator with level characteristic or from a natural or forced draught cooled three-phase or one-phase transformer and rectifier arranged to give output voltages of approximately 14–50 V and ranges of current according to the output of the unit.

The voltage–current characteristic curve, which should be flat or level in a true constant voltage supply, is usually designed to have a slight droop as shown in Fig. 2.3a. Evidently this unit maintains an almost constant arc voltage irrespective of the current flowing. The wire feed motor has an adjustable speed control with which the wire feed speed must be pre-set for a given welding operation. Once pre-set the motor feeds the wire to the arc at constant speed. Supplies for the auxiliaries are generally at 110 V a.c. for the wire feed motor and 25 v a.c. for the torch switch circuit, but some units have a 120 V d.c. supply. For the arc to function correctly the rate of wire feed must be exactly balanced by the burn-off rate to keep the arc length constant. Suppose the normal arc length is that with voltage drop V_M

Fig. 2.3. Volt–ampere curves of constant voltage and drooping characteristic sources.

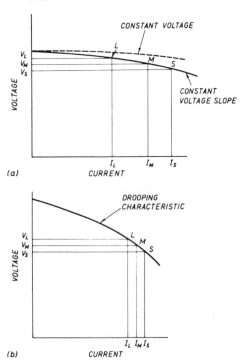

indicated in Fig. 2.3*a* at *M* and the current for this length is I_M amperes. If the arc shortens (manually or due to slight variation in motor speed) to *S* (the volts drop is now V_S) the current now increases to I_S, increasing the burn-off rate, and the arc is lengthened to *M*. Similarly if the arc lengthens to *L*, current decreases to I_L and burn-off rate decreases, and the arc shortens to *M*.

Evidently the gradient of slope of the output curve affects the welding characteristics and slope-controlled units are produced in which the gradient or steepness of the slope can be varied as required and the correct slope selected for given welding conditions.

Drooping characteristic d.c. supply

With this system the d.c. supply is obtained from a welding generator with a drooping characteristic or more usually from a transformer–rectifier unit. If a.c. is required it is supplied at the correct voltages from a transformer.

The characteristic curve of this type of supply (Fig. 2.3*b*) shows that the voltage falls considerably as the current increases, hence the name. If normal arc length *M* has volts drop V_M and if the arc length increases to *L*, the volts drop increases substantially to V_L. If the arc is shortened the volts drop falls to V_S while the current does not vary greatly, hence the name constant current which is often given to this type of supply. The variations in voltage due to changing arc length are fed through control gear to the wire feed motor, the speed of which is thus varied so as to keep a constant arc length, the motor speeding up as the arc lengthens and slowing down as the arc shortens. With this system, therefore, the welding current must be selected for given welding conditions and the control circuits are more complicated than those for the constant voltage method.

Power source–dip transfer (or short-circuit transfer)

In order to keep stable welding conditions with a low voltage arc (17–20 V) which is being rapidly short-circuited, the power source must have the right characteristics. If the short-circuit current is low the electrode will freeze to the plate when welding with low currents and voltages. If the short-circuit current is too high a hole may be formed in the plate or excessive spatter may occur due to scattering of the arc pool when the arc is re-established. The power supply must fulfil the following conditions:

(1) During short-circuit the current must increase enough to melt the wire tip but not so much that it causes spatter when the arc is re-established.

(2) The inductance in the circuit must store enough energy during short-circuit to help to start the arc again and assist in maintaining it during the decay of voltage and current. If an inductive reactor or choke connected in the arc circuit when the arc is short-circuited the current does not rise to a maximum immediately, so the effect of the choke is to limit the rate of rise of current, and the amount by which it limits it depends upon the inductance of the choke. This limitation is used to prevent spatter in CO_2 welding. When the current reaches its maximum value there is maximum energy stored in the magnetic field of the choke. When the droplet necks off in dip transfer and the arc is re-struck, the current is reduced and hence the magnetic field of the choke is reduced in strength, the reduction in energy being fed into the circuit helping to re-establish the arc. If the circuit is to have variable inductance so that the choke can be adjusted to given conditions the coil is usually tapped to a selector switch and by varying the number of turns in circuit the inductive effect is varied. The inductance can also be varied by using a variable air gap in the magnetic circuit of the choke.

To summarize: the voltage of a constant voltage power source remains substantially constant as the current increases. In the case of a welding power unit the voltage drop may be one or two volts per hundred amperes of welding current, and in these circumstances the short-circuit current will be high. This presents no problem with spray transfer where the current adjusts to arc length and thus prevents short-circuiting, but in the case of short-circuiting (dip) transfer, excessive short-circuit currents would cause much spatter.

The steeper the slope of the power unit volt–ampere characteristic curve the less the short-circuit current and the less the 'pinch effect' (which is the resultant inward magnetic force acting on the molten metal in transfer) so that spatter is reduced. Too much reduction of the short-circuit current, however, may lead to difficulty in arc initiation and stubbing. Power units are available having slope control so that the slope can be varied to suit welding conditions, and the control can be by tapped reactor or by infinitely variable reactor, the power factor of the latter being better than that of the former.

Three variables can thus be provided on the power unit – slope control, voltage and inductance. Machines with all three controls give the most accurate control of welding conditions but are more expensive and require more setting than those with only two variables, slope–voltage or voltage–inductance. In general, units with voltage–inductance control offer better characteristics for short-circuit transfer than those with slope–voltage control. For spray transfer conditions all types perform well, with the proviso that for aluminium welding the unit should have sufficient slope.

Transformer–rectifier power unit

Fig. 2.4*a* shows a transformer–rectifier power unit with output voltages by tapped transformer giving various wire speeds and feeds. A variable choke is fitted.

Fig. 2.4*b* shows an engine driven transformer–rectifier unit.

Fig. 2.5 shows a general method of connection for 3 phase input.

Wire feed and control cabinet

In the most popular system the wire is pushed through the guide tube to the gun and is particularly suitable for hard wires such as steel and stainless and heat-resistant steel. In the second method the motor is in the handle of the gun (Fig. 2.7*f*) and the wire is pulled from the reel through the guide tube to the gun. This allows a greater distance to be worked from power unit to welding position and is particularly useful for aluminium as the roll pressure is not so great but the reel capacity is smaller. In some cases a combination of these two methods, the push–pull type, is used applicable to both hard and soft wires but higher in initial cost.

In the push type, the most popular, the wires of diameter 0.8, 1.0, 1.2, 1.6, 2.4 mm hard and 1.2, 1.6, 2.4 soft, and 3.2 mm flux cored are supplied on reels of approximately 15, 25, 30 kg, etc., and are supplied for steel, low-alloy steel, creep-resistant and weathering steels, stainless heat-resistant steel, hard facing, aluminium, bronze and copper. The wire passes between motor-driven feed and pressure rollers which may have serrations or grooves to provide grip and which drive the wire at speeds between 2.5 and 15 m per min. The pressure on this drive can be varied and care must be taken that there is enough pressure to prevent slipping, in which case the arc lengthens and may burn back to the contact tube, and on the other hand that the pressure is not so great as to cause distortion of the wire or the flaking off of small metal particles with consequent increased wear on the guide tubes and possibility of jamming. Some units have a steel channel between feed and pressure rollers through which the wire passes and which helps prevent kinking. Other machines have a small removable magnet fixed after the wire drive to pick up such particles when ferrous wires are used. The flexible outer cables through which the wire is fed to the gun may have nylon liners for smoother feeding of fine wire sizes. The cabinet houses the wire drive motor and assembly, gas and cooling water valves and main

Fig. 2.4. (*a*) NBC 400 d.c. power unit for dip or spray transfer, with wire feed unit, welding gun and gas gauges, etc. 400 A at 60% duty cycle, fan cooled. Input 380–440 V, 20 A, 3-phase 50 Hz. Tapped input, star connected for 14–24 V output range and delta connected for 24–34 V range in 32 OCV steps. Delta connected secondary with silicon diode bridge rectifier. Overload protected. Variable choke for dip transfer. 115 V supply for CO_2 heater, etc., and 24 V for torch trigger control. Wire feed unit has variac or solid state module control. Wire speed 1.5–20 m/minute. Wire diameters 0.6–1.6 mm hard and 1.2–1.6 mm soft. Gas purge and wire inch control. Friction drum brake, geared top roller on swing arm. Quick release torch connection.

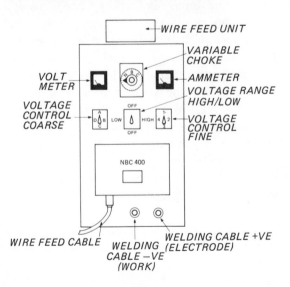

Fig. 2.4. (*b*) Mega-arc diesel engine driven, 1800 rpm, d.c. power unit. Output 400 A at 40 V and 100% duty cycle for MMA, MIG, TIG and plasma arc welding. SCR control with two current ranges, 2–250 A and 40–500 A with 3 kW at 115 V or 230 V for auxiliaries. Arc 'force' control gives 'soft' or 'hard' arc and the unit is easily serviced.

Fig. 2.5. 3-phase rectification (capacitors not shown).

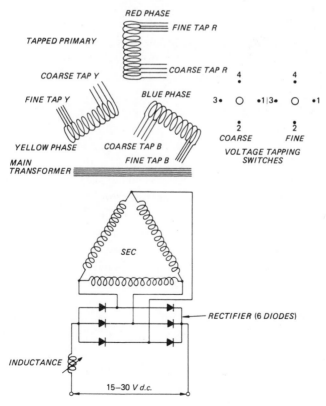

current contactor (controlled from the switch on the gun or automatic head) and the wire speed control, which is also the current control, is pre-set manually or automatically (Figs. 2.6*a*, *b* and *c*). There are also gas purging and inching switches, and in some units regenerative braking on the wire drive motor prevents over-run at the end of a weld.

Sets are now also available with programmable power sources. Using known quantities such as amperes, seconds, metres per minute feed, the welding program is divided into a chosen number of sections and the welding parameters as indicated previously are used to program the computer which controls the welding source. The program can be stored in the computer memory of up to say 50 numbered welding programs or it can be stored on a separate magnetic data card for external storage or use on another unit. By pressing the correct numbers on the keyboard of the unit any programs can be selected and the chosen program begins, controlling welding current, shielding and backing gas, gas pre-flow, wire feed speed, arc length, pulsed welding current and slope control, etc. All safety controls are fitted and changes in the welding program can be made without affecting other data.

Torch (Fig. 2.7*a*, *b*, *c*, *d*, *e*, *f* and *g*)

To the welding torch of either gooseneck or pistol type or an automatic head are connected the following supplies:

(1) Flexible cable through which the wire electrode is fed.
(2) Tubes carrying shielding gas and cooling water flow and return (if water cooled).
(3) Cable carrying the main current and control wire cable.

A centrally placed and replaceable contact tube or tip screws into the torch head and is chosen to be a sliding fit in the diameter of wire being. Contact from power unit to welding wire is made at the contact tube, which must be removed and cleaned at intervals and replaced as required. A metal shield or nozzle surrounds the wire emerging from the tube through which the shielding gas flows and surrounds arc and molten pool. Air-cooled torches are used up to 400 A and water-cooled up to 600 A. In the latter type the cooling water return flows around the copper cable carrying the welding current and thus this cable can be of smaller cross-sectional area and thus lighter and more flexible. A water-cooled fuse in the circuit ensures that the water-cooling flow must be in operation before welding commences and thus protects the circuit.

The wire feed can be contained in the head (Fig. 2.7*f*) only when the feed is by 'pull' or there can be a 'pull' torch unit, a 'push' unit, and a control box. These units handle wires of 0.8, 1.0, 1.2 or 1.6 mm diameter in soft

Fig. 2.6. (*a*) LAH 500, d.c. power unit with wire feed unit, remote control unit and counterbalanced welding arm. 500 A at 60% duty cycle. Supply 3-phase, 50 Hz at 220, 380, 415 and 500 V, primary current 38 A at 415 V and 60% duty cycle. Thyristor control (SCR) gives one knob stepless control of voltage and current. Basic version 150–500 A and the other version 30–500 A divided into two ranges, one high amperage and the other low amperage. Wire feed unit (40 V) has stepless adjustment of wire feed and automatic control for temperature, voltage and friction variation. All feed rollers exert equal pressure on wire. The four driving rollers have different tracks for different diameters of wire. Wire change by reversible motors and lever action. Remote control unit is connected to power unit and wire feed unit giving remote control of current, voltage and pulsing. Facilities for spot and interval welding (the latter gives a colder weld for sheet metal and for bridging larger gaps). Shielding gases: argon, CO_2 or mixtures. Wire diameters: steel 1.2–2.0 mm, tubular wire 1.6–2.4 mm, aluminium 1.6–2.0 mm.

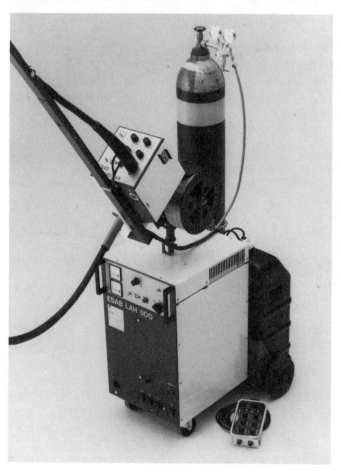

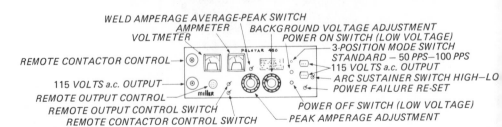

WELD AMPERAGE AVERAGE-PEAK SWITCH
AMPMETER   BACKGROUND VOLTAGE ADJUSTMENT
VOLTMETER   POWER ON SWITCH (LOW VOLTAGE)
                                            3-POSITION MODE SWITCH
REMOTE CONTACTOR CONTROL   STANDARD — 50 PPS—100 PPS
                                            115 VOLTS a.c. OUTPUT
115 VOLTS a.c. OUTPUT   ARC SUSTAINER SWITCH HIGH—LO
REMOTE OUTPUT CONTROL   POWER FAILURE RE-SET
REMOTE OUTPUT CONTROL SWITCH   POWER OFF SWITCH (LOW VOLTAGE)
REMOTE CONTACTOR CONTROL SWITCH   PEAK AMPERAGE ADJUSTMENT

◄ Fig. 2.6. (*b*) Pulstar 450 power unit and wire feed. Solid state d.c. constant potential power unit and wire feed unit. Input: 350–415 V, 3-phase 50 Hz, with line voltage compensation 10%. Output: Voltage 14–38 current 450 A maximum and 100% duty cycle. Single knob current control. 3-position switch for standard operation on 50 or 100 pulses per second. Wire feed unit: Feed speed 1.3–19.8 m/min with 0.8–3.2 mm diameter hard or soft wires. Power supply 115 V with permanent magnet field feed motor. Inch control and gas purge, digital wire speed display. 4 roller feed system. Remote hand voltage control, burnback control and wire straightener if required. Control of background voltage and peak amperage in pulsed mode. Dip transfer (short circuiting arc) spray and flux cored methods can be used.

Fig. 2.6. (*c*) Controls for solid state wire feed unit.

Fig. 2.7. (*a*) Straight-necked, air-cooled torch, variable in length by 38 mm.
Current: at 75% duty cycle, 300 A with argon-rich gases, 500 A with CO_2; at
50% duty cycle, 350 and 550 A respectively. Wires: hard and soft, 1.2 and
1.6 mm, flux cored 1.6, 2.0 and 2.4 mm. (*b*) air-cooled torch, 45° neck angle.
Current: at 75% duty cycle 300 A with argon-rich gases, 350 A with CO_2; at
50% duty cycle 350 A and 400 A respectively. Wires: hard 0.8, 1.0, 1.2 and
1.6 mm; soft 1.2 and 1.6 mm, flux cored, 1.6 mm. (*c*) Air-cooled torch, 60°
neck angle. Current: at 75% duty cycle 300 A with argon-rich gases, 350 A
with CO_2; at 50% duty cycle 350 and 400 A respectively. Wires: hard 0.8, 1.0,
1.2 and 1.6 mm.

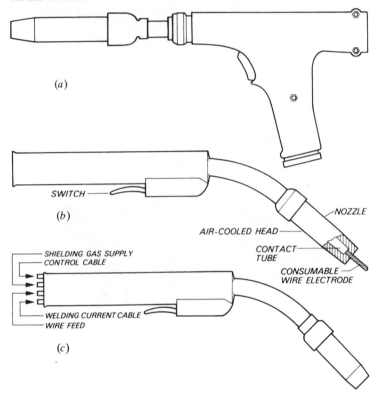

(*a*)

SWITCH

(*b*)

NOZZLE

AIR-COOLED HEAD

SHIELDING GAS SUPPLY
CONTROL CABLE

CONTACT
TUBE

CONSUMABLE
WIRE ELECTRODE

WELDING CURRENT CABLE
WIRE FEED

(*c*)

Fig. 2.7. (*d*) MIG welding guns: 600 A, 400 A, 300 A.

Fig. 2.7. (*e*) 400 A gun showing 'quick disconnect'.

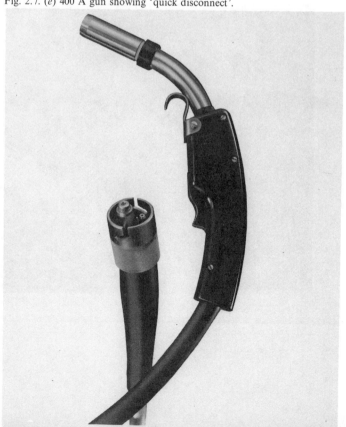

Fig. 2.7. (*f*) Welding torch with air motor for wire feed drive built into the handle. Speed of motor is set by means of a knob which controls an air valve. Guns can be fitted with swan necks rotatable through 360°. Available in sizes up to 400 A, these latter have a heat shield fitted to protect the hand. A fume extraction unit is also fitted.

Fig. 2.7. (*g*) Torch with fume extraction unit.

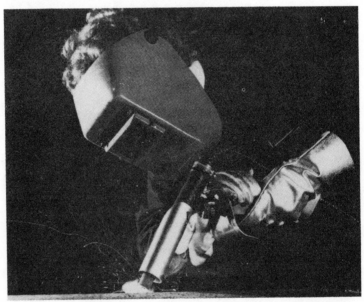

Fig. 2.7. (*h*) Welding torch or gun.

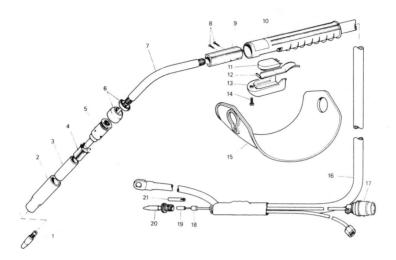

1. Contact tip for 1.0, 1.2 and 1.6 mm hard and soft wire; 1.6, 2.0 and 2.4 mm tubular wire.
2. Nozzle.
3. Nozzle insulator.
4. Nozzle spring clip.
5. Torch head.
6. Head insulator and   clip.
7. Neck.
8. Self-tapping screw.
9. Spider.
10. Handle mounting.
11. Microswitch.
12. Switch lever.
13. Switch housing.
14. Screw.
15. Heat shield.
16. Integrated cable.
17. Plug: 7 pin.
18. Basic liner.
19. Liner; 1.2, 1.6 mm soft wire, 1.2, 1.6 mm hard wire, 1.6, 2.0 mm tubular wire.
20. Outlet guide; 1.2, 1.6 mm soft wire, 1.2, 1.6 mm hard wire, 1.6, 2.0, 2.4 mm tubular wire.
21. Collet (for soft wire outlet guides).

aluminium or 0.8, 1.0 and 1.2 mm in hard steel wire, the unit being rated at 300 A. Pre-weld and post-weld gas flows are operated by the gas trigger and operate automatically. They operate from the standard MIG or CO_2 rectifier units and are suitable for mixed gas or CO_2 shielding, the very compact bulk greatly adding to their usefulness since they have a very large working radius.

Gases

Since CO_2 and oxygen are not inert gases, the title metallic inert gas is not true when either of these gases is mixed with argon or CO_2 is used on its own. The title *metallic active gas* (MAG) is sometimes used in these cases.

Argon, Ar. Commercial grade purity argon (99.996%) is obtained by fractional distillation of liquid air from the atmosphere, in which it is present to about 1% (0.932%) by volume. It is supplied in blue-painted cylinders containing 8500 litres at a pressure of 17.5 N/mm^2 (175 bar) or from bulk supply. It is used as a shielding gas because it is chemically inert and forms no compounds.

Carbon dioxide, CO_2. This is produced as a by-product of industrial processes such as the manufacture of ammonia, from the burning of fuels in an oxygen-rich atmosphere or from the fermentation processes in alcohol production, and is supplied in black-painted steel cylinders containing up to 35 kg of liquid CO_2.* To avoid increase of water vapour above the limit of 0.015% in the gas as the cylinder is emptied, a dip tube or syphon is fitted so that the liquid CO_2 is drawn from the cylinder, producing little fall in temperature. Fig. 2.8 shows a CO_2 vaporizer or heater with pressure reducing regulator and ball type flowmeter. A cartridge type electric heating element at 110 V is in direct contact with the liquid CO_2 to vaporize it. The 150 W version gives a flow rate of 21 litres/min while the 200 W version gives 28 litres/min. A neon warning light connected via a thermostat indicates when the element is heating and is extinguished when the heater has warmed up sufficiently. The syphon cylinder has a white longitudinal stripe down the black cylinder while the non-syphon cylinder, used when the volume of CO_2 to be taken from it is less than 15 litres/min, is all black. Manifold cylinders can be fed into a single vaporizer, and if the supply is in a bulk storage tank, this is fed into an evaporator, and then fed to the welding points at correct pressure as with bulk argon and oxygen supplies.

----

* The cylinder pressure depends upon the temperature being approximately 33 bar at 0 °C and 50 bar at 15 °C.

Helium, He. Atomic weight 4, liquifying point $-268\,^{\circ}$C, is thus much lighter than argon, atomic weight 40. It is present in the atmosphere in extremely small quantities but is found in association with natural gas in Texas, Oklahoma, Kansas, Alberta, etc., which are the main sources of supply. There is little associated with North Sea gas. Being lighter it requires a greater flow rate than argon and has a 'hotter' arc which may give rise to certain health hazards. Cylinder identifying colour is brown.

Oxygen, O$_2$. Atomic weight 16, boiling point $-183\,^{\circ}$C. Obtained by distillation from liquid air. Used in connection with the inert gas in small percentages to assist in wetting and stabilizing the arc. Wetting is discussed on page 305 under the heading Surface tension.

Application of gases

(1) Argon. Although argon is a very suitable shielding gas for the non-ferrous metals and alloys, if it is used for the welding of steel there exists an unstable negative pole in the work-piece (the wire being positive) which produces an uneven weld profile. Mixtures of argon and oxygen are selected to give optimum welding conditions for various metals.

Fig. 2.8

(2) Argon + **1% or 2% oxygen.** The addition of oxygen as a small percentage to argon gives higher arc temperatures and the oxygen acts as a wetting agent to the molten pool, making it more fluid and stabilizing the arc. It reduces surface tension and produces good fusion and penetration. Argon-rich mixtures such as argon and up to 25% CO_2 with the addition of oxygen make the dip transfer process applicable to positional steel work and to thin sheet. Argon + 1–2% O_2 is used for stainless steel and helium, argon, CO_2 can be used on thin sections similarly and has good weld profile. The argon + 1% mixture is used for stainless steels with spray and pulse methods. Dip transfer is better with argon +2% oxygen as long as the increased oxidation can be tolerated. The addition of 5% hydrogen is the maximum for titanium-stabilized stainless steel and larger amounts than this increase porosity.

(3) Helium is nearly always found in mixed gases. Because of the greater arc temperature, mixing it with argon, oxygen or CO_2 controls the pool temperature, increases wetting and stabilizes the arc. The higher the helium content the higher the arc voltage and the greater the heat output. It is used in gas mixtures for aluminium, nickel, cupro-nickels, etc., and is particularly applicable to stainless steels with the helium–argon–oxygen or CO_2 mixtures. Helium–argon mixtures are also used for the welding of 9% nickel steels.

(4) Carbon dioxide. Pure CO_2 is the cheapest of the shielding gases and can be used as a shield for welding steel up to 0.4% C and low-alloy steel. Because there is some dissociation of the CO_2 in the arc resulting in carbon monoxide and oxygen being formed, the filler wire is triple deoxidized to prevent porosity, and this adds somewhat to its cost and results in some small areas of slag being present in the finished weld. The droplet rate is less than that with pure argon, the arc voltage drop is higher, and the threshold value for spray transfer much higher than with argon. The forces on the droplets being transferred across the arc are less balanced than with argon–oxygen so that the arc is not as smooth and there is some spatter, the arc conditions being more critical than with argon–oxygen.

Using spray transfer there is a high rate of metal deposition and low-hydrogen properties of the weld metal. The dip transfer process is especially suited for positional work. The process has not displaced the submerged arc and electro-slag methods for welding thick steel sections, but complements them and competes in some fields with the manual metal arc process using iron powder electrodes. It offers the most competitive method for repetitive welding operations, and the use of flux cored wire greatly increases the scope of the CO_2 process. Thickness up to 75 mm can be welded in steel using fully automatic heads.

With stainless steel, because of the loss of stabilizers (titanium and niobium) in the CO_2 shielded arc, there is some carbon pick-up resulting in some precipitation of chromium carbide along the grain boundaries and increased carbon content of the weld, reducing the corrosion resistance. Multi-pass runs result in further reduction in corrosion resistance, but with stabilized filler wire and dip transfer on thinner sections satisfactory single-pass welds can be made very economically. Non-toxic, non-flammable paste is available to reduce spatter problems.

(5) Argon + 5% CO_2, Argon + 20% CO_2. The addition of CO_2 to argon for the welding of steel improves the 'wetting' action, reduces surface tension and makes the molten pool more fluid. Both mixtures give excellent results with dip and spray transfer, but the 20% mixture gives poor results with pulse while the 5% mixture gives much better results. The mixtures are more expensive than pure CO_2 but give a smoother, less critical arc with reduced spatter and a flatter weld profile, especially in fillets (Fig. 2.9b). The current required for spray transfer is less than for similar conditions with CO_2 and the 5% mixture is suitable for single-run welding of stainless steel with the exception of those with extra low carbon content (L), below 0.06%. Both mixtures contain a small amount of oxygen.

If the CO_2 content is increased above 25% the mixture behaves more and more like pure CO_2. If argon and CO_2 are on bulk supply, mixers enable the percentage to be varied as required. Figs 2.9a and 2.9b show penetration beads and weld profiles with differing gas mixtures.

Fig. 2.9. (a) Stainless steel weld bead profiles: dip transfer.

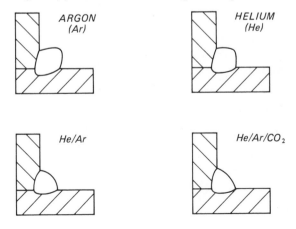

Filler wires*

Note. BS 2901 (1970) gives specifications for *Filler rods and wires for inert gas welding.* Part 1, *Ferritic steels;* Part 2, *Austenitic stainless steels;* Part 3, *Copper and copper alloys;* Part 4, *Aluminium and aluminium alloys and magnesium alloys;* Part 5, *Nickel and nickel alloys.*

Filler wires are supplied on convenient reels of 300 mm or more diameter and of varying capacities with wire diameters of 0.6, 0.8, 1.0, 1.2, 1.6, and 2.4 mm. The bare steel wire is usually copper coated to improve conductivity, reduce friction at high feed speeds and minimize corrosion while in stock. Manganese and silicon are used as deoxidizers in many cases but triple deoxidized wire using aluminium, titanium and zirconium gives high-quality welds and is especially suitable for use with CO_2. The following are examples of available steel wires and can be used with argon–5% CO_2 argon–20% CO_2 and CO_2, the designation being to BS 2901 Part 1 (1970).

A 18. General purpose mild steel used for mild and certain low alloy steels. Analysis: 0.12% C; 0.9–1.6% Mn; 0.7–1.2% Si. 0.04% max S and P.

Fig. 2.9. (*b*) Penetration beads: spray transfer.

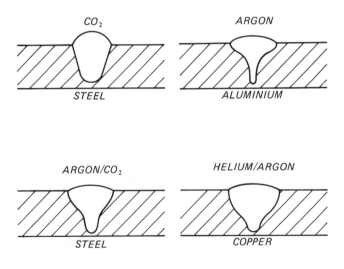

* American classification of mild steel welding wires AWS A5 18–69. *Example* E 70S–3 E = electrode, S = bare solid electrode, 3 is a particular classification based on the as manufactured chemical composition.

Metal arc gas shielded process. Recommended gases and gas mixtures for various metals and alloys

Metal type	Gas shield	Remarks
Carbon and low-alloy steels	CO_2	For dip transfer, and spray transfer
		Spatter problems. Use deoxidized wire
	Ar–15/20% CO_2	For dip or spray transfer. Minimum spatter
	Ar–5% CO_2	For dip and spray transfer
	Ar–5% O_2	Spray transfer. High impact properties
	Ar–5% CO_2–O_2 2%	For pulsed arc and thin sections
Stainless steels	Ar–1/2% O_2	Spray transfer
	75 He 23.5% Ar 1.5% CO_2	High quality dip transfer. For thin sections and positional work. Good profile
	He 75%–Ar 24%–O_2 1%	
Aluminium and its alloys	Argon	Stable arc with little spatter
	Helium	Hotter arc, less pre-heat, more spatter
	He 75% Ar 25%	Stable arc, high heat input. Good penetration. Recommended for thicknesses above 16 mm
Magnesium and its alloys	Argon	Stable arc
	He 75% Ar 25%	Hotter arc. Less porosity
Copper and its alloys	Argon	For sections up to 9.5 mm thickness
	Helium	
	He 75% Ar 25%	For medium and heavy sections. High heat input
Nickel and its alloys	Argon	Sections up to 9.5 mm thickness. Pulsed arc
	Ar 70% He 30%	High heat input, less cracking in thicker sections of 9% Ni
	Ar 25% He 75%	
Cupro-nickel	Argon	Stable arc
	Ar 70% He 30%	Stable arc with less cracking risk
Titanium, Zirconium and alloys	High purity argon	Very reactive metals. High purity shielding gases are essential

Note. O_2 increases the wetting action.
See appendix 3 for proprietary gases.

The following colour codes are used for cylinders (BS 349, BS 381 C)

Gas or mixture	Colour code
Acetylene, dissolved	Maroon; also with name ACETYLENE.
Argon	Peacock blue.
Argon–CO_2	Peacock blue. Light brunswick green band round middle of cylinder.
Argon–hydrogen	Peacock blue. Signal red band round middle of cylinder.
Argon–nitrogen	Peacock blue. French grey band round middle of cylinder.
Argon–oxygen	Peacock blue. Black band round middle of cylinder. % of oxygen indicated.
CO_2, commercial vapour withdrawal	Black.
CO_2, commercial liquid withdrawal	Black. Narrow white band down length of cylinder (syphon tube).
Helium	Brown.
Hydrogen	Signal red.
Hydrogen–high purity	Signal red. White circle on shoulder.
Hydrogen–nitrogen	Signal red. French grey band round middle of cylinder.
Nitrogen	French grey. Black cylinder shoulder.
Nitrogen (oxygen free)	French grey. Black cylinder shoulder with white circle.
Nitrogen–CO_2	French grey. Light brunswick green band round middle of cylinder and black cylinder shoulder.
Nitrogen–oxygen	French grey. Black band round middle of cylinder and black cylinder shoulder.
Oxygen, commercial	Black.
Propane	Signal red with name PROPANE.
Air	French grey.
Air–CO_2	French grey. Light brunswick green band round middle of cylinder.

Note. Colour bands on shoulder of cylinder denote hazard properties. Cylinders having a red band on the shoulder contain a flammable gas; cylinders having a golden yellow band on the shoulder contain a toxic (poisonous) gas. Cylinders having a red band on the upper shoulder and a yellow band on the lower shoulder contain a flammable and toxic gas. Carbon monoxide has a red cylinder with a yellow shoulder.

A 15. Triple deoxidized steel wire recommended for pipe welds and root runs in heavy vessel construction. Analysis: 0.12% max. C; 0.9–1.6% Mn; 0.3–0.9% Si; 0.04–0.4% Al; 0.15% max. Ti; 0.15% max. Zr; 0.04% max. S and P.

A31. Molybdenum bearing mild steel wire. Used for most mild-steel applications requiring extra strength; high-tensile and quench and tempered steel; and suitable for offshore pipeline application and root runs in thick joints. Analysis: 0.14% max. C; 1.6–2.1% Mn; 0.5–0.9% Si; 0.4–0.6% Mo; 0.03% max. S and P.

Flux cored and metal powder cored filler wire*

Flux cored and metal powder cored wires, used with a gas shield of CO_2 or argon–5% CO_2 to argon–20% CO_2 give rapid deposition of metal and high-quality welds in steel. Metal powder wire used with argon–20% CO_2 gives a smooth arc with little spatter and is used with wire negative polarity. It can be used for positional work and because of the small amount of slag, which is produced intermittently, there is no need to deslag between runs.

The wire is manufactured from a continuous narrow flat steep strip formed into a U shape which is then filled with flux and formed into a tube. It is then pulled through reducing dies which reduce the diameter and compress the flux uniformly and tightly into a centre core (Fig. 2.10a). It is supplied on reels as for MIG welding in diameters 1.6, 2.0, 2.4, 3.2 mm and

Fig. 2.10. (*a*) Flux cored process with gas shield.

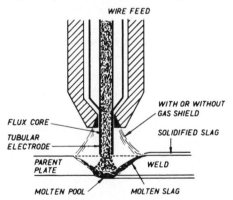

---

* American classification of mild steel tubular welding wires; flux cored arc welding (FCAW), with and without external gas shield. *Example* E 70 T–1: E = electrode; 70 = tensile strength of weld metal in 1000lb/sq in. (ksi); T = tubular continuous electrode with powdered flux in core; 1 = gas type and current. (1, 2 – CO_2 deep; 3, 4, 6, 8 – none, deep: 5 – CO_2 or none, deep; 7 – none, deep; 9 – miscellaneous).

fed to the feed rollers of a MIG type unit. The welding gun can be straight or swan-necked and should be water cooled for currents above 400 A, and the power unit is the same as for MIG welding.

There is deep penetration with smooth weld finish and minimum spatter, and deposition rate is of the order of 10 kg/h using 2.4 mm diameter wire. The deep penetration characteristics enable a narrower V preparation to be found for butt joints resulting in a saving of filler metal, and fillet size can be reduced by 15–20%.

Some of the filler wires available are:

(1) Rutile type, steel, giving a smooth arc and good weld appearance with easy slag removal. Its uses are for butt and fillet welds in mild and medium-tensile steel in the flat and horizontal–vertical position.

(2) Basic hydrogen-controlled type, steel. This is an all-positional type and gives welds having good low-temperature impact values.

(3) Basic hydrogen-controlled type, steel, with $2\frac{1}{2}\%$ Ni for applications at $-50\,^{\circ}$C.

(4) Basic hydrogen-controlled type for welding 1% Cr, 0.5% Mo steels.

(5) Basic hydrogen-controlled type for welding 2.25% Cr, 1% Mo steels.

Self-shielded flux cored wire

Self-shielded flux cored wires are used without an additional gas shield and can be usefully employed in outdoor or other on site draughty situations where a cylinder-supplied gas shield would be difficult to establish.

The core of these wires contains powdered metal together with gas-forming compounds and deoxidizers and cleaners. The gas shield formed protects the molten metal through the arc and slag-forming compounds form a slag over the metal during cooling, protecting it during solidification. To help prevent absorption of nitrogen from the atmosphere by the weld pool, additions of elements are made to the flux and electrode wire to effectively reduce the soluble nitrogen.

This process can be used semi- or fully automatically and is particularly useful for on-site work (Fig. 2.10*b*).

Metal cored filler wire

The wire has a core containing metallic powders with minimal slag-forming constituents and there is good recovery rate (95%) with no interpass deslagging. It is used with CO_2 or argon–CO_2 gas shield to give welds with low hydrogen level (less than 5 ml/100 g weld metal). The equipment is similar to that used for MIG welding, and deposition rates are higher than with stick electrodes especially on root runs.

Safety precautions

The precautions to be taken are similar to those when metal arc welding, given on pp. 29–30. The BS 679 recommended welding filters are up to 200 A, 10 or 11 EW (electric welding); over 200 A, 12, 13, or 14 EW. When welding in dark surroundings choose the higher grade number, and in bright light the lower shade number. Because there is greater emission of infra-red energy in this process a heat absorbing filter should be used. The student should consult the following publications for further information: BS 679, *Filters for use during welding*, British Standards Institution; *Electric arc welding*, new series no. 38, *Safety, health and welfare*. Published for the Department of Employment by HMSO.

Techniques

There are three methods of initiating the arc. (1) The gun switch operates the gas and water solenoids and when released the wire drive is switched on together with the welding current. (2) The gun switch operates the gas and water solenoids and striking the wire end on the plate operates the wire drive and welding current (known as 'scratch start'). (3) The gun switch operates gas and water solenoids and wire feed with welding current, known as 'punch start'.

The table on p. 107 indicates the various gases and mixtures at present in use. As a general rule dip transfer is used for thinner sections up to 6.4 mm

Fig. 2.10. (*b*) Flux cored process (self-shielded).

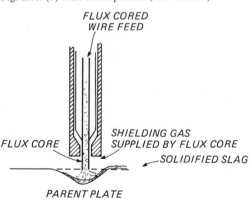

and for positional welding, whilst spray transfer is used for thicker sections. The gun is held at an angle of 80° or slightly less to the line of the weld to obtain a good view of the weld pool, and welding proceeds with the nozzle held 6–12 mm from the work (see Fig. 2.11a). Except under special conditions welding takes place from right to left. If welding takes place for special reasons from left to right the torch has to be held at almost 90° to the line of travel and care must be taken that the gas shield is covering the work.

The further the nozzle is held from the work the less the efficiency of the gas shield, leading to porosity. If the nozzle is held too close to the work spatter may build up, necessitating frequent cleaning of the nozzle, while arcing between nozzle and work can be caused by a bent wire guide tube allowing the wire to touch the nozzle, or by spatter build-up short-circuiting wire and nozzle. If the wire burns back to the guide tube this may be caused by a late start of the wire feed, fouling of the wire in the feed conduit or the feed rolls being too tight. Intermittent wire feed is generally due to insufficient feed roll pressure or looseness due to wear in the rolls. Excessively sharp bends in the flexible guide tubes can also lead to this trouble.

Root runs are performed with no weave and filler runs with as little weave as possible consistent with good fusion since excessive weaving tends to promote porosity. The amount of wire projecting beyond the contact tube is important because the greater the projection, the greater the I^2R effect and the greater the voltage drop, which may reduce the welding current and affect penetration. The least projection commensurate with

Fig. 2.11. (*a*) Angle of torch (flat position). (*b*) Stubbing. (*c*) Dip transfer on thin sheet.

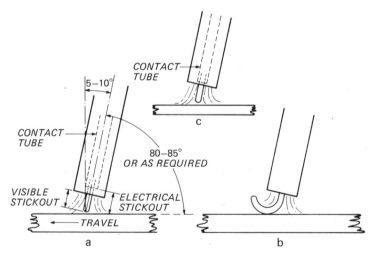

accessibility to the joint being welded should be aimed at. Backing strips which are welded permanently on to the reverse side of the plate by the root run are often used to ensure sound root fusion. Backing bars of copper or ceramics with grooves of the required penetration bead profile can be used and are removed after welding. It is not necessary to back-chip the roof run of the light alloys but with stainless steel this is often done and a sealing run put down. This is a more expensive way compared with efficient back purging. The importance of fit-up in securing continuity and evenness of the penetration bead cannot be over-emphasized.

Flat welds may be slightly tilted to allow the molten metal to flow against the deposited metal and thus give a better profile. If the first run has a very convex profile poor manipulation of the gun may cause cold laps in the subsequent run.

Run-on and run-off plates can be used as in TIG welding to obviate cold start and crater finishing leading to cracks. Slope in and out controls obviate much of this.

Stubbing

If the wire speed is too high the rate of feed of the wire is greater than the burn-off rate and the wire stubs. If a reduction in the wire speed feed rate does not cure the stubbing, check the contact tube for poor current pick-up and replace if necessary (Fig. 2.11b).

Burn back control

This operates a variable delay so that the wire burns back to the correct 'stick-out' beyond the nozzle before being switched off. It is then ready for the next welding sequence. Excessive burn back would cause the wire to burn to the contact tip. Varying the stick-out length (Fig. 2.11a) varies the temperature of the molten pool. Increasing the length gives a somewhat 'cooler' pool and is useful in cases of poor fit-up, for example.

The nearer that the contact tube end is to the outer end of the torch nozzle, the greater the likelihood of the contact tube being contaminated with spatter. With dip transfer, in which the spatter is least, the distance is about 2–3 mm, but for spray transfer it should be in the region of 5–12 mm. The contact tube may even be used slightly projecting from the gas nozzle when being used on thin sheet (e.g. body panels) using dip transfer. This gives a very clear view of the arc but is not recommended as the contact tube becomes rapidly contaminated even when treated with anti-spatter paint or paste and must be cleaned frequently (Fig. 2.11c.) Whatever the stick-out, the flow of shielding gas should be such that molten pool and immediate surrounding areas are covered.

Positional welding

This is best performed by the dip transfer method since the lower arc energy enables the molten metal to solidify more quickly after deposition. Vertical welds in thin sections are usually made downwards with no weave. Thicker sections are welded upwards or with the root run downwards and subsequent runs upwards, weaving as required. Overhead welding, which is performed only when absolutely necessary, is performed with no weave.

Fillet welds are performed with the gun held backwards to the line of welding, as near as possible to the vertical consistent with a good view of the molten pool, bisecting the angle between the plates and with a contact tip-to-work distance of 16–20 mm (Fig. 2.12). On unequal sections the arc is held more towards the thicker section. The root run is performed with no weave and subsequent runs with enough weave to ensure equal fusion on the legs. Tilted fillets give better weld profile and equal leg length more easily.

Some of the filler wires available are:

Note. Except where indicated they can be used with CO_2, argon–5% CO_2 and argon–20% CO_2. CO_2 gives good results with dip transfer whereas the argon mixtures give better weld profile. L indicates 0.08% C or lower.

(1) 1.5% Mn, L. Smooth arc, little slag, so no deslagging, Argon–20% CO_2, cored wire −ve.

(2) 1.8% Mn, 0.5% Mo, 0.14% C. High strength mild steels and for highet tensile steels, solid wire +ve.

(3) 1.4% Mn, 0.5% Mo, L. Similar wire to (1) but molybdenum gives extra strength. Argon–20% CO_2, cored wire −ve.

(4) 1.4% Mn, 0.45% Cu, L. A similar wire to (1) but with copper added for weathering steels such as Corten. Use argon–5% or argon–20% CO_2, cored wire −ve.

Fig. 2.12. Angle of torch: horizontal–vertical fillet.

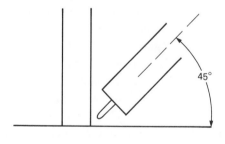

(5) 1.3% Cr, 1.1% Mn, 0.2% Mo, 1.0% Ni, L. High strength welds, tensile strengths of 750 N/mm². Rutile basic flux cored wire +ve.

(6) 2.0% Mn, 1.0% Mo, L. Basic flux core with iron powder, creep-resistant steels, cored wire −ve.

(7) 2.2% Cr, 1.0% Mo, 0.12% C. Basic flux core for creep-resistant steels to temperatures of 580°C, solid wire − ve.

(8) 2.2% Ni, L. For low temperatures to −50°C. Good resistance to cracking cored wire −ve, use argon–CO_2 mixtures.

Stainless steels

Preparation angle is usually similar for semi-automatic and automatic welding for butt welds in stainless steel, being 70–80° with a 0–1.5 mm gap (Fig. 3.22). Torch angle should be 80–90° with an arc length long enough to prevent spatter but not long enough to introduce instability. All oxides should be removed by stainless steel wire-brushing. Back chipping is usually performed by grinding. With spray transfer using high currents there is considerable dilution effect and welds can only be made flat, but excessive weaving should be avoided as this increases dilution. Direct current with the torch (electrode) positive is used with a shield of argon + 1% oxygen.

When welding stainless steel of any thickness it is imperative to obtain good penetration and this is only possible if there is a good back purge. Without this the penetration has to be back chipped and a sealing run made which adds considerably to the cost of the joint.

Back purging can be obtained simply by tack welding a strip on the underside of the weld and removing after welding (Fig. 2.13). This uses argon from the torch and reduces oxide formation on the penetration bead. Back purging of tubes can often be done using soluble paper dams or piped bladders which are blown up in position inside the pipe and serve to contain the inert gas on the underside of the weld, thus effecting considerable saving.

Fig. 2.13. Butt welding stainless steel MIG, 99% A–1% O_2, d.c. wire positive. Automatic or semi-automatic.

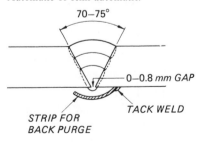

70–75°

0–0.8 *mm* GAP

TACK WELD

STRIP FOR
BACK PURGE

In the welding of dissimilar metals to stainless steel the shortcircuiting arc (dip transfer) gives much less dilution with lower heat input and is generally preferred. The arc should be kept on the edge of the parent metal next to the molten pool and not in the pool itself, to reduce the tendency to the formation of cold laps. Argon 1–2% oxygen is used as the shielding gas and filler wires should be chosen to match the analysis of the parent metal while allowing for some loss in the transfer across the arc. Examples are given in the table. Fig. 3.22 gives suitable methods of preparation.

Filler wires available include the following

BS 2901 Pt 2	AISI	AWS	
308S92	304, 304L	A59 ER308L	Low carbon (0.03% max.) 19.5–22% Cr, 9–11% Ni, suitable for welding 18% Cr, 8% Ni steels including low carbon types.
	321, 347		18% Cr, 8 Ni, Nb stabilized at temperatures up to 400°C
316S92	316, 316L 318	A59 ER 316L	Low carbon 0.03% max., 18–20% Cr, 11–14% Ni, 2–3% Mo, for welding 18% Cr, 8% Ni, Mo and 18% Cr, 8% Ni, Mo, Nb types
316S96	316, 318		Carbon 0.08% max., 18–20% Cr, 11–14% Ni for steels of similar composition to give high impact strength at low temperatures
347S96	347, 321, 304		Carbon 0.08% max., 19–21.5% Cr, 9–11% Ni niobium stabilized. For steels of 20% Cr, 10% Ni type, giving high corrosion resistance
309S94	347		Carbon 0.12% max., 25% Cr, 12% Ni for welding mild and low-alloy steel to stainless steel. for buttering, when welding type 347 to mild steel and for temperatures up to 300°C

Note. Argon 1 or 2% oxygen is used as the shielding gas.
These wires are also used for submerged arc welding of stainless steels.
AISI: American Iron and Steel Institute.

Nickel alloys

The welding of nickel and nickel alloys can be done using spray transfer and short circuit (dip transfer) arc methods.

Spray transfer with its higher heat input gives high welding speeds and deposition rate and is used for thicker sections, usually downhand because of the large molten pool. Argon is used as the shielding gas but 10–20%

helium can be added for welding Nickel 200 and Inconel 600, giving a wider and flatter bead with reduced penetration. The gun should be held as near to 90° to the work as possible, to preserve the efficiency of the gas shield, and the arc should be kept just long enough to prevent spatter yet not long enough to affect arc control.

Short circuiting arc conditions are used for thinner sections and positional welding, the lower heat input giving a more controllable molten pool and minimum dilution. The shielding gas is pure argon but the addition of helium gives a hotter arc and more wetting action so that the danger of cold laps is reduced. To further reduce this danger the gun should be held as near as possible to 90° to the work and moved so as to keep the arc on the plate and not on the pool. High-crowned profile welds increase the danger of cold laps.

The table in chapter 3 in TIG welding of nickel alloys indicates materials and filler metal suitable for welding by the MIG process and Fig. 3.28 the recommended methods of preparation.

Aluminium alloys

When fabricating aluminium alloy sections and vessels, the plates and sections are cut and profiled using shears, cold saw, band saw or arc plasma and are bent and rolled as required and the edges cut to the necessary angle for welding where required. (*Note*. As the welding currents employed in the welding of aluminium are high, welding lenses of a deeper than normal shade should be used to ensure eye protection.)

The sections are degreased using a degreaser such as methyl chloride and are tack welded on the reverse side using either TIG or MIG so as to give as far as possible a penetration gap of close tolerance. Where this is excessive it can be filled using the TIG process, and pre-heating may be performed for drying. The areas to be welded are stainless steel wire brushed to remove all oxide and the root run is made. Each successive run is similarly wire brushed and any stop–start irregularities should be removed. It is important that the tack welds should be incorporated into the overbead completely so that there is no variation in it. Back-chipping of the root run, where required for a sealing run, should be performed with chisels, routers or saws and not ground, as this can introduce impurities into the weld.

Back-chipping and a sealing run add substantially to the cost of a welded seam. In many cases it can be avoided by correctly backpurging the seam and welding from one side only. Fig. 2.14 and table indicate typical butt weld preparation in the flat position using 1.6 mm filler wire. Fig. 2.15a shows the section of a MIG weld in aluminium and Fig. 2.15b a TIG weld on the preparation bead of Fig. 2.15a.

Plate thickness (mm)	Number of runs	Approximate current (A)		Root face (mm)
		Root	Subsequent	
6 and 8	2	220	250–280	1.6
9.5 and 11	3	230	260–280	1.6
12.7 and 16	4	230–240	270–290	3.2
19	5	240–250	280–310	3.2
22 and 25	6	240–260	280–330	4.8

Refer also to BS 3571 *General recommendations for manual inert gas metal arc welding,* Part 1, *Aluminium and aluminium alloys.*
Suitable filler wires are given on p. 161.

Fig. 2.14. Flat butt weld preparation, aluminium plate MIG, semi-automatic process argon shield. Work negative.

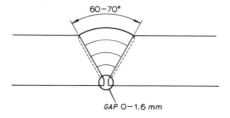

Fig. 2.15. (*a*) MIG weld in aluminium 5083 (NS8) plate 8 mm thick with 5556 A (NG61) wire. Stainless steel backing bar.
1st run: 250–270 A, 26–27 V, speed 9–10 mm/second.
2nd run: 260–280 A, 28–29 V, speed 8–10 mm/second.
Preparation: 90° V, 0–1 mm root gap, 3–4 mm root face.

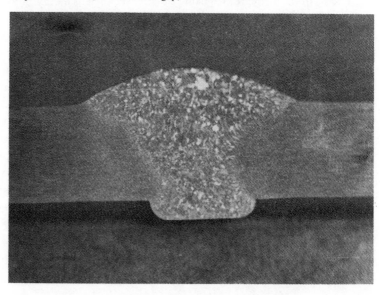

Copper and copper alloys

Many of the problems associated with the welding of copper and copper alloys are discussed in the chapter on the TIG process (Chapter 3) and apply equally to the MIG process. Because of the high thermal conductivity of copper and to reduce the amount of pre-heating, it is usual in all but the thinnest sections to use high currents with spray transfer conditions, which are obtainable with argon and argon–helium mixtures. The addition of nitrogen to argon destroys the spray transfer conditions but arc conditions are improved with up to 50% helium added to argon and this results in an increased heat output. For thin sections, fine feed wires can be used giving spray transfer conditions with lower current densities, thus preventing burn-through. Fig. 2.16 shows edge preparation for TIG and MIG butt welds in copper.

SG cast irons

Pearlitic and ferritic cast irons are very satisfactorily welded using dip transfer conditions (e.g. 150 A, 22 V with 0.8 mm diameter wire) to give low heat input using filler wire of Nickel 6 (93% Ni) or Monel 60 (62/69 Ni 21–28% Cu). Carbide precipitation in the HAZ is confined to thin envelopes around some of the spheroids, unlike the continuous film associated with MMA welding. SG iron can be welded to other metals and Monel 60 can be used to give a corrosion-resistant surface on SG iron castings or as a buffer layer for other weld deposits. Cleaning and

Fig. 2.15. (*b*) TIG weld on penetration bead of the above weld. No back chipping, no filler wire (can be added if required). 350 A, 32 V, speed 3–4 mm/second. Section etched with cupric chloride $CuCl_2$ followed by a 50% nitric acid wash.

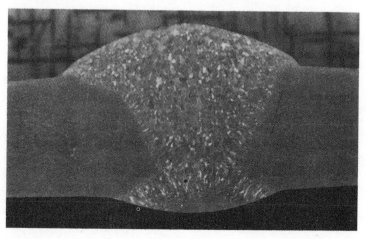

degreasing should be performed before welding, and pre-heat is only required for heavy pearlitic section or joints under heavy restraint when pre-heat of 200 °C is suitable. Minimum-heat input compatible with adequate fusion should always be used (Fig. 2.17).

MIG Process. Recommended filler wires for copper welding

| Type (BS 2870–2875) | Grade | Filler wire | |
		Argon or helium shield	Nitrogen shield
C106	Phosphorus deoxidized, non-arsenical	C7, C8, C21	Not recommended
C107	Phosphorus deoxidized, arsenical		
C101	Electrolytic tough pitch, high conductivity	C7, C8, C21	Not recommended
C102	Fire refined tough pitch, high conductivity		
C103	Oxygen-free, high conductivity	C7, C21	Not recommended

Fig. 2.16. Edge preparation for TIG and MIG butt welds in copper.

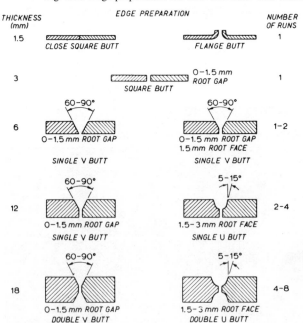

CO₂ welding of mild steel

There are four controls to enable optimum welding conditions to be achieved: (1) wire feed speed which also controls the welding current, (2) voltage, (3) choke or series inductance and (4) gas flow.

For a given wire diameter the wire feed rate must be above a certain minimum value to obtain a droplet transfer rate of above about 20 per second, below which transfer is unsatisfactory. With increasing wire feed rate the droplet transfer rate and hence the burn-off increases and the upper limit is usually determined by the capacity of the wire feed unit. The voltage setting also affects the droplet frequency rate and determines the type of transfer, about 15–20 V for short-circuit or dip, and 27–45 V for spray, with an intermediate lesser used zone of about 22–27 V for the semi-short-circuiting arc.

The choke, which limits the rate of current rise and decay, is also important because too low a value can give a noisy arc with much spatter and poor weld profile, while too high a value can give unstable arc conditions with more difficult start and even occasional arc extinguishing. Between these limits there is a value which with correct arc voltage and wire feed rate gives a smooth arc with minimum spatter and good weld profile. Penetration is also affected by the choke value, see p. 90.

As stated before, the short-circuiting arc is generally used for welding thinner sections, positional welding, tacking and on thicknesses up to 6.5 mm. In positional welding the root run may be made downwards with no weave and subsequent runs upward. The lower heat output of this type of arc reduced distortion on fabrications in thinner sections and minimizes

Fig. 2.17. Preparation for welding of SG cast irons.

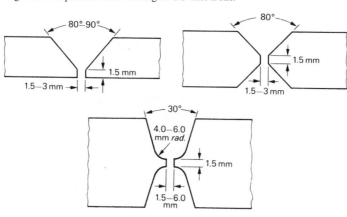

over-penetration. The spray-type arc is used for flat welding of thicker sections and gives high deposition rates.

Gas flow rate can greatly affect the quality of the weld. Too low a flow rate gives inadequate gas shielding and leads to the inclusion of oxides and nitrides, while too high a rate can introduce a turbulent flow of the CO_2 which occurs at a lower rate than with argon. This affects the efficiency of the shield and leads to a porosity in the weld. The aim should be to achieve an even non-turbulent flow and for this reason spatter should not be allowed to accumulate on the nozzle, which should be directed as nearly as possible at 90° to the weld, again to avoid turbulence.

The torch angle is, in practice, about 70–80° to the line of travel consistent with good visibility and the nozzle is held about 10–18 mm from the work. If the torch is held too close, excess spatter build-up necessitates frequent cleaning, and in deep U or V preparation the angle can be increased to obtain better access. Weaving is generally kept as low as convenient to preserve the efficiency of the gas shield and reduce the tendency to porosity. Wide weld beads can be made up of narrower 'stringer' runs, and tilted fillets compared with HV fillets give equal leg length more easily, with better profile.

Economic considerations

Although filler wire for the CO_2 process, together with the cost of the shielding gas, is more expensive than conventional electrodes, other factors greatly affect the economic viability of the process. The deposition rate governs the welding speed which in turn governs the labour charge on a given fabrication.

The deposition rate of the filler metal is a direct function of the welding current. With metal arc welding the upper limit is governed by the overheating of the electrode. The current I amperes flows through the electrode, the wire of which has an electrical resistance R ohms, so that the heating effect is $\propto I^2 R$. The resistance of any metallic conductor increases with the rise in temperature, so that, as the electrode becomes hotter, the resistance and hence the $I^2 R$ loss increases so that with excessive currents, when half of the electrode has been consumed, the remaining half has become red hot and the coating ruined. Iron powder electrodes have a greater current-carrying capacity due to the conductivity of the coating, the electrical resistance being reduced. With the CO_2 process the distance from contact tube to wire tip is of the order of 20 mm so that the electrical resistance is greatly reduced even though the wire diameter is smaller. The current can thus be increased greatly, resulting in higher deposition rates, greater welding speeds and reduced labour charges. In addition the duty cycle is increased since there is no constant electrode change and need for deslagging, but there may be greater spatter loss. Economically therefore the CO_2 process shows an advantage in very many applications, though the final choice of process is governed by the application and working conditions.

By using the argon–CO_2 gas mixtures, certain advantages are obtained over CO_2 which may offset the greater price of the mixtures. These are faster welding speed and reduced spatter, with an improved weld profile.

When welding thin steel sections with a dip transfer arc, argon–CO_2 mixtures are often preferred because there is a lower voltage drop across the arc which thus gives a cooler molten pool and prevents burn-through, especially if the fit-up is poor.

Details of plate preparation are given as a guide only, since there are too many variables to give generalized recommendations (see Figs. 2.18 and 2.19).

Automatic welding

Automatic welding, by MIG, pulse and CO_2 processes, now plays an important part in welding fabrication practice. It enables welds of consistently high quality and accuracy to radiographic standard to be

Fig. 2.18. Various preparations for thicker plate (varying with applications). CO_2 welding.

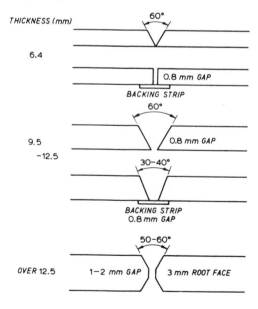

performed at high welding speeds because of the close degree of control over the rate of travel and nozzle-to-work distance. It is less tolerant than semi-automatic welding to variations of root gap and fit-up but reduces the number of start–stop breaks in long sequences. The choice between semi-automatic and automatic process becomes a question of economics, involving the length of runs, number involved, volume of deposited metal if the sections are thick, method of mechanization and set-up time. The torches are now usually air cooled even for currents up to 450 A and are carried on welding heads fitted with controls similar to those used for semi-automatic welding, and may be remote controlled (Fig, 2.20).

The head may be: (1) fixed, with the work arranged to move or be rotated beneath it, (2) mounted on a boom and column which can either be of the

Mild steel sheet, butt welds, CO_2 shielding, flat, 0.8 mm diam. wire (approximate values)

Thickness (mm)	Gap (mm)	Wire feed (m/minute)	Arc (volts)	Current (A)
1	0	2.8–3.8	16–17	65–80
1.2	0	3.2–4.0	18–19	70–85
1.6	0.5	4.0–4.8	19–20	85–95
2.0	0.8	5.8–7.0	19–20	110–125
2.5	0.8	7.0–8.4	20–21	125–140
3.0	1.5	7.0–8.4	20–21	125–140

Fig. 2.19. (*a*) Unprepared fillets. (*b*) multi-run prepared fillets in thick plate, (*c*) deep preparation fillets, (*d*) multi-run unprepared fillet. CO_2 welding.

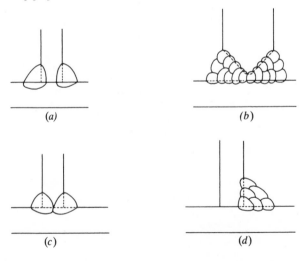

(*a*)   (*b*)

(*c*)   (*d*)

positioning type in which the work moves, or the boom can traverse over the work (Figs. 2.21a and b), (3) gantry mounted so as to traverse over the stationary work, (4) tractor mounted, running on guide rails to move over the fixed work, (5) mounted on a special machine or fixture designed for a specific production. A head may carry two torches arranged to weld simultaneously, thus greatly reducing the welding time.

For the CO_2 process in steel a typical example would be with wire of 1.2 mm diameter using 150–170 A on thinner sections and multiple passes up to 30 on thicknesses up to 75 mm with current in the 400–500 A range and 2.4 mm wire. With automatic surge (pulse) arc welding on stainless

Fig. 2.20. MIG head showing wire feed.

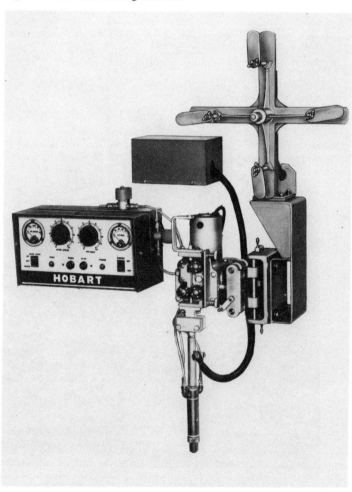

steel, accurate control of the underbead is achieved, obviating the necessity for back-chipping and sealing run. In the case of aluminium welding on plate above 10 mm thickness the accuracy of the underbead produced with the MIG automatic process results in more economical welding than by the double-operator vertical TIG method. In general the full automation of these processes results in greater productivity with high-quality welds.

Magnetic arc control systems

Magnetic probes with electronic solid state control and with water cooling if required can be fitted onto or near to the welding head and are applicable to spray, MIG, TIG, plasma and submerged arc systems (Fig. 2.22a and b).

Fig. 2.21. (a) A ram-type boom with welding head.

Fig. 2.21. (*b*) Extra heavy duty column and boom and roller bed.

Fig. 2.22. (*a*) Types of probe. (*b*) Magnetic arc control.

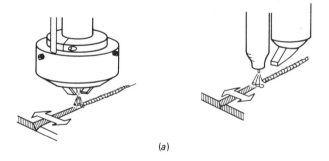

(*a*)

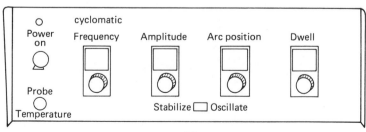

(*b*)

The arc can be stabilized or oscillated by the applied magnetic field and magnetic and non-magnetic materials are equally well welded when using this system. Controls on the unit enables the sweep frequency to be varied from 0–100 oscillations per second while sweep amplitude and arc position are controllable and proportional to the arc length in an approximately 1:1 ratio. There is controllable variable dwell on each side of the oscillation in order to obtain good fusion, arc blow is counteracted and the heat imput can be balanced when welding thin to thick sections, and undercut is minimized. Porosity is decreased and the system is applicable, for example, to the welding of thicker stainless steel sections resulting in reduced danger from cracking, reduced porosity and greater weld reliability. This method of arc stabilization and control is now becoming increasingly popular as MIG and submerged arc welding become more automated.

Seam tracker

This unit bolts on to the head of the welding torch and enables the torch to follow accurately the line of the joint to be welded.

The torch head is placed over the seam or joint and the tracker keeps the arc over the seam with an accuracy of less than 0.25 mm.

One type has an electro-mechanical sensor which responds to the movement of the probe tip which moves along the seam. The signals are sent by the probe to the solid state control which operates two linear slides whose axes are at right angles to each other, one horizontal and the other vertical. These are driven by small d.c. motors and keep the welding head in place exactly over the seam to be welded, which keeps the torch-to-work distance constant (Fig. 2.23).

Fig. 2.23. Programmable seam tracker.

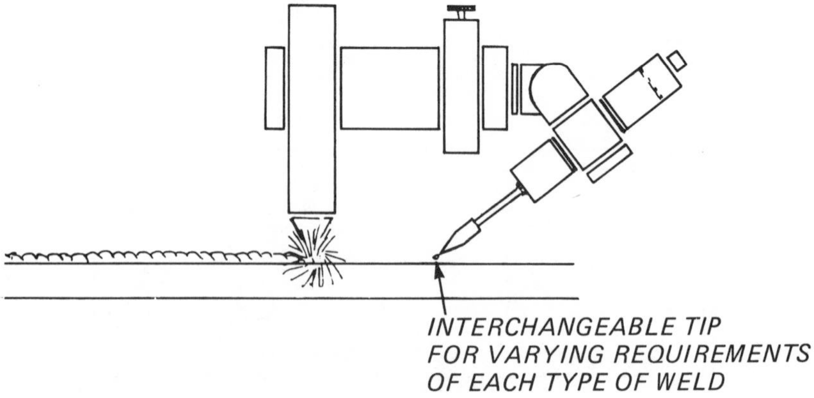

INTERCHANGEABLE TIP
FOR VARYING REQUIREMENTS
OF EACH TYPE OF WELD

The optical seam tracker using flexible fibre optics for operation and with microcircuit control operating the cross slides as in the previous unit keeps the torch head accurately over the joint, which should be a tight machined edge butt weld, although this system can tolerate a gap of 2 mm and still keep the torch central to the joint.

MIG spot welding

Spot welding with this process needs access to the joint on one side only and consists of a MIG weld, held in one spot only, for a controlled period of time. Modified nozzles are fitted to the gun, the contact tip being set 8–12 mm inside the nozzle, and the timing unit can either be built into the power unit or fitted externally, in which case an on–off switch does away with the necessity of disconnecting the unit when it is required for continuous welding. The timer controls the arcing time, and welds can be made in ferrous material with full or partial penetration as required. A typical application of a smaller unit is that of welding thin sheet as used on car bodies. The spot welding equipment timer is built into a MIG unit and is selected by a switch. Wire of 0.6 mm diameter is used with argon + CO_2 5% + O_2 2% as the shielding gas. Currents of 40–100 A are available at 14–17 V for continuous welding with a maximum duty cycle of 60%. The spot welding control gives a maximum current of 160 A at 27 V, enabling spot welds to be made in material down to 0.5 mm thickness, the timing control varying from 0.5–2.0 seconds.

On larger units arcing time can be controlled from 0.3 to 4.5 seconds with selected heavier currents enabling full penetration to be achieved on ferrous plate from 0.7 to 2.0 mm thick and up to 2.5 mm sheet can be welded on to plate of any thickness.

Cycle arc welding is similar to spot welding except that the cycle keeps on repeating itself automatically as long as the gun switch is pressed. The duration of the pause between welds is constant at about 0.35 seconds while the weld time can be varied from 0.1 to 1.5 seconds. This process is used for welding light-section components which are prone to burn-through. In the pause between the welding period, the molten pool cools and just solidifies, thus giving more accurate control over the molten metal.

MIG pulsed arc welding

Pulsed arc welding is a modified form of spray transfer in which there is a controlled and periodic melting off of the droplets followed by projection across the arc. A pulse of current is applied for a brief duration at regular frequency and thus results in a lower heat output than with pure spray transfer, yet greater than that with dip transfer. Because of this, thinner sections can be welded than with spray transfer and there is no danger of poor fusion in a root run as sometimes occurs with dip transfer and positional welding is performed much more easily. There is regular and even penetration, no spatter and the welds are of high quality and appearance.

To obtain these conditions of transfer it is necessary to have two currents fed to the arc: (*a*) a background current which keeps the gap ionized and maintains the arc and (*b*) the pulsed current which is applied at 50 or 100 Hz and which melts off the wire tip into a droplet which is then projected across the arc gap. These two currents, which have critical values of satisfactory welding conditions are to be obtained, are supplied from two sources, a background source and a pulse source contained in one unit, and their voltages are selected separately. The background current, of much lower value than the pulse, is half a cycle out of phase with it (see Figs. 2.24*a* and *b*). A switch enables the pulse source to be used for dip and spray transfer methods as required. The power supply is a silicon rectifier with constant voltage output and a maximum current value of about 350 A at open-circuit voltages of 11–45 V similar to those already described.

To operate the unit, the 'pulse height' (the value of pulse current) and the background current are selected on separate switches, the wire is adjusted to protrude about 10 mm beyond the nozzle and welding is commenced

Fig. 2.24. (*a*) Wave shape of pulse supply.

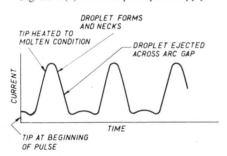

Fig. 2.24. (*b*) Modified pulse supply with background pulse.

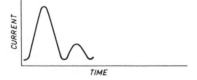

moving from right to left down the line of the weld with the gun making an angle of 60–70° with the line of weld. When welding fillets the gun is usually held at right angles to the line of weld with the wire pointing directly into the joint, and thus an excellent view is obtained of the degree of fusion into the root; welding is again performed from right to left. The process gives good root fusion with even penetration and good fill-in and is especially efficient when used fully automatically. As in all other welding processes, welding is best performed flat but positional welding with pulsed arc is very satisfactory and relatively easy to perform. Thicknesses between 2 mm and 6.5 mm which fall intermediate in the ranges for dip and spray transfer are easily welded. One of the drawbacks to pulse arc is the necessity to ensure accurate fit-up. If there are any sizeable gaps a keyhole effect is produced and it is impossible to obtain a regular underbead. The gap should be of the order of from 0 to about 1.0 mm max. The shielding gas is argon with 1% or 2% oxygen or argon with 5% CO_2 and 2% oxygen for welding mild and low-alloy steels, stainless steel and heat-resistant steel, and pure argon for aluminium and its alloys, 9% nickel steel and nickel alloys. Pulse arc is especially useful when used automatically for stainless steel welding, since the accurately formed under- or penetration bead obviates the expensive operation of back-chipping and a sealing run and there is little carbon pick-up and thus little increase of carbon content in the weld. Aluminium requires no back purge and for the back purge on stainless steel a thin strip of plate can be tack welded on the underside of the point (see Fig. 2.13) and removed after welding. The argon from the torch supplies the back purge and prevents oxidation of the underside of the weld.

There is good alloy recovery when welding alloy steel and because of the accurate heat control, welding in aluminium is consistently good without porosity and a regular underbead so that it can be used in place of double argon TIG with a saving of time and cost.

Fabrications and vessels can be fully tack weld fabricated with TIG on the underside of the seam and pulse arc welded, greatly reducing the tendency to distortion. The torch should be held at 75–80° to the line of the weld and good results are also obtained fully automatically by welding from left to right with the torch held vertically to the seam. This method allows greater tolerance in fit-up and preparation with better penetration control. Because this process is sensitive to the accuracy of fit-up and preparation, care should be taken to work to close tolerances and excessive gaps should be made up from the reverse side with, for example, the TIG process, and then carefully cleaned.

Fig. 2.25 and the table give a typical preparation for automatic flat butt welds in stainless steel using 99% A, 1% O_2 as shielding gas and 1.6 mm diameter filler wire. For the alloys of nickel the recommended shielding gas is pure argon. A similar technique is used as for the MIG welding of stainless steel, with a slight pause each side of the weave to avoid undercut.

Plate thickness (mm)	Runs	Current (A) Roots and subsequent	Volts (approx.)
4.8, 6.4, 8	2	180–200	27–28
9.5	3	180–210	27–28

There appears to be no advantage in using the pulse method over the normal shielded metal arc with unpulsed wave when using CO_2 as the shielding gas. (See Appendix 10).

Plasma MIG process. A process is being developed in which MIG filler wire is fed through a contact tube situated centrally in the torch head. A tungsten electrode is set at an angle to this contact tube and projects nearly to the mouth of the shielding nozzle through which the plasma gas issues. An outer nozzle provides a gas shield. The current through the wire assists the melting and gives good starting characteristics and a stable arc, so that thin plate is weldable at high welding speeds.

Repair of cast iron by automated MIG process using tubular electrodes. With the advent of flux cored wire (2.10*a*) carbon can be added to the core and high speed welding performed using high nickel iron wire. As an example, nickel FC 55 is a tubular wire, the core being filled with carbon, slagging constituents and deoxidizers. It can be used with or without a shielding gas such as CO_2 and no post-weld heat treatment is required. Standard gas–metal arc equipment is used, the joints have high strength and the HAZ is free from the carbide complex which can cause embrittlement. It makes possible the automatic welding of cast irons to themselves or to stainless

Fig. 2.25. Automatic pulsed arc welding of stainless steel, flat, 99% argon–1% oxygen.

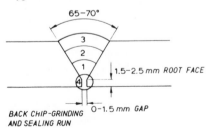

steel or high nickel alloys and can be used for overlaying and metal arc spot welding.

Liquid (gas) storage and gas mixing systems

The gas mixer (Fig. 2.26) mixes the gases in proportions suitable for welding. The proportion of each gas can be adjusted and the panel maintains this proportion over a wide range of flow requirements. A typical example is the mixing of argon and carbon dioxide as in the Coogar or Argoshield range (see appendix 3).

Fig. 2.26. Gas mixing system for mixing shielding gases.

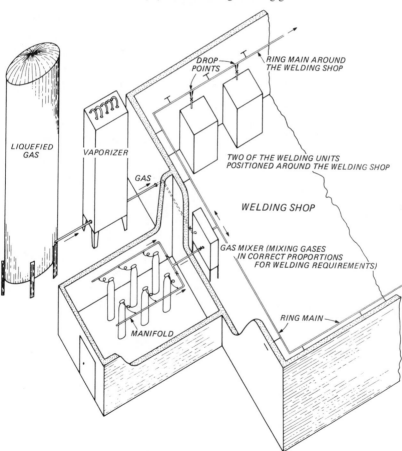

3

Tungsten electrode, inert gas shielded welding processes (TIG), and the plasma arc process*

Technology and equipment

The welding of aluminium and magnesium alloys by the oxy-acetylene and manual metal arc processes is limited by the necessity to use a corrosive flux. The gas shielded, tungsten arc process (Fig. 3.1) enables these metals and a wide range of ferrous alloys to be welded without the use of the flux. The choice of either a.c. or d.c. depends upon the metal to be welded. for metals having refractory surface oxides such as aluminium and its alloys, magnesium alloys and aluminium bronze, a.c. is used whilst d.c. is used for carbon and alloy steels, heat-resistant and stainless steels, copper and its alloys, nickel and its alloys, titanium, zirconium and silver.

Fig. 3.1. Connexions for inert gas welding using air-cooled torch.

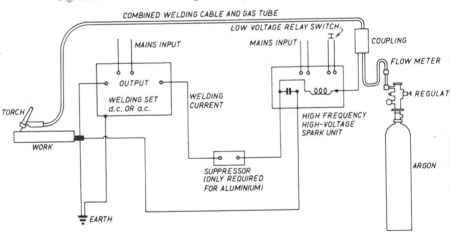

*American designation: gas tungsten arc welding (GTAW).

(See also BS 3019, *General recommendations for manual inert gas tungsten arc welding*. Part 1, *Wrought aluminium, aluminium alloys and magnesium alloys*; Part 2, *Austenitic stainless and heat resisting steels*.)

The arc burns between a tungsten electrode and the workpiece within a shield of the inert gas argon, which excludes the atmosphere and prevents contamination of electrode and molten metal. The hot tungsten arc ionizes argon atoms within the shield to form a gas plasma consisting of almost equal numbers of free electrons and positive ions. Unlike the electrode in the manual metal arc process, the tungsten is not transferred to the work and evaporates very slowly, being classed as 'non-consumable'. Small amounts of other elements are added to the tungsten to improve electron emission. It is, however, a relatively slow method of welding.

Gases

Argon in its commercial purity state (99.996%) is used for metals named above, but for titanium extreme purity is required. Argon with 5% hydrogen gives increased welding speed and/or penetration in the welding of stainless steel and nickel alloys; nitrogen can be used for copper welding on deoxidized coppers only. Helium may be used for aluminium and its alloys and copper, but it is more expensive than argon and, due to its lower density, a greater volume is required than with argon to ensure adequate shielding, and small variations in arc length cause greater changes in weld conditions. A mixture of 30% helium and 70% argon is now used, and gives fast welding speeds. The mechanized d.c. welding of aluminium with helium gives deep penetration and high speeds.

The characteristics of the arc are changed considerably with change of direction of flow of current, that is with arc polarity.

Electrode positive

The electron stream is from work to electrode while the heavier positive ions travel from electrode to work-piece (Fig. 3.2a). If the work is of aluminium or magnesium alloys there is always a thin layer of refractory oxide of melting point about 2000 °C present over the surface and which has to be dispersed in other processes by means of a corrosive flux to ensure weldability. The positive ions in the TIG arc bombard this oxide and, together with the electron emission from the plate, break up and disperse the oxide film. It is this characteristic which has made the process so successful for the welding of the light alloys. The electrons streaming to the tungsten electrode generate great heat, so its diameter must be relatively large and it forms a bulbous end. It is this overheating with consequent vaporization of the tungsten and the possibility of tungsten being transferred to the molten pool (pick-up) and contaminating it that is the drawback to the use of the process with electrode positive. Very much less heat is generated at the molten pool and this is therefore wide and shallow.

Electrode negative

The electron stream is now from electrode to work with the zone of greatest heat concentrated in the workpiece so that penetration is deep and the pool is narrower. The ion flow is from work to electrode so that there is no dispersal of oxide film and this polarity cannot be used for welding the light alloys. The electrode is now near the zone of lesser heat and needs be of reduced diameter compared with that with positive polarity. For a given diameter the electrode, when negative, will carry from four to eight times

Fig. 3.2. Electron streams between electrode and work: (*a*) d.c., tungsten electrode +ve of large diameter tends to overheat; (*b*) d.c., tungsten electrode −ve of small diameter; (*c*) a.c., electrode diameter between that of electrode +ve and −ve electrode.

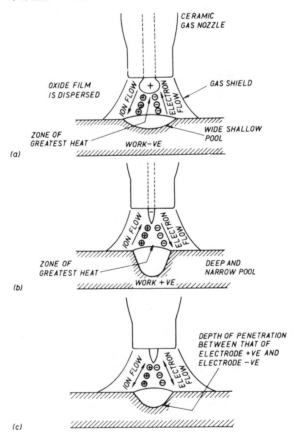

the current than when it is positive and twice as much as when a.c. is used. (Fig. 3.2*b*.)

Alternating current

When a.c. is used on a 50 Hz supply, voltage and current are reversing direction 100 times a second so that there is a state of affairs between that of electrode positive and electrode negative, the heat being fairly evenly distributed between electrode and work (Fig. 3.2*c*). Depth of penetration is between that of electrode positive and electrode negative and the electrode diameter is between the previous diameters. When the electrode is positive it is termed the positive half-cycle and when negative the negative half-cycle. Oxide removal takes place on the positive half-cycle.

See note on square wave equipment and wave balance control (pp. 145–9).

Inherent rectification in the a.c. arc

In the a.c. arc the current in the positive half-cycle is less than that in the negative half-cycle (Fig. 3.3*b*). This is known as inherent rectification and is a characteristic of arcs between dissimilar metals such as tungsten and aluminium. It is due to the layer of oxide acting as a barrier layer to the current flowing in one direction and to the greater emission of electrons from the tungsten electrode when it is of negative polarity. The result of this imbalance is that an excess pulsating current flows in one direction only and the unbalanced wave can be considered as a balanced a.c. wave, plus an excess pulsating current flowing on one direction only on the negative half-cycle. This latter is known as the d.c. component and can be measured with

Fig. 3.3. Alternating current: (*a*) balanced wave; (*b*) unbalanced wave, inherent rectification (current in +ve half-cycle less than that in −ve half-cycle).

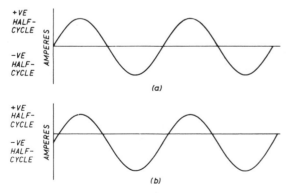

a d.c. ammeter. (The suppression of this d.c. component is discussed later.) The reduction of current in the positive half-cycle due to the inherent rectification results in a reduction of oxide removal.

Partial rectification

A greater voltage is required to strike the arc than to maintain it and re-ignition on the negative half-cycle requires a lower voltage than for the positive half-cycle, partly due to the greater electron emission from the tungsten when it is negative polarity, but actual re-ignition depends upon many factors including the surface condition of the weld pool and electrode, the temperature of the pool and the type of shielding gas. There may be a delay in an arc re-ignition on the positive half-cycle until sufficient voltage is available and this will result in a short period of zero current (Fig. 3.4a) until the arc ignites. This delay reduces the current in the positive half-cycle and this state is known as partial rectification. If the available voltage is not sufficient, ignition of the arc may not occur at all on the positive half-cycle, the arc is extinguished on the one half-pulse and continues burning on the uni-directional pulses of the negative half-cycle, and we have complete rectification with gradual extinguishing of the arc (Fig. 3.4b).

Re-ignition voltages

To ensure re-ignition of the arc on the positive half-cycle, the available voltage should be of the order of 150 V, which is greater than that of the supply transformer. To ensure re-ignition, auxiliary devices are used which obviate the need for high open-circuit transformer voltages.

Fig. 3.4. (a) Partial rectification; (b) half-wave rectification (+ ve half-cycle misssing).

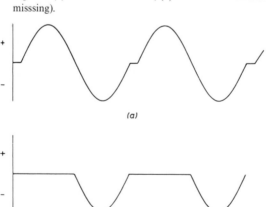

(a)

(b)

Ignition and re-ignition equipment

High-frequency, high-voltage, spark gap oscillator. This device enables the arc to be ignited without touching down the electrode on the work and thus it prevents electrode contamination. It also helps arc re-ignition at the beginning of the positive half-cycle.

The oscillator consists of an iron-cored transformer with a high voltage secondary winding, a capacitor, a spark gap and an air core transformer or inductive circuit, one coil of which is in the high voltage circuit and the other in the welding circuit (Fig. 3.5). The capacitor is charged every half cycle to 3000–5000 V and discharges across the spark gap. The discharge is oscillatory, that is, it is not a single spark but a series of sparks oscillating across the spark gap during discharge. This discharge, occurring on every half-cycle, sets up oscillatory currents in the circuit and these are induced and superimposed on the welding current through the inductance of the coils *L* (Fig. 3.6).

The spark discharge is phased to occur at the beginning of each half-cycle (although for re-ignition purposes it is required only for the positive half-cycle) and is of about 5 milliseconds or less duration compared with the

Fig. 3.5. High-frequency spark oscillator.

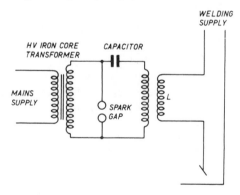

Fig. 3.6. Arc voltage with superimposed HF spark main for re-ignition and stabilization; the HF is injected on both +ve and −ve half-cycles, but is only required at the beginning of the former.

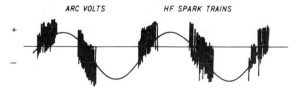

half-cycle duration of 10 milliseconds (Fig. 3.7). To initiate the arc, the electrode is brought to about 6 mm from the work with the HF unit and welding current switched on. Groups of sparks pass across the gap, ionizing it, and the welding current flows in the form of an arc without contamination of the electrode by touching down. The HF unit can give rise to considerable radio and TV interference and adequate suppression and screening must be provided to eliminate this as far as possible. The use of HF stabilization with the a.c. arc enables this method to be used for aluminium welding, although inherent rectification is still present, but partial rectification can be reduced to a minimum by correct phasing of the spark train.

Scratch start. Some units do not employ an HF unit for starting the arc without touch down. The tungsten is scratched momentarily onto the work and the arc is ionized. Naturally there will be some very small tungsten contamination but this will not greatly interfere with the mechanical properties of the joint in non-critical conditions.

Surge injection unit. This device supplies a single pulse surge of about 300 V phased to occur at the point when the negative half-cycle changes to the positive half-cycle (Fig. 3.7), so that the pulse occurs at 20 millisecond intervals and lasts for only a few microseconds. This unit enables transformers of 80 V open-circuit voltage to be used and it does not produce the high frequency radiation of the spark oscillator and thus does not interfere to any extent with radio and TV apparatus.

The unit consists of a rectifier which supplies d.c. to a circuit containing resistance and capacitance, a surge valve which supplies the short pulses and a trigger valve which releases the pulses into the welding circuit. The

Fig. 3.7. Surge voltage superimposed on arc voltage.

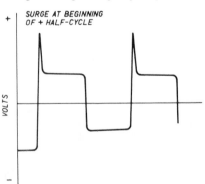

trigger valve, which is sensitive to change in arc voltage, releases the pulse at the end of the negative half-cycle, just when the positive half-cycle is beginning. As the pulse is usually unable to initiate the arc from cold, a spark oscillator is also included in the unit and is cut out of circuit automatically when the arc is established so that the surge injection unit is an alternative method to the HF oscillator for re-igniting the arc on the positive half-cycle (Fig. 3.8).

Suppression of the d.c. component in the a.c. arc

In spite of the use of the HF unit, the imbalance between the positive and negative half-cycles remains and there is a d.c. component flowing. This direct current flows through the transformer winding and saturates the iron core magnetically, giving rise to high primary currents with such heating effect that the rating of the transformer is lowered, that is, it cannot supply its rated output without overheating. The insertion of banks of electrolytic capacitors in series with the welding circuit has two effects:

(1) They offer low impedence to the a.c. which flows practically uninterrupted, but they offer very high impedance to the d.c. which is therefore suppressed or blocked.

(2) During the negative half-cycle the capacitors receive a greater charge (because of the imbalance) and during the following positive half-cycle this excess adds to the positive half-cycle voltage so that it is increased, and if the open-circuit voltage is greater than about 100 V the arc is re-ignited on each half-cycle without the aid of high-frequency currents, which can then be used for starting only. The effect of this increase in voltage on the positive half-cycle is to improve the balance of the wave (Fig. 3.8), so that the heating effect between electrode and plate is more equal and disposal of the oxide film is increased.

Fig. 3.8. Current and arc voltage wave-forms showing re-ignition.

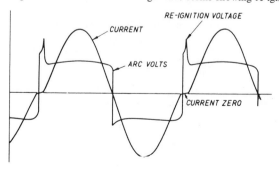

The value of the capacitor must be chosen so that there is no danger of electrical resonance in the circuit which contains resistance, inductance and capacitance, which would result in dangerous excessive currents and voltages irrespective of the transformer output (Fig. 3.9*a*). If external capacitors are used they may affect the current output of the transformer by altering the power factor, and this must be taken into account since the current calibrations on the set will be increased.

The cost of large banks of capacitors however is considerable and is accompanied by a high voltage across them, so that this method has been largely superceded in the units which use solid state technology and thyristor control (see later).

Power sources a.c. and d.c.

Equipment can be chosen to give a.c. or d.c. or both a.c. and d.c. from one unit and may even be designed for specific industries: (1) d.c. output for the fabrication of a variety of steels and special steels such as stainless, heat resistant and 9% Ni, etc., (2) a.c. output for fabrication aluminium and its alloys, (3) a.c. and d.c., which includes the fabrication of both the above ferrous and non-ferrous metals and it is this type that covers most types of fabrication.

a.c. power unit. For the light alloys of aluminium and magnesium a transformer similar to that used for MMA welding can be used. Cooling can be by forced draught or oil, usually the former, and primary tappings are provided on the input side for single- or three-phase 50 Hz supply, with an output voltage of about 80 V. Current control can be by tapped choke, saturable reactor and auxiliary units such as HF oscillator, d.c. component suppressor, and surge injector can be built in or fitted externally. Scratch start obviates the use of the HF oscillator but has certain disadvantages.

d.c. power unit. These units consist of a step-down transformer with input from single- or three-phase 50 Hz mains which feeds into a silicon bridge

Fig. 3.9. (*a*) Series capacitor used to block d.c. component.

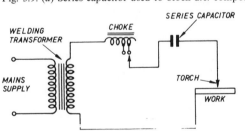

rectifier (SCR) which gives a stepless control of current and obviates the use of other methods of current control. On larger units there may be a two position switch for high or low current ranges, and these are usually constant voltage units.

a.c. and d.c. power unit. These units, which supply either a.c. or d.c., are connected to single- or three-phase 50 Hz mains, fed into a step-down transformer and then into a silicon controlled rectifier (SCR) which may also act as a contactor so that there are no mechanical parts to open and close when welding is begun and ended.

The SCR's, which can be switched under load, are fitted to a heat sink and the whole is forced draught cooled. Current can be supplied for MMA and TIG and output can be for electrode +ve or −ve or a.c. by means of the controlling printed circuit.

The auxiliary supply is at a lower voltage of about 110 V and there is automatic regulation for variation of the mains voltage. In addition there can be burn back control, soft start, pulse and spot welding facilities and the units are suitable for using hard or soft solid core wires and flux cored wire using either short circuit or spray transfer.

Duty cycle

Over a period of 10 minutes, a 60% duty cycle means that the unit can be used at that particular current (about 275 A in Fig. 3.9*b*) for 6

Fig. 3.9. (*b*) Illustrating how the welding current decreases as the % duty cycle of a power unit increases.

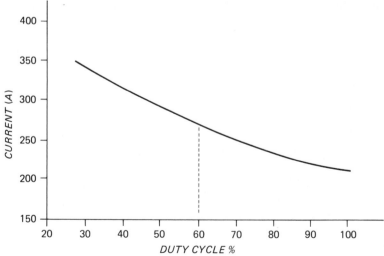

minutes and then given 4 minutes to cool down. Increasing the welding current above that recommended for the particular duty cycle increases the quantity of heat evolved and the excessive temperature rise can lead to failure. Decreasing the welding current increases the period within the 10 minutes for which the unit can be operated. Thermal cut-outs may be fitted which trip and prevent serious damage while SCR's are bolted to a heat sink and the whole unit force draught cooled.

Exceeding the duty cycle value of current can cause overheating and may damage the unit.

Slope-in (slope-up) and soft start. The use of the normal welding current at the beginning of the weld often causes burn-through and increases the risk of tungsten contamination. To prevent this a slope-in control is provided which gives a controlled rise of the current to its normal value over a preselected period of time, which can be from 0 to 10 seconds. The soft start control is available on some equipment and performs the same function but it has a fixed time of operation and can be switched in or out as required.

Slope-out (slope-down) (crater fill). The crater which would normally form at a weld termination can be filled to the correct amount with the use of this control. The crater-filling device reduces the current from that used for the welding operation to the minimum that the equipment can supply in a series of steps. The slope-out control performs the same function, but the current is reduced over a period of about 20 seconds depending upon the setting of the control (not steps). The tendency to cracking is reduced by the use of this control (Fig. 3.10).

Fig. 3.10

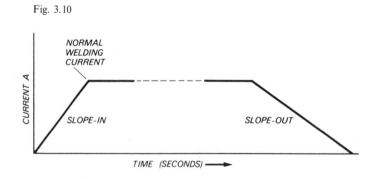

Thyristor or silicon controlled rectifier (SCR) control of welding current

The value of welding current required is set by the single dial or knob current control on the front panel of the unit. This value is compared by the control circuit (p.c.b.) with the value of output current received from the sensor in the outgoing welding line and the firing angle of the SCR's is altered to bring the output current to the value set by the operator on the control dial. Hence stepless control is achieved (Fig. 3.11a, b). A simple explanation of the operation of an SCR, with diagrams, is given in appendix 2.

Square wave output units (a.c./d.c.)

Power units are now available in which the voltage and current waves are not sinusoidal, but have been modified by solid state technology

Fig. 3.11. (a) Heavy duty, a.c., d.c., 400 A power source for TIG welding. Input, single-phase 220, 380–415, 440 or 550 V with two current ranges. OCV's, a.c. 69 V, d.c. 94 V. Load at 35% duty cycle, a.c. 400 A, 26 V.; d.c. 370 A, 25 V. Load at 60% duty cycle, a.c. 340 A, 24 V; d.c. 320 A, 23 V. Single knob control of welding current by SCR's giving infinitely variable current. Built in HF unit. Pulsed current, slope-up and slope-down. TIG spot welding MMA and remote control facilities. Pre- and post-flow gas. Water cooling has built in cooling unit.

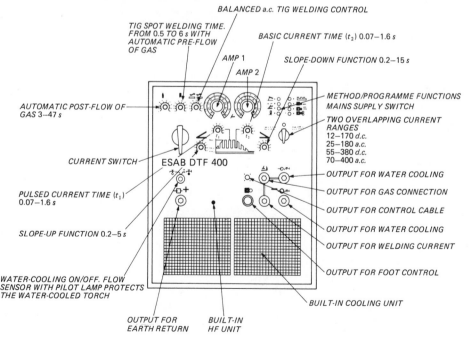

Fig. 3.11. (*b*) TR250–HF, TIG and MMA power unit a.c. and d.c. Input 220, 380, 415 or 550 V single phase 50 Hz a.c. Maximum output at 80 OCV, 250 a.c. or d.c. at 40% duty cycle. Three current ranges from 5 A to 310 A at 30 V. SCR control. Start-continuous HF switch with intensity control. Gas and water valves and overload protection.

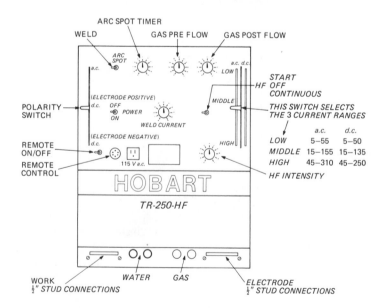

and using printed circuit boards (p.c.b.'s) to give a very rapid rise of the a.c. wave from zero value to maximum value to give a 'square wave' output (Fig. 3.12).

When TIG welding aluminium using a sinusoidal wave form current the arc tends to become unstable and the electrode is easily overloaded. This gives tungsten inclusions in the weld (spitting) and leads to faults in the weld bead and more rapid consumption of the tungsten electrode. Square wave current overcomes these drawbacks and the arc is greatly stabilised and risk of inclusions greatly reduced. In addition MMA welding characteristics are greatly improved and the arc is smooth with reduced spatter.

The units are designed for precision a.c. TIG welding of aluminium, etc., and for d.c. TIG and manual metal arc welding. They have a transformer and a silicon bridge rectifier, SCR or thyristor with a square wave output. A memory core stores energy proportional to the previous half-cycle and then injects it into the circuit just as the wave passes through zero at the beginning of the next half-cycle. The rapid rise of the wave from zero to maximum, of about 80 microseconds from peak to peak (Figs. 3.13 and 14) means that the high frequency and high voltage at the beginning of the cycle needed to initiate the arc and often to keep it ionized without touchdown, need only be applied when the arc is first initiated. After this first initiation the HF is switched off automatically and the rapid rise of voltage from peak to peak keeps the arc ionized without application of the HF. A switch enables this HF to be used continuously if required although its continuous use may interfere with radio frequency apparatus in the vicinity.

Variation in mains voltage is compensated automatically, there is provision for MMA, pulse and spot welding and often there may be two

Fig. 3.12. a.c. square wave output.

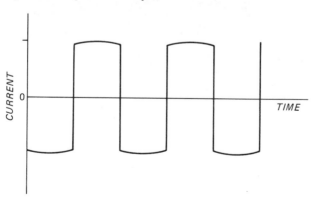

ranges of current, either of which can be controlled by the one knob control (Fig. 3.13).

Soft start (slope-up) and crater fill (slope-down) are provided with pre- and post-gas and water supply controls. Pulse duration, height and background and wave balance controls are also fitted.

a.c. wave balance control.

On square wave equipment the wave balance control enables either tungsten electrode or work to be biased as required. Fig. 3.15 indicates how this is used for controlling penetration and cleaning action. (1) is a balanced wave. Moving the control to one side gives a wave balance (2) in which the tungsten electrode is positive polarity for a larger period,

Fig. 3.13. Syncrowave 300 and 500 A square wave a.c., d.c., power unit for TIG and MMA welding. Input 220–576 V as required, mains voltage compensated. 500 A model gives 500 A at 100% duty cycle. Solid state contactors, slope-in and slope-out. Pulsed mode for improved penetration and control of weld pool. Rectification by SCR's with one knob control of welding current. 300 A model has dual current range, 5–75 A and 15–375 A, 500 A model has single range 25–625 A. Square wave output has wave balance control. Post timer for gas and water. Slope-up and slope-down (crater fill).

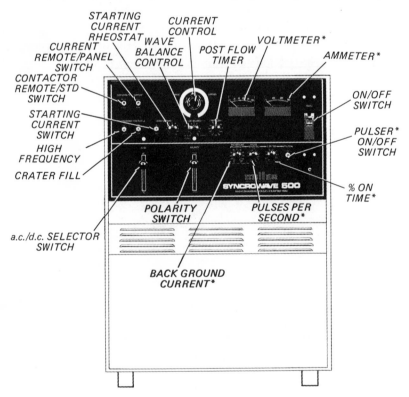

giving maximum cleaning but with minimum penetration. With the control in the other direction the greatest heat is in the work so that we have maximum penetration but with minimum cleaning. With d.c. TIG welding as with MMA the control is set for a balanced wave.

Electrical contactors

The contactors control the various circuits for welding, inert gas, water flow and ancillary equipment. The contactor control voltage is of the order of 25–45 V and if the TIG head is machine mounted, a 110 V supply for the wire feed and tractor is provided. The post-weld argon flow

Fig. 3.14. DTA 200 a.c., d.c. power unit, 200 A with square wave a.c. wave form. Forced draught cooling. Single phase input 50–60 Hz with input voltages 220, 380, 415 V with mains compensation of 10% o.c. volts a.c. 59 V, d.c. 52 V. Mains voltage stabilisation. Slope-up and slope-down, built-in HF generator, pre- and post-flow shielding gas, control by SCR's giving one knob control of welding current. Remote control and pulse unit facilities.

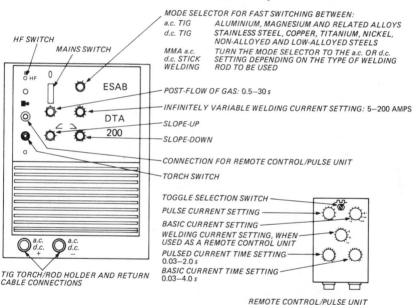

Fig. 3.15. Wave balance control.

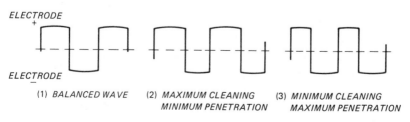

contactor allows the arc to be extinguished without removing the torch with its argon shield from the hot weld area, thus safeguarding the hot electrode end and weld area from contamination whilst cooling. An argon–water purge is also provided and contactors may be foot-switch operated for convenience. Where the ancillary equipment is built into one unit an off-manual metal arc–TIG switch enables the unit to be used for either process.

With the advanced SCR method of control the SCR's act as a contactor so that there are no mechanical contactors in the welding current circuit. Cooling is by heat sink and forced draught and this method is very reliable since there are no moving parts.

Gas regulator, flowmeter and economizer

The gas regulator reduces the pressure in the argon cylinder from 17.5 N/mm^2 (175 bar)* down to 2 bar for supply to the torch (Fig. 3.16*a* and *b*). The flowmeter, which has a manually operated needle valve, controls the argon flow from 0–600 litres/hour to 0–2100 litres/hour according to type (Fig. 3.16*c*).

The economizer may be fitted in a convenient position near the welder and when the torch is hung from the projecting lever on the unit, argon gas and (if fitted) water supplies are cut off. A micro-switch operated by the lever can also be used to control the HF unit.

Fig. 3.16. (*a*) Single-stage argon regulator and flowmeter.

---

* Cylinders are now filled to 200 bar maximum.

Fig. 3.16. (*b*) Two-stage argon regulator.

Fig. 13.16. (*c*) Argon flowmeter.

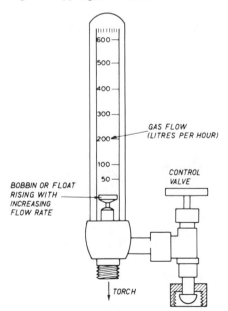

Additional equipment

Add-on equipment can be used with existing transformers and rectifiers. The equipment shown in Fig. 3.17 is fitted with a pulse generator and can be connected to a thyristor controlled rectifier power unit or to a.c. or d.c. arc welding sources. When connected to a d.c. source the pulse generator is used for arc ignition only, being automatically cut off after ignition, but it cuts in again if required to establish the arc. When used with an a.c. source the pulse generator ensures arc ignition without touchdown and ensures the re-ignition of the arc when welding. Soft start, crater fill, current control with one knob and remote control facilities are also fitted.

Torch

There is a variety of torches available varying from lightweight air cooled to heavy duty water cooled types (Fig. 3.18a and b). The main factors to be considered in choosing a torch are:

(1) Current-carrying capacity for the work in hand.
(2) Weight, balance and accessibility of the torch head to the work in hand.

The torch body holds a top-loading compression-type collet assembly which accommodates electrodes of various diameters. They are securely gripped yet the collet is easily slackened for removal or reposition of the electrode. As the thickness of plate to be welded increases, size of torch and electrode diameter must increase to deal with the larger welding currents required.

Small lightweight air cooled torches rated at 75 A d.c. and 55 A a.c. are ideal for small fittings and welds in awkward places and may be of pencil or

Fig. 3.17. Add-on unit for TIG welding.

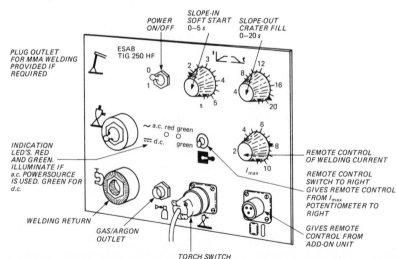

Fig. 3.18. (*a*) 160 A air-cooled torch (manual). Torches are air cooled for lower currents and water cooled for higher currents.

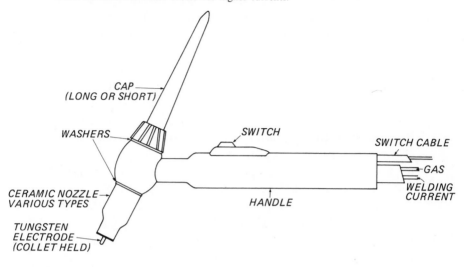

Fig. 3.18. (*b*) A water-cooled torch.

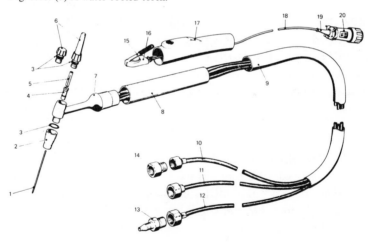

Key:

1. Thoriated or zirconiated tungsten electrode (0.8, 1.2, 1.6, 2.4, 3.2 mm diameter).
2. Ceramic nozzle.
3. O ring.
4. Collet holder.
5. Collet (sizes as above to take various diameters of electrodes).
6. Electrode cap (long and short).
7. Body assembly.
8. Sheath.
10. Argon hose assembly.

11. Water hose assembly.
12. Power cable assembly.
13. Adapter power/water; required only in certain cases.
14. Adapter argon; required only in certain cases.
15. Switch actuator.
16. Switch.
17. Switch-retaining sheath.
18. Cable, 2 core.
19. Insulating sleeve.
20. Plug.

swivel head type. Collet sizes on these are generally 0.8 mm, 1.2 mm and 1.6 mm diameter. Larger air cooled torches of 75 A d.c. or a.c. continuous rating or 100 A intermittent usually have a collet of 1.6 mm diameter. Air or water cooled torches rated at 300 A intermittent may be used with electrodes from 1.6 to 6.35 mm diameter and can be fitted with water cooled shields while heavy duty water cooled torches with a water cooled nozzle of 500 A a.c. or d.c. continuous rating and 600 A intermittent employ larger electrodes. A gas lens can be fitted to the torch to give better gas coverage and to obtain greater accessibility or visibility.

Normally, because of turbulence in the flow of gas from the nozzle, the electrode is adjusted to project up to a maximum of 4–9 mm beyond the nozzle (Fig. 3.20). By the use of a lens which contains wire gauzes of coarse and fine mesh, turbulence is prevented and a smooth even gas stream is obtained, enveloping the electrode which, if the gas flow is suitably increased, can be used on a flat surface projecting up to 19 mm from the nozzle orifice, greatly improving accessibility. The lens is screwed on to the torch body in place of the standard nozzle and as the projection of electrode from nozzle is increased the torch must be held more vertically to the work to obtain good gas coverage.

The ceramic nozzles (of alumina or silicon carbide), which direct the flow of gas, screw on to the torch head and are easily removable for cleaning and replacement. Nozzle orifices range from 9.5 to 15.9 mm in diameter and they are available in a variety of patterns for various applications. Ceramic nozzles are generally used up to 200 A a.c. or d.c. but above this water cooled nozzles or shields are recommended because they avoid constant replacement.

Electrodes

The electrode may be of pure tungsten but more generally is of tungsten alloyed with thorium oxide (thoria ThO_2) or zirconium oxide (zirconia ZrO_2). 1% thoriated is used for d.c. welding; 2% thoriated gives good arc striking characteristics on low d.c. values; 1% zirconiated is used for welding aluminium and its alloys, reducing the risk of tungsten contamination.

Tungsten, with a melting point of 3380°C, has a boiling point of 5950°C so there is only little vaporization in the welding arc and it retains its hardness when red hot.

The electrodes are supplied with a ground grey finish to ensure good collet contact and standard diameters are 1.2, 1.6, 2.4, 3.2, 4.0, 4.8, 6.4 and 8 mm to a tolerance of 0.075 mm.

Pure tungsten electrodes are generally used for ordinary quality welds.

They give a stable a.c. arc when used with balanced wave or HF stabilization and can be used with d.c. and with argon, helium, or argon–helium mixtures. The thoriated tungsten electrodes give easier starting, a more stable arc and less possibility of weld contamination with tungsten particles, and in addition they have a greater current-carrying capacity for a given diameter than pure tungsten. Difficulty is, however, encountered when they are used on a.c. in maintaining a hemispherical end, and as a result zirconiated electrodes are preferred for a.c. welding because of the high resistance to tungsten contamination and good arc starting characteristics with reduced tungsten contamination. They are used therefore for high quality welds in aluminium and magnesium, and like pure tungsten they produce a hemispherical or ball end. Selection of electrode size is usually made by choosing one near the maximum current range for the electrode and work. Too small an electrode will result in overheating and thus contamination of the weld while too large an electrode results in difficult control of the arc. Aim for a shining hemispherical end on the electrode.

Electrode grinding

Usually electrodes need grinding to a point only when thin materials are to be welded. They should be ground on a fine grit, hard abrasive wheel used only for this purpose to avoid contamination, ground with the electrode in the plane of the wheel and rotated while grinding (Fig. 3.19).

Electrode current ratings

Electrode diameter (mm)	d.c.					a.c.						
	1.2	1.6	2.4	3.2	4.0	1.2	1.6	2.4	3.2	4.0	4.8	6.0
Max. current (A):						Thoriated may be used but zirconiated is preferable						
Thoriated	70	150	240	380	400	30	60	90	140	195	250	320
Zirconiated	—	—	—	—	—	30	60	90	150	210	275	350

Pulsed current TIG welding (see also Chapter 2, pulsed MIG welding)

In MMA welding the operator chooses the current for a given joint and then, when making the weld, uses his skill to balance the heat input by the arc with the heat output due to the melting of the metal and conduction, convection and radiation by the work. He keeps this balance by electrode angle, amount of weave and rate of travel and achieves correct penetration.

In manual TIG welding this balance is achieved by weaving and the addition of filler metal. With pulsed current TIG welding the melting of the metal is controlled by pulsing the current at a higher value and at regular intervals.

Fig. 3.19. Tungsten electrode preparation.

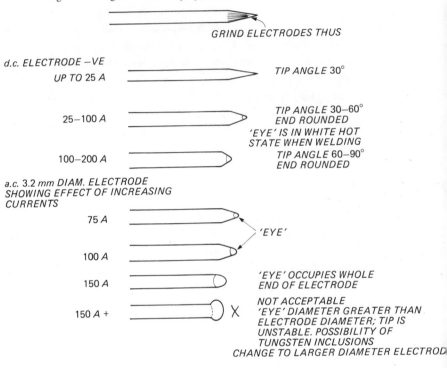

GRIND ELECTRODES THUS

d.c. ELECTRODE –VE
UP TO 25 A — TIP ANGLE 30°

25–100 A — TIP ANGLE 30–60°
END ROUNDED
'EYE' IS IN WHITE HOT
STATE WHEN WELDING

100–200 A — TIP ANGLE 60–90°
END ROUNDED

a.c. 3.2 mm DIAM. ELECTRODE
SHOWING EFFECT OF INCREASING
CURRENTS

75 A — 'EYE'

100 A —

150 A — 'EYE' OCCUPIES WHOLE
END OF ELECTRODE

150 A + — X NOT ACCEPTABLE
'EYE' DIAMETER GREATER THAN
ELECTRODE DIAMETER; TIP IS
UNSTABLE. POSSIBILITY OF
TUNGSTEN INCLUSIONS
CHANGE TO LARGER DIAMETER ELECTRODE

Fig. 3.20. Electrode protrusion. No gas lens. In awkward places protrusion can be increased. Gas shield *must* cover area being welded.

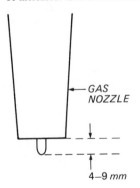

GAS
NOZZLE

4–9 mm

There are two current values, the background current which is steady throughout and the pulsed current which is intermittent but at regular intervals. This pulse increases the energy in the arc and melts off an amount of filler rod to allow correct penetration, and then when the pulse ceases there is left only the background current to keep the arc ionized and the weld metal freezes.

The weld thus progresses as a series of overlapping spot welds and by judicious selection of background current and pulse height (amount of current) and duration the operator has complete control of welding parameters. It can be seen that this method is very advantageous for thin and medium sections, positional welding and in mechanized TIG welding. Pulse facilities are now incorporated on most power units.

Firstly the background current is set at about 5–12 A, then the pulse is set at a value which depends upon the metal to be welded, say 140 A, and the duration is varied until the correct amount of penetration is achieved (remember that the heat input depends upon (pulse current × duration). The pulse heats up the molten pool, more filler wire is melted off,

Fig. 3.21. Pulsed current. Pulses are variable in frequency (pulses per second) and width (peak current).

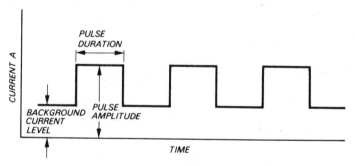

PULSED CURRENT d.c.

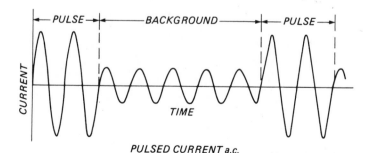

PULSED CURRENT a.c.

penetration is achieved and the pulse ceases; the metal then solidifies, cooling slightly ready for the next pulse.

Fig. 3.21 shows a basic pulse for d.c. and a.c. and since there is such control over the molten pool, fit-up tolerances are not so severe as in welding with non-pulsing current. Pulsed TIG is used on mild and low-alloy steels, stainless steels, heat-resistant steels, titanium, inconel, monel, nickel and aluminium and its alloys and for the orbital welding of thin pipes.

The pulse frequency can be increased from about 1 to 10 pulses per second by the use of the pulse control, while the width control varies pulse width and the background current control varies this as a constant percentage of the weld current as shown in Fig. 3.22. Using these controls enables the operator to control the heat input into the weld and thus the root penetration.

Safety precautions

The precautions to be taken are similar to those when manual metal arc welding. In addition however there is the high voltage unit used for arc initiation without touchdown. This unit is similar in output to that of the car ignition coil and care should be taken when it is switched on. Do

Fig. 3.22

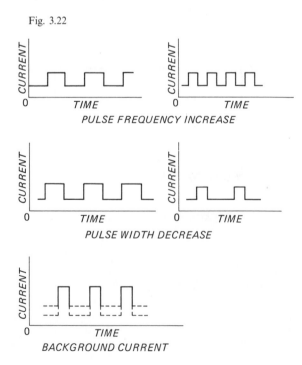

PULSE FREQUENCY INCREASE

PULSE WIDTH DECREASE

BACKGROUND CURRENT

not let the electrode touch the body when switched on or a severe shock will be felt. BS 639 recommended EW filters are graded according to welding current thus: up to 50 A, no. 8; 15–75 A, no. 9; 75–100 A, no. 10; 100–200 A, no 11; 200–250 A, no, 12; 250–300 A, no. 13 or 14.

It will be noted that these filters are darker than those used for similar current ranges in MMA welding. The TIG (and MIG) arcs are richer in infra-red and ultra-violet radiation, the former requiring some provision for absorbing the extra associated heat, and the latter requiring the use of darker lenses.

It is recommended that the student reads the booklet *Electric arc welding, safety, health and welfare*, new series no. 38; published for the Department of Employment by HMSO; and also BS 679, *Filters for use during welding*; British Standards Institution.

Welding techniques

Carbon steel

TIG welding can be used with excellent results on carbon steel sheet and pipework, and for this and other steel welds the angle of torch and rod are similar to that for aluminium, namely torch 80–90° to line of travel and filler wire at 20–30° to the plate (Fig. 3.23). The filter wire is chosen to match the analysis of the plate or pipe.

Fig. 3.23. Angles of torch and rod showing typical weld: aluminium alloy plate 11 mm thick 70–80° prep., 1.5 mm root face, no gap, a.c. volts 102, arc volts drop, 17–18 volts: electrode 4.8 mm diameter, current 255 A, arc length approx. 4 mm.

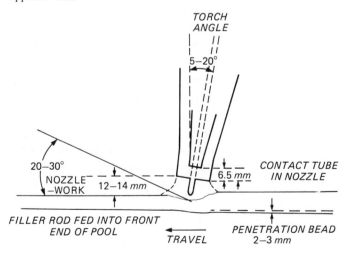

TORCH ANGLE

5–20°

20–30°

NOZZLE –WORK   12–14 mm

6.5 mm   CONTACT TUBE IN NOZZLE

FILLER ROD FED INTO FRONT END OF POOL

TRAVEL

PENETRATION BEAD 2–3 mm

The welding of aluminium and its alloys

The table gives details of the filler wire types for welding the various alloys. Angles of torch and filler wire are shown in Fig. 3.23.

Casting alloys. Those with high silicon content, LM2, LM6, LM8 and LM9, can be welded to the Al–Mg–Si alloys with Al–Si 4043 wire. Similarly the Al–Mg casting alloys LM5 and LM10 can be welded to the wrought alloys using 5356 wire. Serious instability of LM10 may occur due to local heating and the alloy must be solution treated after welding.

Technique. The supply is a.c. and the shielding gas pure argon. Remember that, with TIG and MIG processes using shielding gases, the welding position must be protected from stray draughts and winds which will destroy the argon shield. This is particularly important when site welding and (waterproof) covers or a tent to protect welder and gas shields from wind and rain must be erected if they are to be fully protected. Pre-heating is required for drying only to produce welds of the highest quality. All surfaces and welding wire should be degreased and the area near the joint and the welding wire should be stainless steel wire brushed or scraped to remove oxide and each run brushed before the next is laid. After switching on the gas, water, welding current and HF unit, the arc is struck by bringing the tungsten electrode near the work (without touching down). The HF sparks jump the gap and the welding current flows.

Arc length should be about 3 mm. Practice starting by laying the holder on its side and bringing it to the vertical position using the gas nozzle as a fulcrum, or use a striking plate and get the tungsten hot before starting the weld. (This does not apply to scratch start.) The arc is held in one position on the plate until a molten pool is obtained and welding is commenced, proceeding from right to left, the rod being fed into the forward edge of the molten pool and always kept within the gas shield. It must not be allowed to touch the electrode or contamination occurs. A black appearance on the weld metal indicates insufficient argon supply. The flow rate should be checked and the line inspected for leaks. A brown film on the weld metal indicates presence of oxygen in the argon while a chalky white appearance of the weld metal accompanied by difficulty in controlling the weld indicates excessive current and overheating. The weld continues with the edge of the portion sinking through, clearly visible, and the amount of sinking, which determines the size of the penetration bead, is controlled by the welding rate.

Run-on and run-off plates are often used to prevent cold starts and craters at the end of a run, which may lead to cracking. Modern sets have

Aluminium and its alloys. Filler rod guide

Parent metal	6061 6063 6082	5083	5454	5154A 5251	3103	1050A
1050A	4043 5356(3)	5356 4043	4043 5183 5356	4043 5183 5356	4043 5356 1050A(1)	4043 1050A(1) (2)
3103	4043 5356(3)	5356 5183	5154A 5183 5356	5154A 5183 5356	4043 1050A 5356	
5154A 5251	5356 5183 4043	5356 5183 5556A	5356 5183 5154A	5356 5183	(1) corrosive conditions (2) elevated temperatures	
5454	5356 5183 4043	5356 5183 5556A	5154A 5154A 5554(2) 5356		(3) colour matching (after anodizing)	
5083	5356 5183 5556A	5183 5356 55556A			Old designations 1050A 1B 5154A NG5 5356 NG6	
6061 6063 6083	4043 5356(3) 5183				5183 NG8 4043 NG21 5554 NG 52 5556 NG 61	

Examples of preparation: Flatt butt joints, no backing

Plate thickness (mm)	No. of runs	Current (A)	Filler rod diam. (mm)
2.0	1	100–110	2.4
2.4	1	120	3.2
3.2	1	130–160	3.2
4.8	2	230	4.8
6.4	2	240	4.8

Plate thickness (mm)		
2	Butt no gap	
2.4	Butt 0.8 mm gap	
3.2	Butt 0.8 mm gap	

slope-up and slope-down controls which greatly assist the welding operation but it is as well to practice using these plates.

Preparations for single V butt joints are shown in Fig. 3.24a while 3.24b shows a backing strip in position.

Tack welding and the use of jigs for line-up is similar to that used for steel. The tacks should be reduced in size by chiseling or grinding before welding the seam and care should be taken, when breaking any jigs away that were temporarily welded in, that damage does not occur to the structure or casting.

TIG tack welding is used for line-up of plate and tube before finally MIG welding the seam. Temporary backing bars are grooved to the shape of the underbead and made of mild or stainless steel. These help to form and shape the penetration bead when root runs cannot be made and control of penetration is difficult. When ungrooved bars are used they should be removed after welding, the root run back-chipped and a sealing run made. Any backing strips to be welded in place should be of aluminium alloy, tack welded in place and fused in with the root run.

The vertical double-operator method (Fig. 3.25), using filler wire on one side only, gives sound non-porous welds with accurate penetration bead controlled by one operator 'pulling through' on the reverse side. There is excellent argon shielding from both sides but the method is expensive in man-hours of work and needs more skill. It however reduces residual stresses as there is less total heat input and welding speed is increased.

Fig. 3.24. (a) Manual TIG process: preparation of aluminium and aluminium alloy plate. Single V butt joints. Flat. Plate thickness 4.8 mm–9.5 mm, root run may be back chipped and a sealing run made, though this adds to the cost of the weld.

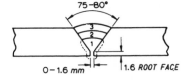

Fig. 3.24. (b) Use of backing strip and preparation.

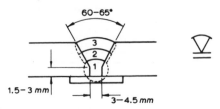

Magnesium and magnesium alloys

Equipment is similar to that for the aluminium alloys and the technique similar, welding from right to left with a short arc and with the same angles of torch and filler rod, with pure argon as the shielding gas and an a.c. supply. Little movement of the filler rod is required but with material over 3 mm thick some weaving may be used. For fillet welding the torch is held so as to bisect the angle between the plates and at sufficient angle from the vertical to obtain a clear view of the molten pool, with enough weave to obtain equal leg fusion. Tilted fillets are used to obtain equal length.

The material is supplied greased or chromated. Degreasing removes the grease and it is usual to remove the chromate by wire brushing from the side to be welded for about 12 mm on each side of the weld and to leave the chromate on the underside where it helps to support the penetration bead. No back purge of argon is required. The surface and edges should be wire brushed and the filler rod cleaned before use. Each run should be brushed before the next is laid.

Backing plates can be used to profile the underbead and can be of mild steel, or of aluminium or copper 6.4–9.5 mm wide, with grooves 1.6–3.2 mm deep for material 1.6–6.4 mm thick. (Fig. 3.26.)

Jigs, together with correct welding sequence, can be used to prevent distortion, but if it occurs the parts can be raised to stress relief

Fig. 3.25. (*a*) Angles of torches and filler wire. (*b*) Double operator vertical welding aluminium alloys.

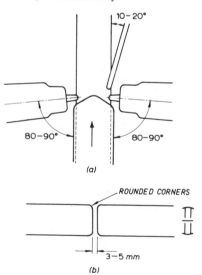

temperatures of approximately 250°–300°C. Ensuring sufficient flow of argon will prevent any oxidized areas, porosity, and entrapped oxides and nitrides.

Welding rods are of similar composition to that of the plate and the table indicates the relative weldability of similar and dissimilar alloys. It is not recommended that the Mg–Zr alloys should be welded to alloys containing Al or Mn.

For fillet welds the torch is held at about 90° to the line of travel and roughly bisecting the angle between the plates, so that the nozzle is clear of either plate, and it is often necessary to have more projection of the wire beyond the nozzle to give good visibility, but it should be kept to a minimum. Tilted fillets give a better weld profile and equal leg length.

Fig. 3.26. Suitable edge preparations for magnesium alloy butt welds (dimensions in mm).

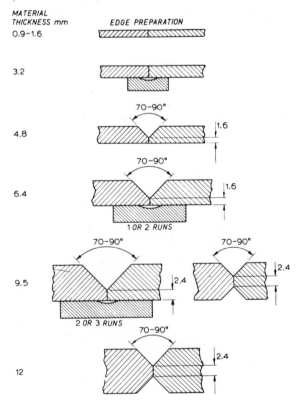

Typical magnesium alloy welding rod, composition and suitability

Elektron alloy	Zn%	Zr%	RE%	Others%	MEL welding rod type
RZE 1	2.0–3.0	0.5–1.0	2.5–4.0		W 6
RZ5	3.5–5.0	0.4–1.0	1.2–1.75		W 7
ZT1	1.7–2.5	0.5–1.0	0.1	2.5–4.0 Th	W 1
TZ6	5.2–6.0	0.5–1.0	0.2	1.5–2.2 Th	W 12
MSR B	0.2	0.4–1.0	2.2–2.7	2.0–3.0 Ag	W 13
AM503	0.03	0.01	—	1.3–1.7 Mn, 0.02 Ca	W 2
AZ31	0.7–1.3	—	—	2.5–3.5 Al, 0.04 Ca	W 15
A8	0.4–1.0	—	—	0.1 Sn, Cu, Si, Fe, Ni, to 0.4 max, 7.5–8.1 Al	W 14

Note. RE – Rare earths.

The table gives recommended filler wires for the various alloys.

Stainless and heat-resistant steel

The production welding of stainless steel by this process is apt to be rather slow compared with MMA, MIG or pulsed arc. (See the section on the MMA welding of stainless steel.)

Areas adjacent to the weld should be thoroughly cleaned with stainless steel wire brushes and a d.c. supply with the torch negative used. The shielding gas is pure argon or argon–hydrogen mixtures (up to 5%) which give a more fluid weld pool, faster rate of deposition (due to the higher temperature of the arc), better 'wetting' and reduction of slag skin by the hydrogen.

The addition of oxygen, CO_2 or nitrogen is not recommended. Joint preparation is given in Fig. 3.27.

Technique. The torch is held almost vertically to the line of travel and the filler rod fed into the leading edge of the molten pool, the hot end never being removed from the argon shield. Electrode extension beyond the nozzle should be as short as possible: 3.0–5.0 mm for butt and 6.0–11.0 for fillets. To prevent excessive dilution, which occurs with a wide weave, multi-run fillet welds can be made with a series of 'stringer beads'. The arc length should be about 2 mm when no filler wire is used and 3–4 mm with filler. Gas flow should be generous and a gas lens can be used to advantage to give a non-turbulent flow. Any draughts should be avoided.

Fig. 3.27. Preparation of stainless steel and corrosion- and heat-resistant steels. (TIG and MIG.)

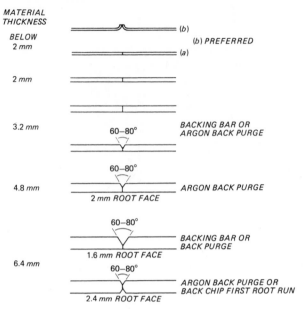

Nickel alloys

Although this process is very suitable for welding nickel and its alloys it is generally considered rather slow, so that it is mostly done on thinner sections of sheet and tube.

Shielding gas should be commercially pure argon and addition of hydrogen up to 10% helps to reduce porosity and increases welding speeds especially for Monel 400 and Nickel 200. Helium has the same advantages, greatly increasing welding speed. Great care should be taken to avoid disturbance of the protective inert gas shield by draughts, and the largest nozzle diameter possible should be used, with minimum distance between nozzle and work. The use of a gas lens increases the efficiency of the shield and gives increased gas flow without turbulence. Argon flow is of the order of 17–35 litres per hour for manual operation and higher for automatic welding. Helium flow should be about $1\frac{1}{2}$-3 times this flow rate.

The torch is held as near 90° to the work as possible – the more acute the angle the greater the danger of an aspirating effect causing contamination of the gas shield. Electrodes of pure, thoriated or zirconiated tungsten can be used, ground to a point to give good arc control and should project 3–4.5 mm beyond the nozzle for butt welds and 6–12 mm for fillets.

Stray arcing should be avoided as it contaminates the parent plate, and there should be no weaving or puddling of the pool, the hot filler rod end being kept always within the gas shield, and fed into the pool by touching it on the outer rim ahead of the weld. The arc is kept as short as possible – up to 0.5 mm maximum when no filler is used.

To avoid crater shrinkage at the end of a weld a crater-filling unit can be used or failing this, extension tabs for run-off are left on the work and removed on completion. Fig. 3.28 gives suitable joint preparations.

For vertical and overhead joints the technique is similar to that for steel, but downhand welding gives the best quality welds.

Filler metals for inert-gas shielded metal-arc welding

Material	Filler metal
Nickel 200	Nickel 61
Nickel 201	Nickel 61
Monel 400	Inconel 82
Monel K 500	Inconel 64
Inconel 600	Inconel 82
Inconel 625	Inconel 625
Inconel 718	Inconel 718
Inconel 750	Inconel 69
Incoloy DS	NC 80/20
Incoloy 800	Inconel 82
Incoloy 825	Incoloy 65
Nimonic 75	NC 80/20
Nimonic 80A	Nimonic 90
Nimonic 90	Nimonic 90
Nimonic 263	Nimonic 263
Nimonic PE13	Nimonic PE13
Nimonic PE16	Nimonic PE16
Brightray alloys	NC 80/20
Nilo alloys	Inconel 82 or 92
	Nickel 61

Copper and copper alloys

Direct current, electrode negative is used with argon as the shielding gas for copper and most of its alloys with currents up to 400 A. and angle of torch and rod roughly as for aluminium welding. With aluminium bronze and copper–chromium alloys however, dispersal of the oxide film is difficult when using d.c., and for these alloys a.c. is generally used, but d.c. can be used with helium as the shielding gas. The weld areas should be bronze wire brushed and degreased and each run brushed to remove oxide film. Jigs or tack welds can be used for accurate positioning and to prevent distortion.

Mild steel or stainless steel backing bars, coated with graphite or anti-spatter compound to prevent fusion, can be used to control the penetration bead and also to prevent heat dissipation, or backing may be welded into the joint.

Because of the high coefficient of thermal expansion of copper the root gap has a tendency to close up as welding proceeds, and due to its high thermal conductivity pre-heating from 400–700°C according to thickness is essential on all but the thinnest sections to obtain a good molten pool. The thermal conductivity of most of the copper alloys is much lower than copper so that pre-heating to about 150°C is usually sufficient. In the range 400–700°C a reduction in ductility occurs in most of the alloys so that a cooling period should be allowed between runs. The main grades of copper are (1) tough pitch (oxygen containing) high conductivity, (2) oxygen free, high conductivity, (3) phosphorus deoxidized, and in most cases it is the last type that is used for pressure vessels, heat exchangers and food processing equipment, etc. If tough pitch copper, in the cast form, is to be welded, a deoxidizing filler rod should be used to give a deoxidized weld and in all cases the weld should be performed as quickly as possible to prevent overheating.

Argon, helium and nitrogen can be used as shielding gases. Argon has a lower heat output than helium and nitrogen because it has a lower arc voltage. Argon can be mixed with helium and the mixture increases the heat output proportionally with the helium percentage, so that using the

Fig. 3.28

Nickel alloys. Examples of joint design for fusion welding. Type of joint	Material thickness (mm) T	Width of groove (top) (mm) W	Root space (mm) S
Square butt	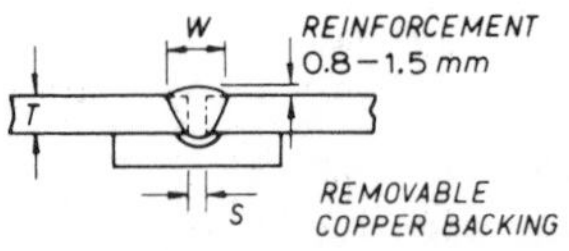 1.0	3.2	0
	1.2	4.0	0
	1.6	4.8	0
	2.4	4.8–6.4	0–0.8
	3.2	6.4	0–0.8
V groove	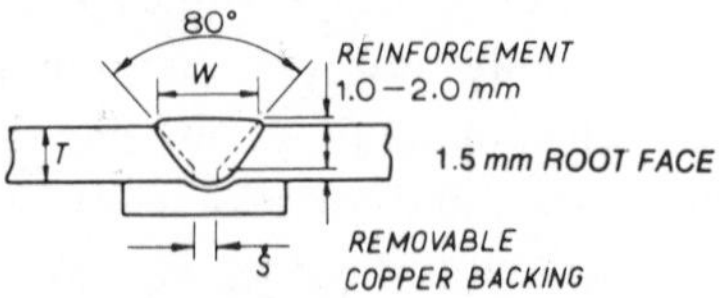 4.8	8.9	3.2
	6.4	13.0	4.8
	8.0	15.5	4.8
	9.5	18.0	4.8
	12.7	23.0	4.8
	16.0	29.5	4.8
V groove	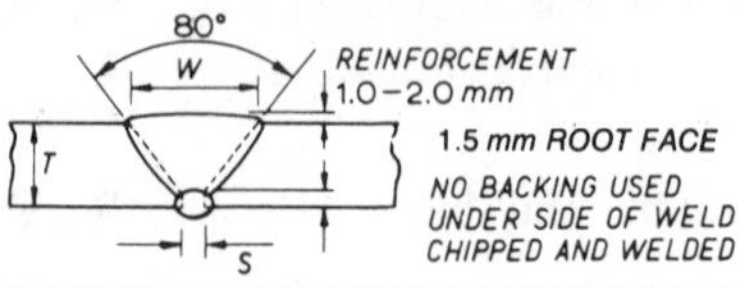 6.4	10.4	2.4
	8.0	13.0	2.4
	9.5	16.5	3.2
	12.7	21.6	3.2
	16.0	27.0	3.2

mixture for copper welding a lower pre-heat input is required and penetration is increased. Nitrogen gives the greatest heat output but although the welds are sound they are of rough appearance.

Recommended filler alloys are:

C7 0.05–0.35% Mn, 0.2–0.35% Si, up to 1.0% Sn, remainder Cu.
C8 0.1–0.3% Al, 0.1–0.3% Ti, remainder Cu.
C21 0.02–0.1% B, 99.8% Cu.

Note. Filler alloys for copper and its alloys are to BS 2901, Pt 3, 1970.

Copper–silicon (silicon bronzes). d.c. supply, electrode negative and argon (or helium) shield. These alloys contain about 3% silicon and 1% manganese. They tend to be hot short in the temperature range 800–950°C so that cooling should be rapid through this range. Too rapid cooling, however, may promote weld cracking.

Filler alloy C9, 96% Cu, 3% Si, 1% Mn is recommended.

Copper–aluminium (aluminium bronzes). a.c. with argon shield or d.c. with helium shield. The single phase alloys contain about 7% Al and the two phase alloys up to 11% Al with additions of Fe, Mn and Ni. Welding may affect the ductility of both types. Weld root cracking in the 6–8% Al, 2–2.3% Fe alloy used for heat exchangers can be avoided by using a non-matching filler rod. Due to a reduction in ductility at welding temperatures difficulty may be experienced in welding these alloys. The alloy containing 9% Al and 12% Mn with additions of Fe and Ni has good weldability but requires heat treatment after welding to restore corrosion resistance and other mechanical properties.

Recommended filler alloys are:

C12 6.0–7.0% Al, remainder Cu.
C13 9.0–11.0% Al, remainder Cu.
C20 8.0–10.0% Al, 1.5–3.5% Fe, 4.0–7.0% Ni, remainder Cu.

Copper–nickel (cupro–nickels). d.c. with argon shield and electrode negative. These alloys, used for example for pipe work and heat exchangers, contain from 5–30% Ni, often with additions of Fe or Mn to increase corrosion resistance. They are very successfully welded by both TIG and MIG processes but since they are prone to contamination by oxygen and hydrogen, an adequate supply of shielding gas must be used including a back purge in certain cases, if possible.

Recommended filler alloys are:

C17 19–21% Ni, 0.2–0.5% Ti, 0.2–1.0% Mn, remainder Cu.
C19 29–31% Ni, 0.2–0.5% Ti, 0.2–1.0% Mn, remainder Cu.

Copper–tin (phosphor bronzes and gunmetal). d.c. with argon shield. The wrought phosphor bronzes contain up to 8% Sn with phosphorus additions up to 0.4%. Gunmetal is a tin bronze containing zinc and often some lead. Welding of these alloys is most often associated with repair work using a phosphor bronze filler wire but to remove the danger of porosity, non-matching filler wires containing deoxidizers should be used.

Filler alloys recommended are:

C10 4.5–6.0% Sn, 0.4% P, remainder Cu.
C11 6.0–7.5% Sn, 0.4% P, remainder Cu.

Copper–zinc (brass and nickel silvers). a.c. with argon shield, d.c. with helium shield. The copper–zinc alloys most frequently welded are Admiralty brass (70% Cu, 29% Zn, 1% Sn), Naval brass (62% Cu, 36.75% Zn, 1.25% Sn) and aluminium brass (76% Cu, 22% Zn, 2% Al). Porosity, due to the formation of zinc oxide, occurs when matching filler wire is used so that it is often preferable to use a silicon–bronze wire which reduces evolution of fumes. This may, however, induce cracking in the HAZ. Post-weld stress relief in the 250–300 °C range reduces the risk of stress corrosion cracking. The range of nickel silver alloys is often brazed.

Filler alloys recommended are:

C14 70–73% Cu, 1.0–1.5% Sn, 0.02–0.06% As.
C15 76–78% Cu, 1.8–2.3% Sn, 0.02–0.06% As.

TIG process. Recommended filler wires for copper welding

Type (BS 2870–2875)	Grade	Argon or helium shield	Nitrogen shield
		Filler wire	
C106	Phosphorus deoxidized non-arsenical	C7, C21	C8
C107	Phosphorus deoxidized arsenical		
C101	Electrolytic tough pitch high conductivity	C7, C21	Not recommended
C102	Fire-refined tough pitch high conductivity		
C103	Oxygen-free high conductivity	C7, C21	Not recommended

Work-hardening and precipitation-hardening copper-rich alloys

The work-hardening alloys are not often welded because mechanical properties are lost in the welding operation. Heat-treatable alloys such as copper–chromium and copper–beryllium are welded with matching filler rods using an a.c. supply to disperse the surface oxides. They are usually welded in the solution-treated or in the over-aged condition and then finally heat-treated.

Hardfacing

The tungsten arc is a good method for deposition of stellite and other hard surfacing materials on a base metal. The heat is more concentrated than that of the oxy-acetylene flame and although there may be some base metal pick-up this can be kept to a minimum by using correct technique and there is not a great deal of dilution. It produces a clean, sound deposit and is very useful for reactive base metals.

The electrode is connected to the −ve pole (straight polarity) with the largest diameter filler rod possible to reduce tungsten contamination. Deposition is similar to that when the oxy-acetylene flame is used and deposition moves from right to left. Filler rods of 3.2, 4.0, 4.8, 6.4 and 8 mm are available and alloys include stellite and nickel-based alloys, and deposition proceeds in the normal way for hard surfacing – do not puddle the pool as this increases dilution.

Easily hardfaced

Low and medium carbon steel to 0.4% C. Above 0.4% C, oxy-acetylene only.

Low-alloy steels.

Nickel and nickel–copper alloys.

Chrome–nickel stainless steels, niobium stabilized. (Not free machining nor titanium stabilized).

11–14% manganese steels.

More difficult to hardface

Cast iron.

Stainless steel, straight and titanium stabilized.

Tool and die steels, water-hardening, oil-hardening and air-hardening types and hot work grade.

Straight chromium stainless steels.

Automatic welding

For automatic welding the TIG torch is usually water cooled and may be carried on a tractor moving along a track or mounted on a boom so as to move over the work or for the work to move under the head. The head has a control panel for spark starter, water and gas and current contactor, and the torch has lateral and vertical movement, the arc length being kept constant by a motor-driven movement controlled by circuits operating from the arc length. Filler wire is supplied from the reel to the weld pool by rollers driven by an adjustable speed motor. Heavier currents can be used than with manual operation resulting in greater deposition rates, and the accurate control of speed of travel and arc length results in welds of high quality. Plate up to 9 mm thick can be welded in one run, the welding being of course downhand. Fig. 3.29 illustrates a typical tractor-driven head.

Orbital pipe welding unit

This can operate from a standard TIG power source and enables circumferential joints on pipes and tubes to be made automatically. It is of caliper type enabling rapid adjustments to be made, and rotational drive is by a small motor mounted in the handle of the unit, rotational speed being controlled by transistor regulator. A wire feed unit feeds wire to the arc, the feed speed of the wire being controlled by a thyristor regulator. Three sizes are available for pipes with outside diameters of 18–40, 36–80 and 71–160 mm (Fig. 3.30).

Mechanized TIG hard facing

In this process the argon shielded TIG arc oscillates over a width from 6 to 25 mm and is operated by variable controls for speed and width. Rod and torch also have controls for vertical height and horizontal positioning and for feeding the hard surfacing rod.

Fig. 3.29. Automatic TIG welding head mounted on electrically driven tractor; 500 A water cooled torch.

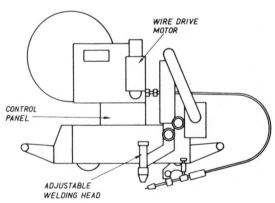

Small quantities of hydrogen are added to the argon by means of a gas blender to give a more consistent deposit. The stellite alloy, in rod form, can be laid on large circular components of 100–600 mm diameter and this is ideal for valves and valve seatings, rings and sleeves, etc., the work being mounted on a manipulator.

A smaller system enables components from 25–220 mm diameter to be hard faced. The unit can be adapted for flat straight working if required and the usual power unit for TIG welding is required. A complete range of hard facing alloys is available.

TIG robot welding

TIG welding has now been applied to the robot system with two manipulators. In this method of welding the workpieces are held in jigs on the plates of the manipulators, one being welded while the other is being loaded. The six axes of the robot enable it to handle the filler wire and place it in the best position without limitation of the wrist movement of the robot.

Since the HF necessary for the arc initiation may cause unwanted signals to the microprocesser all conductors have been screened. The control computer selects and records the particular welding program and together

Fig. 3.30

with the robot achieves synchronized movements, and welding operations are commenced when the operator has loaded one of the two manipulators of the system. (See Chapter 5, Welding with robots, for illustration of this system.)

Various parameter sequences for various operations can be pre-recorded and recalled as required.

Spot welding

Mild steel, stainless and alloy steel and the rarer metals such as titanium can be spot welded using a TIG spot welding gun. The pistol grip water cooled gun carries a collet assembly holding a thoriated tungsten electrode positioned so that the tip of the electrode is approximately 5 mm within the nozzle orifice, thus determining the arc length. Interchangeable nozzles enable various types of joints to be welded flat, or vertically. A variable timing control on the power unit allows the current to be timed to flow to ensure fusion of the joint and a switch on the gun controls gas, water, HF unit and current contractor. To make a weld, the nozzle of the gun is pressed on to the work in the required position and when the gun switch is pressed the following sequence occurs: (1) gas and water flow to the gun, (2) the starter ignites the arc and the current flows for a period controlled by the timer, (3) current is automatically switched off and post-weld argon flows to cool electrode tip and weld.

Note. Refer also to BS 3019, *General recommendations for inert gas, tungsten arc welding,* Part 1: *Wrought aluminium, aluminium alloys and magnesium alloys,* Part 2: *Austenitic stainless and heat resisting steels.*

Titanium

Titanium is a silvery coloured metal with atomic weight 47.9, specific gravity 4.5 (cf. Al 2.7 and Fe 7.8) melting point 1800°C and UTS 310 N/mm^2 in the pure state. It is used as deoxidizer for steel and sometimes in stainless steel to prevent weld decay and is being used in increasing amounts in the aircraft and atomic energy industries because of its high strength-to-weight ratio.

Titanium absorbs nitrogen and oxygen rapidly at temperatures above 1000°C and most commercial grades contain small amounts of these gases; the difficulty in reducing the amounts present makes titanium expensive. Because of this absorption it can be seen that special precautions have to be taken when welding this metal.

Where seams are reasonably straight a TIG torch with tungsten electrode gives reasonable shrouding effect on the upper side of the weld, and jigs may be employed to concentrate the argon shroud even more over the weld. In addition the underbead must be protected and this may be done by welding over a grooved plate, argon being fed into this groove so that a uniform distribution is obtained on the underside. It can be seen therefore that the success of welding in these conditions depends upon the successful shrouding of the seam to be welded.

For more complicated shapes a vacuum chamber may be employed. The chamber into which the part to be welded is placed is fitted with hand holes, and over these are bolted long-sleeve rubber gloves into which the operator places his hands, and operates the TIG torch and filler rod inside the chamber. An inspection window on the sloping top fitted with welding glass enables a clear view to be obtained. All cable entries are sealed tightly and the air is extracted from the chamber as completely as possible by pumps, and the chamber filled with argon, this being carried out again if necessary to get the level of oxygen and nitrogen down to the lowest permissible level. The welding may be performed with a continuous flow of argon through the chamber or in a static argon atmosphere. Since the whole success of the operation depends upon keeping oxygen and nitrogen levels down to an absolute minimum great care must be taken to avoid leaks as welding progresses and to watch for discoloration of the weld indicating absorption of the gases.

Tantalum

Tantalum is a metallic element of atomic weight 180.88, HV 45, melting point 2910 °C, specific gravity 16.6, it is a good conductor of heat and is used in sheet form for heat exchangers, condensers, small tubes, etc. Like titanium it readily absorbs nitrogen and oxygen at high temperatures and also readily combines with other metals so that it is necessary to weld it with a TIG torch in a vacuum-purged argon chamber as for titanium.

Beryllium

Beryllium is a light, steely coloured metallic element, atomic weight 9, melting point 1280–1300 °C, specific gravity 1.8, HV 55–60. It is used as an alloy with copper (beryllium bronze) to give high strength with elasticity. As with tantalum and titanium it can be welded in a vacuum-purged argon chamber using a TIG torch. It can also be pressure welded and resistance welded.

Electron beam welding is now being applied to the welding of the rare metals such as the foregoing, in the vacuum chamber. A beam of electrons is concentrated on the spot where the weld is required. The beam can be focused so as to give a small or large spot, and the power and thus the heating effect controlled. Successful welds in a vacuum-purged argon chamber are made in beryllium and tantalum.

Plasma-arc welding*

Plasma welding is a process which complements, and in some cases is a substitute for, the TIG process, offering, for certain applications, greater welding speed, better weld quality and less sensitivity to process variations. The constricted arc allows lower current operation than TIG for similar joints and gives a very stable controllable arc at currents down to 0.1 A, below the range of the TIG arc, for welding thin metal foil sections. Manual plasma welding is operated over a range 0.1–100 A from foil to 3–4 mm thickness in stainless steel, nickel and nickel alloys, copper, titanium and other rare earth metals (not aluminium or magnesium).

Ions and plasma

An ion is an atom (or group of atoms bound into a molecule) which has gained or lost an electron or electrons. In its normal state an atom exhibits no external charge but when transference of electrons takes place an atom will exhibit a positive or negative charge depending upon whether it has lost or gained electrons. The charged atom is called an ion and when a group of atoms is involved in this transference the gas becomes ionized.

All elements can be ionized by heat to varying degrees (thermal ionization) and each varies in the amount of heat required to produce a given degree of ionization, for example argon is more easily ionized than helium. In a mass of ionized gas there will be electrons, positive ions and neutral atoms of gas, the ratio of these depending upon the degree of ionization.

A plasma is the gas region in which there is practically no resultant charge, that is, where positive ions and electrons are equal in number; the region is an electrical conductor and is affected by electric and magnetic fields. The TIG torch produces a plasma effect due to the shield of argon and the tungsten arc but a plasma jet can be produced by placing a tungsten electrode centrally within a water-cooled constricted copper nozzle. The tungsten is connected to the negative pole (cathode) of a d.c. supply and the nozzle to the positive pole (anode). Gas is fed into the nozzle and when an arc is struck between tungsten electrode and nozzle, the gas is ionized in its passage through the arc and, due to the restricted shape of the nozzle orifice, ionization is greatly increased and the gas issues from the nozzle orifice, as a high-temperature, high-velocity plasma jet, cylindrical in shape

* American designation: plasma-arc welding (PAW).

and of very narrow diameter realizing temperatures up to 10000 °C. This type is known as the non-transferred plasma (Fig. 3.31*a*). With the transferred arc process used for welding, cutting and surfacing, the restricting orifice is in an inner water-cooled nozzle within which the tungsten electrode is centrally placed. Both work and nozzle are connected to the anode and the tungsten electrode to the cathode of a d.c. supply (in American terms, d.c.s.p., direct current straight polarity). Relatively low plasma gas flow (of argon, argon–helium or argon–hydrogen) is necessary to prevent turbulence and disturbance of the weld pool, so a further supply of argon is fed to the outer shielding nozzle to protect the weld (Fig. 3.31*b*). In the lead from work to power unit there is a contactor switch as shown.

A high-frequency unit fed from a separate source from the mains supply initiates the pilot arc and the torch nozzle is positioned exactly over the work. Upon closing the contactor switch in the work-to-power unit

Fig. 3.31. (*a*) Non-transferred plasma arc. (*b*) Transferred plasma arc.

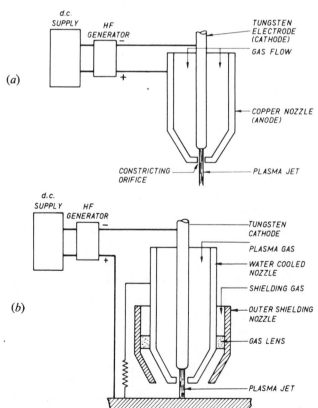

connexion, the arc is transferred from electrode to work via the plasma. Temperatures up to 17000 °C can be obtained with this arc. To shape the arc two auxiliary gas passages on each side of the main orifice may be included in the nozzle design. The flow of cooler gas through these squeezes the circular pattern of the jet into oval form, giving a narrower heat-affected zone and increased welding speed. If a copper electrode is used instead of tungsten as in the welding of zirconium, it is made the anode. The low-current arc plasma with currents in the range 0.1–15 A has a longer operating length than the TIG arc, with much greater tolerance to change in arc length without significant variation in the heat energy input into the weld. This is because it is straight, of narrow diameter, directional and cylindrical, giving a smaller weld pool, deeper penetration and less heat spread whereas the TIG arc is conical so that small changes in arc length have much more effect on the heat output. Fig. 3.32 compares the two arcs. Since the tungsten electrode is well inside the nozzle (about 3 mm) in plasma welding, tungsten contamination by touchdown or by filler rod is avoided, making welding easier. Fig. 3.32 shows the comparison between plasma and TIG arcs.

Equipment. In the range 5–200 A a d.c. rectifier power unit with drooping characteristic and an OCV of 70 V can be used for argon and argon–hydrogen mixtures. If more than 5% hydrogen is used, 100 V or more is required for pilot arc ignition. This arc may be left in circuit with the main arc to give added stability at low current values. Existing TIG power sources such as that in Fig. 3.11 may be used satisfactorily, reducing capital cost, the extra equipment required being in the form of a console placed on or near the power unit. Input is 380–440 V, 50 Hz, a.c. single-phase with approximately 3.5 A full load current. It houses relays and solenoid valves controlling safety interlocks to prevent arc initiation unless gas and water pressures are correct; flowmeters for plasma and shielding gases, gas purge and post-weld gas delay and cooling water controls. For low-current welding in the range 0.1–15 A, often referred to in Europe as micro-plasma and in America as needle plasma welding, the power unit is about 3 KVA, 200–250 V, 50 Hz single-phase input, fan cooled with OCV of 100 V nominal and 150 V peak d.c., for the main arc, and output current ranges 0.1–2.0 A and 1.0–15 A. The pilot arc takes 6A at 25 V start and 2.5 A at 24 V running, the main arc being cylindrical and only 0.8 mm wide. Tungsten electrodes are 1.6, 2.4 and 3.2 mm diameter, depending upon application.

Gases. Very pure argon is used for plasma and shield (or orifice) when welding reactive metals such as titanium and zirconium which have a strong affinity for hydrogen. For stainless steel and high-strength nickel alloys argon, or argon–hydrogen mixtures are used. Argon–5% hydrogen mixtures are applied for this purpose (cylinders coloured blue with a wide red band around the middle) and argon–8% hydrogen and even up to 15% hydrogen are also used. With these mixtures the arc voltage is increased

Fig. 3.32. (*a*) Plasma arc. (*b*) TIG arc.

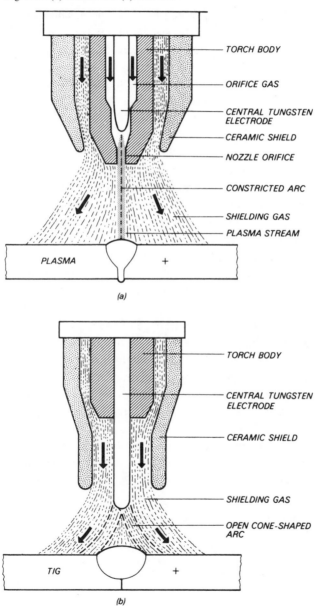

giving higher welding speeds (up to 40% higher) and the thin oxide film present even on stainless and alloy steels is removed by hydrogen reduction giving a clean bright weld, and the wetting action is improved. For copper, nickel and their alloys argon is used in keyhole welding (Fig. 3.33) for thinner sections, for both orifice and shielding gas. Argon and helium is used in the melt-in welding of thinner sections and helium for orifice and shielding gas for sections over 3 mm thick.

De-ionized cooling water should be used, and gas flow rates at 2.0 bar pressure are about 0.25–3.3 litres per minute plasma and 3.8–7 litres per minute shielding in the 5–100 A range. Cleaning and preparation are similar to those for TIG, but when welding very thin sections, oil films or even fingerprints can vary the degree of melting, so that degreasing should be through and the parts not handled in the vicinity of the weld after cleaning.

Technique. Using currents of 25–100 A, square butt joints in stainless steel can be made in thicknesses of 0.8–3.2 mm with or without filler rod, the angle of torch and rod being similar to that for TIG welding. The variables are current, gas flow and welding rate. Too high currents may break down the stabilizing effect of the gas, the arc wanders and rapid nozzle wear may occur. In 'double arcing' the arc extends from electrode to nozzle and then to the work. Increasing plasma gas flow improves weld appearance and increases penetration, and the reverse applies with decreased flow. A turbulent pool gives poor weld appearance.

In 'keyhole' welding of thicknesses of 2.5–6.5 mm a hole is formed in the square-edge butt joint at the front edge of the molten pool, with the arc passing through the section (Fig. 3.33). As the weld proceeds, surface tension causes the molten metal to flow up behind the hole to form the welding bead, indicating complete penetration, and gas backing for the underbead is required. When butt welding very thin sections, the edges of the joint must be in continuous contact so that each edge melts and fuses into the weld bead. Separation of the edges gives separate melting and no weld. Holding clamps spaced close together near the weld joint and a backing bar should be used to give good alignment, and gas backing is recommended to ensure a fluid pool and good wetting action. Flanging is

Fig. 3.33. Surface tension causes molten metal to flow and close keyhole.

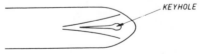

KEYHOLE

recommended for all butt joints below 0.25 mm thickness and allows for greater tolerance in alignment. In the higher current ranges joints up to 6 mm thick can be welded in one run. Lap and fillet welds made with filler rod are similar in appearance and employ a similar technique to those made with the manual TIG process. Edge welds are the best type of joint for foil thickness, an example being a plug welded into a thin-walled tube.

In tube welding any inert gas can be used within the tube for underbead protection.

Faults in welding very thin sections are: excessive gaps which the metal cannot bridge; poor clamping allowing the joint to warp; oil films varying the degree of melting; oxidation or base metal oxides preventing good 'wetting'; and nicking at the ends of a butt joint. This latter can be avoided by using run-off plates or by using filler rod for the start and finish of the weld.

Plasma spray hard facing

In this method the arc plasma melts the hard facing powder particles and the high-velocity gas stream carries the molten particles on to the surface (Fig. 3.34). Although there may be some voids caused, 90–95% densities can be obtained and the method is particularly suitable for applying refractory coatings because this method has a better bond with the parent plate than the spray and fuse method. Because the heat is localized, surfaces can be applied to finished parts with no distortion and the finish is very smooth. The cobalt, nickel- and tungsten-based powders are all supplied for this process and suitable applications are gas turbine parts, such as sealing rings of gas turbines, mixer and feed parts, sleeves, sheets and wear pads and slides, etc.

Fig. 3.34. Plasma surfacing.

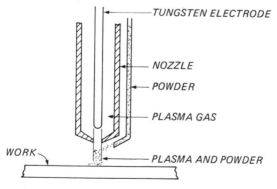

Plasma transferred arc hardfacing

This powder deposition process is fully automated and uses a solid state SCR, the power output being 300 A, 32 V, at 100% duty cycle. The pilot arc is initiated by HF and the unit has pulse facilities. The powder is supplied from a powder feeder operating on the rotating drum principle. Powder feed rate is controlled by varying the feed opening over the knurled rotating wheel upon which a regulated ribbon of powder is applied, and the powder feed can be regulated to provide an upslope and downslope feed rate. This powder is carried from the container to the work in a stream of argon and is melted in the plasma, the argon providing a shield around the plasma heat zone (Fig. 3.35). Torches are of varying size and cover most hard surfacing requirements. As can be seen in the illustration the plasma and powder are supplied through separate passages. Cobalt-based (stellite), nickel-based, iron-based with chrome or molybdenum alloys are available and applications include valve seats, oil drill tool joints, gate valve inserts, diesel engine crossheads, etc. Stellite alloys Nos. 1, 6, 12, 21, 156, 157, 158, F of the stellite range are suitable.

Fig. 3.35. Plasma transferred arc.

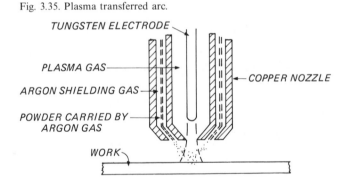

4

Resistance welding and flash butt welding

Spot welding

In this method of welding use is made of the heating effect which occurs when a current flows through a resistance. Suppose two plates A and B (Fig. 4.1) are to be welded together at X. Two copper electrodes are pressed against the plates squeezing them together. The electrical resistance is greatest at the interface where the plates are in contact, and if a large current at low voltage is passed between the electrodes through the plates, heat is evolved at the interface, the heat evolved being equal to $I^2 R t$ Joules, where I is current in amperes, R the resistance in ohms and t the time in seconds. A transformer supplies a.c. at low voltage and high current to the electrodes (Figs. 4.1 and 4.2).

Fig. 4.1

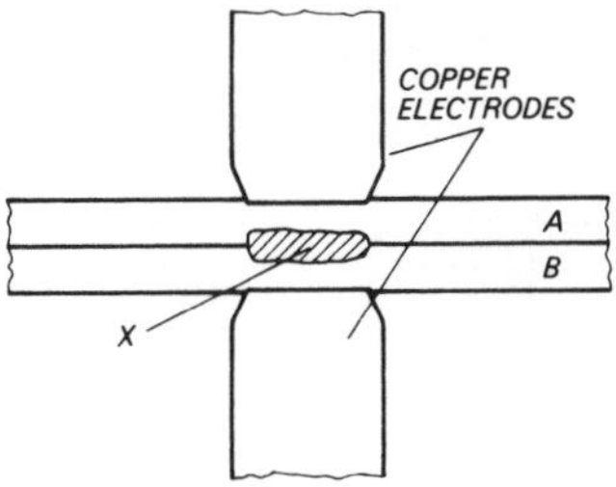

Fig. 4.2

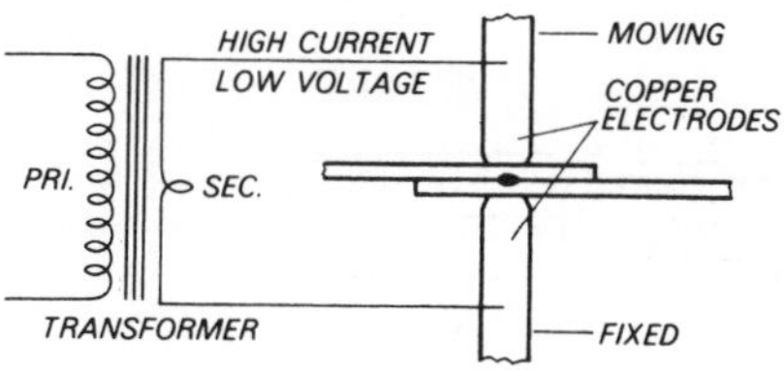

For any given joint between two sheets a suitable time is selected and the current varied until a sound weld is obtained. If welds are made near each other some of the current is shunted through the adjacent weld (Fig. 4.3) so that a single weld is not representative of what may occur when several welds are made. Once current and time are set, other welds will be of consistent quality. If the apparatus is controlled by a pedal as in the simplest form of welder, then pressure is applied mechanically and the time for which the current flows is switch controlled, as for example in certain types of welding guns used for car body repairs. This method has the great disadvantage that the time cannot be accurately controlled, so that if it is too short there is insufficient heat and there is no fusion between the plates, whilst if the time is too long there is too much heat generated and the section of the plates between the electrodes melts, the molten metal is spattered out due to the pressure of the electrodes and the result is a hole in the plates. In modern spot-welding machines the pressure can be applied pneumatically or hydraulically or a combination of both and can be accurately controlled. The current is selected by a tapping switch on the primary winding of the transformer and the time is controlled electronically, the making and breaking of the circuit being performed by thyristors.

When the current flows across the interface between the plates, the heating effect causes melting and fusion occurs at *A* (Fig. 4.4). Around this there is a narrow heat-affected zone (HAZ) since there is a quenching effect

Fig. 4.3

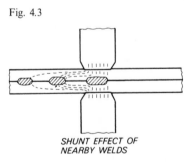

SHUNT EFFECT OF
NEARBY WELDS

Fig. 4.4

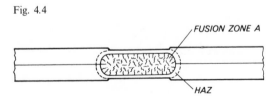

FUSION ZONE A

HAZ

due to the electrodes, which are often water cooled. There are equi-axed crystals in the centre of the nugget and small columnar crystals grow inwards towards the centre of greatest heat.

Types of spot welders

In the pedestal type there is a fixed vertical pedestal frame and integral transformer and control cabinet. The bottom arm is fixed to the frame and is stationary during welding and takes the weight of the workpiece. The top arm may be hinged so as to move down in the arc of a circle (Fig. 4.5a) or it may be moved down in a straight line (Fig. 4.5b). Pivoting arms are adjustable so as to have a large gap between the electrodes, the arms are easily adjusted in the hubs and various length arms are easily fitted giving easy access to difficult joints. The vertical travel machine has arms of great rigidity so that high pressures can be applied, and the electrode tips remain in line irrespective of the length of stroke, so that the machine is easily adapted for projection welding. Additionally the spot can be accurately positioned on the work, and since the moving parts have low inertia, high welding speeds can be achieved without hammering (see Fig. 4.6a and b).

Welding guns

Portable welding guns are extensively used in mass production. The equipment consists of the welding station often with hydro-pneumatic booster to apply the pressures, a water manifold, one or two welding guns

Fig. 4.5. (a) Pivoting arms, (b) vertical travel.

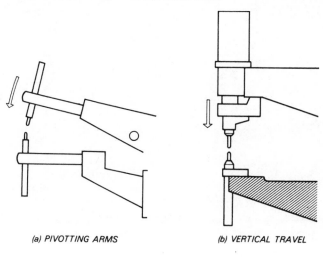

(a) PIVOTTING ARMS   (b) VERTICAL TRAVEL

with balancers, cable between transformer and guns and a control station. In modern machines the composite station with built-in cabinet comprises sequence controls with integrated circuits, thyristor contactors and disconnect switch. Articulated guns (Figs. 4.7a and b) have both arms articulated as in a pair of scissors, giving a wide aperture between electrodes, and are used for welding joints difficult of access. Small

Fig. 4.6. (a) Air-operated spot welding machine with pivoting arms.

articulated guns are used for example in car body manufacture for welding small flanges, and in corners and recesses. C guns (Fig. 4.7c) have the piston-type ram of the pressure cylinder connected to the moving electrode, which thus moves in a straight line. There is great rigidity and a high working speed because of the low inertia of the moving parts. The electrodes are always parallel and the precision motion is independent of

Fig. 4.6. (*b*) Air-operated spot welding machine with vertical action.

Fig. 4.7. (*a*) Articulated type welding head; air operated.

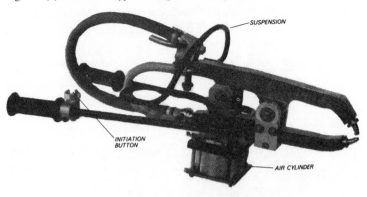

Fig. 4.7. (*b*) Articulated type welding head; hydraulically operated.

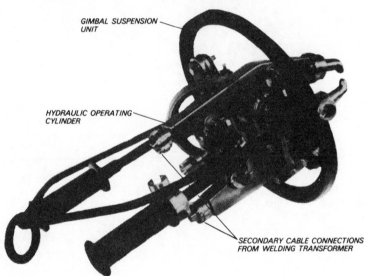

Fig. 4.7. (*c*) Air-operated C type welding head.

the arm length with easily determined point of welding contact. They are used for quality welds where the height does not preclude them. Integral transformer guns are used to reduce handling costs of bulky fabrications, and the manual air-cooled gun and twin spot gun are used in repair work on car bodies and agricultural equipment and for general maintenance, the latter being used where there is access for only one arm as in closed box sections and car side panels.

The welding cycle

The cycle of operations of most modern machines is completely automatic. Once the hand or foot switch is pressed, the cycle proceeds to completion, the latest form having digital sequence controls with printed circuits, semi-conductors, digital counting and air valve operation by relays. The simplest cycle has one function, namely weld time, the electrode force or pressure being pre-set, and it can be best illustrated by two graphs, one the electrode pressure plotted against time and the other the current plotted against the same time axis (Fig. 4.8). The time axis is divided into 100 parts corresponding to the 50 Hz frequency of the power supply. Fig. 4.8 shows a simple four-event control. First the squeeze is applied and held constant (a–b on the graph). The current is switched on at X, held for seven complete cycles (7/50 s) and switched off at Y. The pressure is held until c on the graph, the weld cooling in this period. Finally the squeeze is released and the repeat cycle of operations begins again at a_1. The switches comprise repeat–non-repeat, thyristors on–off and heat control in–out.

For more complex welding cycles, necessary when welding certain metals to give them correct thermal treatment, the graphs are more complex with up to ten functions with variable pressure cycle, and these machines are often operated from three phases. Machines for the aeronautical and space industries are specially designed and incorporate variable pressure heads

Fig. 4.8. Standard four-event sequence for spot welding.

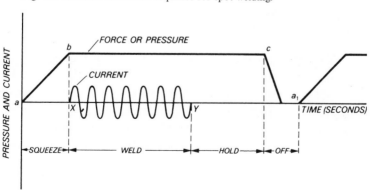

and up to ten functions. Most aluminium alloys lose the properties given to them by work hardening or heat treatment when they are heated. To enable the correct thermal treatment of heating and cooling to be given to them during the welding cycle three-phase machines are often employed.

The transformer

The transformer steps the voltage down from that of the mains to the few volts necessary to send the heavy current through the secondary welding circuit. When the current is flowing the voltage drop across the secondary may be as low as 3–4 volts and in larger machines the current can be up to 35 000 A. Because of these large currents there is considerable force acting between the conductors due to the magnetic field, so transformers must be robustly constructed or movement may cause breakdown of the insulation, and in addition they may be water cooled because of the heating effect of the current. The secondary winding consists of one or two turns of copper strip or plates over which the primary coils fit, the whole being mounted on the laminated iron core. Because any infiltration of moisture may lead to breakdown, modern transformers have the primary winding and secondary plates assembled as a unit which is then vacuum encapsulated in a block of epoxy resin. The primary winding has tappings which are taken to a rotary switch which selects the current to be used.

The current flowing in the secondary circuit of the transformer provides the heating effect and depends upon (1) the open circuit voltage and (2) the impedance of the circuit, which depends upon the gap, throat depth and magnetic mass introduced during welding and the resistance of the metal to be welded. The duty cycle is important, as with all welding machines, as it affects the temperature rise of the transformer. For example, if a spot welder is making 48 spot welds per minute, each of 0.25 seconds duration, the duty cycle is $(48 \times 0.25 \times 100) \div 60 = 20\%$. Evidently knowing the duty cycle and welding time, the number of welds that can be made per minute can be calculated.

Electrodes

The electrode arms and tips (Fig. 4.9a) which must carry the heavy currents involved and apply the necessary pressure must have the following properties: (1) high electrical conductivity so as to keep the I^2R loss (the heating due to the resistance) to a minimum, (2) high thermal conductivity to dissipate any heat generated, (3) high resistance to deformation under large squeeze pressures, (4) must keep their physical properties at elevated temperatures, (5) must not pick up metal from the surface of the workpiece,

Electrode	Properties	Uses
(1) Copper, 1% silver	High conductivity, medium hardness.	Light alloys, coated sheets, scaly steel.
(2) Copper, 0.6% Cr or 0.5% Cr and Be	Best electrical conductivity with greatest hardness.	Clean or lightly oxidized steel, brass and cupro-nickel. Used for the arms of the machine and electrical conductors.
(3) Copper, 2.5% Co and 0.5% Be	Poor conductivity but very hard.	For welding hard metals with high resistivity, e.g. stainless steel, heat-resistant and special steels.
(4) Copper, beryllium.	Poor conductivity, great hardness.	For clamping jaws of flash butt welders.
(5) Molybdenum		Drawn bar or forged buttons used as pressed fit inserts in supports of (2) above. For welding thin sheet or wires or electro-brazing silver-based metals.
(6) Sintered copper-tungsten	Fair conductivity, very hard.	Keeps its mechanical properties when hot.

Note. The addition of various elements increases the hardness but reduces the conductivity.

(6) must be of reasonable cost. The following types of electrodes are chiefly used, and are given in the table. Electrolytic copper (99.95% Cu) has high electrical and thermal conductivity and, when work hardened, resists deformation but at elevated temperatures that part of the electrode tip in contact with the work becomes annealed due to the heating and the tip softens and deforms. Because of this it is usual to water-cool the electrodes to prevent excessive temperature rise.

Water cooling

Adequate electrode cooling is the most essential factor to ensure optimum tip life; the object is to prevent the electrode material from reaching its softening temperature, at which point it will lose its hardness

Fig. 4.9. (*a*) Electrodes and holders.

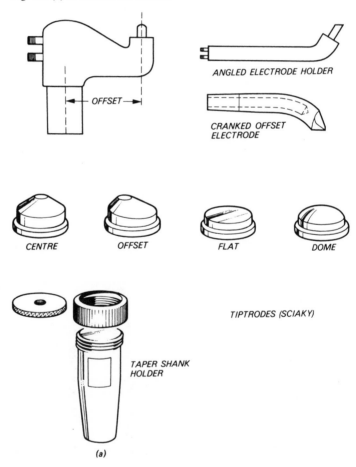

and rapidly deteriorate. The normal cooling method is by internal water circulation where the water is fed via a central tube arranged to direct the water against the end of the electrode cooling hole (Fig. 4.9*b*). In some cases a short telescopic extension tube is used inside the main tube and must be adjusted to suit the length of the electrode used.

Electrodes are available with a taper fit to suit the electrode arms and also as tips to fit on to a tapered shank body, which makes them easily interchangeable. The face that makes contact with the metal to be welded can be a truncated cone with central or offset face, flat or domed (Fig. 4.9) and tip diameters can be calculated from formulae which depend upon the plate thickness, but in all cases these are approximate only and do not replace a test on the actual part. Electrodes are subject to great wear and tear in service due to the constant heating and cooling and varying pressure

Fig. 4.9. (*b*)

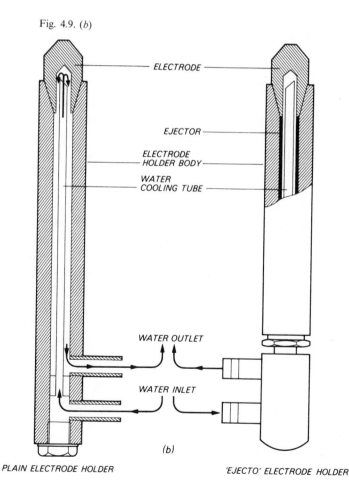

ELECTRODE

EJECTOR

ELECTRODE HOLDER BODY

WATER COOLING TUBE

WATER OUTLET

WATER INLET

(b)

PLAIN ELECTRODE HOLDER

'EJECTO' ELECTRODE HOLDER

cycles. The chief causes of wear are: electrical; wrong electrode material, poor surface being welded, contacts not in line; mechanical; electrode hammering, high squeeze, weld and forge pressures, abrasion in loading and unloading and tearing due to the parting of the electrodes.

Seam welding

In a seam welding machine the electrodes of a spot welder are replaced by copper alloy rollers or wheels which press on the work to be welded (Fig. 4.10a and b). Either one or both are driven and thus the work passes between them. Current is taken to the wheels through the rotary bearings by silver contacts with radial pressure and the drive may be by knurled wheel or the more usual shaft drive which enables various types of wheel to be easily fitted. If the current is passed continuously a continuous seam weld results but, as there is a shunt effect causing the current to flow through that part of the weld already completed, overheating may occur resulting in burning of the sheets. To avoid this the current can be pulsed, allowing sufficient displacement of the already welded portion to take place and thus obviating most of the shunt effect. For materials less than 0.8 mm thick or at high welding speeds (6 m/min) no pulsing is required, the 50 Hz frequency of the supply providing a natural pulse. Above 2×0.8 mm thickness pulsation is advisable, and essential above 2×1.5 mm thickness, while for pressure-tight seams the welds can be arranged to overlap and if the seams are given a small overlap with wide-faced wheels and high pressure a mash weld can be obtained.

By the use of more complex electro-mechanical bearing assemblies, longitudinal and circumferential welds can be made (Fig. 4.10c).

Seam welding guns are extremely useful for fabricating all types of tanks, exhaust systems, barrels, drip-mouldings on car body shells, etc. They have electrode drive which automatically propels the gun along the seam so that

Fig. 4.10. (a)

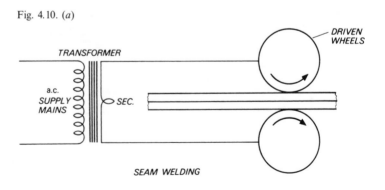

SEAM WELDING

Fig. 4.10. (b) 'Thin wheel' seam welding machine with silver bearings.

Fig. 4.10. (c)

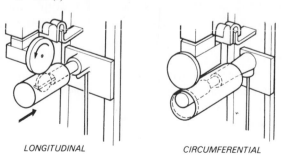

LONGITUDINAL   CIRCUMFERENTIAL

SEAM WELDING

it only requires guidance, and they are operated in the same way as spot welding guns.

Direct current spot welding

We have seen that when iron and steel to be welded is placed in a spot or seam welder the impedance of the secondary circuit is increased and the secondary current varies according to the mass introduced. By using d.c. this loss is obviated, the power consumed is reduced and the electrode life increased because skin effect is eliminated. This enables coated steels, stainless and other special steels in addition to aluminium alloys to be welded to high standards. The machine is similar in appearance to the a.c. machine but has silicon diodes as the rectifying elements.

Three-phase machines

Three-phase machines have been developed to give impulse of current in the secondary circuit at low frequency, with modulated wave form to give correct thermal treatment to the material being welded. These machines have greatly increased the field of application of spot welding into light alloys, stainless, and heat-resistant steels, etc. A typical sequence of operations is: squeeze, which multiplies the number of high spots between the contact faces; welding pressure, which diminishes just before the welding current flows; immediately after the passage of the current, forging or recompression pressure is applied which is above the elastic limit of the material and completes the weld. Setting of the welding current can be done by welding test pieces with increasing current until the diameter of the nugget, found by 'peeling' the joint, is $(2t + 3)$, where t is the thickness of the thinner sheet. This current is noted. It is then increased until spatter of the nugget occurs and the welding current is taken as midway between these two values, with final adjustments made on test pieces.

Nickel and nickel alloys have higher electrical resistivity and lower thermal conductivity than steel and are usually welded in thin sections as lap joints, although the crevice between the sheets may act as a stress raiser and affect the corrosion and fatigue resistance. High pressures are required for the high-nickel alloys so as to forge the solidifying nugget, and the machine should have low inertia of the moving parts so that the electrode has rapid follow-up during welding. Current may be set as in the previous section, and in some cases it is advantageous to use an initial squeeze followed by an increasing (up-sloping) current. When the nugget is just beginning to form with diffusion of the interface, the squeeze force is reduced to about one-third of its initial value and held until the current is switched off; this reduces danger of expulsion of molten metal and gives better penetration of the weld into the two sheets.

Projection welding

Protrusions are pressed on one of the sheets or strips to be welded and determine the exact location of the weld (Fig. 4.11). Upon passage of the current the projection collapses under the electrode pressure and the sheets are welded together. The machines are basically presses, the tipped electrodes of the spot welder being replaced by flat platens with T slots for the attachment of special tools, and special platens are available which allow the machine to be used as a spot welder by fitting arms and electrodes (Fig. 4.12), and automatic indexing tables can be used to give increased output. Projection welding is carried out for a variety of components such as steel radiator coupling elements, brake shoes, tin-plate tank handles and spouts, etc. The press type of machine is also used for resistance brazing in which the joining of the parts is achieved with the use of an alloy with a lower melting point than the parent metal being welded so that there is no melting involved.

Cross wire welding is a form of projection welding, the point of contact of the two wires being the point of location of the current flow. Low-carbon mild steel, brass, 18/8 stainless steel, copper-coated mild steel and galvanized steel wire can be welded but usually the bulk of the work is done with clean mild steel wire, bright galvanized, or copper coated as used for milk bottle containers, cages, cooker and refrigerator grids, etc., and generally several joints are welded simultaneously.

Modern machines are essentially spot welders in ratings of 25, 70, or 150 kVA fitted with platens upon which suitable fixtures are mounted and having a fully controlled pneumatically operated vertical head, and the electrical capacity to weld as many joints as possible simultaneously. Large programmed machines are manufactured for producing reinforcement mesh automatically.

Fig. 4.11. Projection welding.

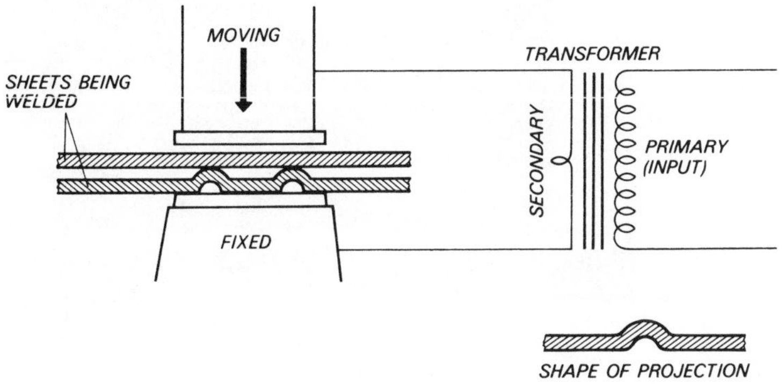

SHEETS BEING WELDED

MOVING

TRANSFORMER

SECONDARY

PRIMARY (INPUT)

FIXED

SHAPE OF PROJECTION

Resistance butt welding

This method is similar to spot welding except that the parts to be butt welded now take the place of the electrodes. The two ends are prepared so that they butt together with good contact. They are then placed in the jaws of the machine, which presses them close together end to end (Fig. 4.13). When a given pressure has been reached, the heavy current is switched on, and the current flowing through the contact resistance

Fig. 4.12. Projection welder.

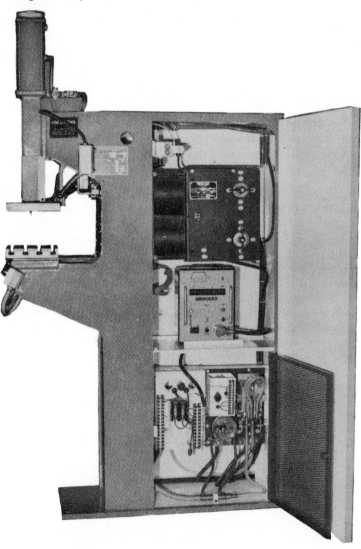

between the ends brings them to welding heat. Extra pressure is now applied and the ends are pushed into each other, the white hot metal welding together and an enlargement of section taking place. The section may be machined to size after the operation if necessary.

*Flash butt welding**

Although this is not a resistance welding process it is convenient to consider it here. Flash butt welding machines must be very robustly made and have great rigidity because considerable pressures are exerted and exact alignment of the components is of prime importance. The clamping dies of copper alloy, which carry the current to the components and hold them during butting up under high pressure, should grip over as large an area as possible to reduce distortion tendencies. The clamping pressure, which is about twice the butting pressure, is usually done pneumatically or hydraulically, and current is of the order of $7-10$ A/mm² of joint area. The following are the stages in the welding cycle.

Pre-heat. The components are butted together and a current passing across the joint heats the ends to red heat.

Flash. The parts are separated and an arc is established between them until metal begins to melt, one of the components being moved to keep the arc length constant.

Upset. The parts are butted together under high pressures with the current still flowing. Impurities are forced out of the joint in the butting process and an impact ridge or flash is formed (Fig. 4.14). Post-heat treatment can be given by a variation of current and pressure after welding. For welding light alloys, pre-heating is generally dispensed with and the flashing is of short duration.

As an example of currents involved, a butt weld in 6 mm thick 18/18/3

Fig. 4.13. Resistance butt welding.

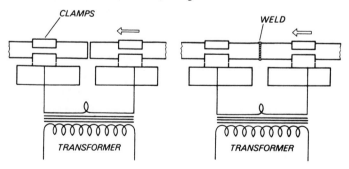

* BS 499 gives this as a resistance process.

stainless steel with a cross-section of 600 mm² involves currents of 20 000 A with a 9 second flashing time.

Flash butt welding is used very extensively by railway systems of the world to weld rails into continuous lengths. British Rail for example weld rails approximately 18.3 m long into continuous lengths from 91 to 366 m and conductor rails are welded in the same way. Fig. 4.15 shows a modern rail welding machine. Hydraulic rams, equally spaced on each side of the rail section, apply the forging load of 200–400 kN to the rails of section approximately 7200 mm². The rail is clamped by two vertically acting cylinders and horizontally acting cylinders align each rail to a common datum and an anti-twist device removes axial twist. When welding long lengths of rail it is more convenient to move the machine to the exact position for welding rather than the rail and for this the machine can be rail mounted. Machines of this type can make 150–200 welds per 8-hour shift. In this case the sequence of operations is pre-flash, pre-heat, flashing and forging.

The upset is removed by a purpose-designed machine with hydraulic power-shearing action usually mounted in line with the welding machine.

The flash butt welded lengths are welded *in situ* on the track by the thermit process.

Fig. 4.14. Flash butt welding.

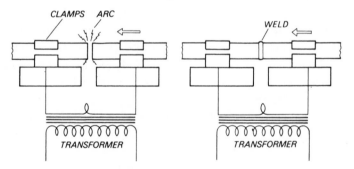

Fig. 4.15. Machine for welding rail sections. 1000 kVA. 3-phase, 440 V. Welding sequence: (1) Both rails securely clamped in welder. (2) Rails aligned vertically and horizontally under full clamping pressure. (3) Weld initiated by push button control. (4) Moving head moves forward on the burn-off or pre-flashing stroke used to square up the ends of the rails if necessary. On completion of this, pre-heating begins until the requisite number of pre-heats have lapsed. At this point the rails should be in a suitably plastic state to allow for straight flashing and finally forging.

5

Additional processes of welding

There has been a great increase in the number of automatic processes designed to speed up welding production. Automatic welding gives high rates of metal deposition because high currents from 400 to 2000 A can be used, compared with the limit of about 600 A with manual arc welding. Automatic arc control gives uniformly good weld quality and finish and the high heat input reduces distortion and the number of runs for a given plate thickness is reduced. Twin welding heads still further reduce welding time, and when used, for example, one on each side of a plate being fillet welded, distortion is reduced. The welding head may be:

(1) Fixed, with the work arranged to move beneath it.
(2) Mounted on a boom and column which can either be of the positioning type in which the work moves or the boom can traverse at welding speed over the fixed work.
(3) Gantry mounted so that it can traverse over the stationary work.
(4) Self propelled on a motor-driven carriage.

The processes which have been described previously in this book, namely TIG, MIG and CO_2 (gas shielded metal arc) with their modifications, are extensively used fully automatically. Heads are now available which, by changing simple components, enable one item of equipment to be used for MIG (inert gas), CO_2 and tubular wire, and submerged arc processes.

Submerged arc welding*

In this automatic process the arc is struck between bare or flux cored wire and the parent plate, the arc, electrode end and the molten pool are submerged or enveloped in an agglomerated or fused powder which

---

* American designation SAW (similar to British).

turns into a slag in its lower layers under the heat of the arc and protects the weld from contamination. The wire electrode is fed continuously to the arc by a feed unit of motor-driven rollers which is voltage-controlled in the same way as the wire feed in other automatic processes and ensures an arc of constant length. The flux is fed from a hopper fixed to the welding head, and a tube from the hopper spreads the flux in a continuous mound in front of the arc along the line of weld and of sufficient depth to completely submerge the arc, so that there is no spatter, the weld is shielded from the atmosphere and there are no radiation effects (UV and IR) in the vicinity. (Fig. 5.1a and b.)

Welding heads

Fully automated welding heads for this process can also be used with modification for gas shielded metal arc welding including CO_2, solid and flux cored, thus greatly increasing the usefulness of the equipment. The head can be stationary and the work moved below it, as for example in the welding of circumferential and longitudinal seams, or the head may be used with positioners or booms or incorporated into custom-built mass production welding units for fabricating such components as brake shoes, axle housings, refrigerator compressor housings, brake vacuum cylinders, etc., and hard surfacing can also be carried out. The unit can also be tractor mounted (cf. Fig. 5.2c) and is self-propelled, with a range of speeds of 100 mm to 2.25 m per minute, and arranged to run on guide bars or rails. Oscillating heads can be used for root runs on butt joints to maintain a constant welding bead on the underside of the joint, and two and even three heads can be mounted together or the heads can be arranged side by side to give a wide deposit as in hard surfacing. Fillet welding can be performed by inclining two heads, one on each side of the joint, with flux feeds and

Fig. 5.1. (a)

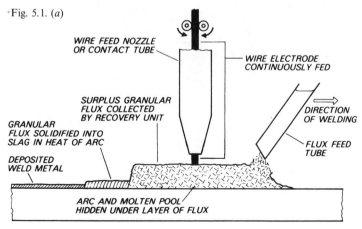

WIRE FEED NOZZLE OR CONTACT TUBE

WIRE ELECTRODE CONTINUOUSLY FED

SURPLUS GRANULAR FLUX COLLECTED BY RECOVERY UNIT

GRANULAR FLUX SOLIDIFIED INTO SLAG IN HEAT OF ARC

DEPOSITED WELD METAL

DIRECTION OF WELDING

FLUX FEED TUBE

ARC AND MOLTEN POOL HIDDEN UNDER LAYER OF FLUX

recovery, the heads being mounted on a carriage which travels along a gantry over the work (Fig. 5.2). Two heads mounted in tandem and travelling either along a guide rail or directly on the workpiece are used for butt joints on thick plate, and both can operate on d.c. or the leading head can operate on d.c. and the trailing head on a.c.

Three electrode heads can be gantry mounted on a carriage, the leading electrode being d.c. operated with the trailing electrodes a.c. This method gives high deposition rates with deep penetration. Special guide units ensure in all cases that the electrode is correctly positioned relative to the joint.

The main components in the control box are: welding voltage and arc current controls, and wire feed controlled by a thyristor regulator which maintains set values of arc voltage. The head is accurately positioned by

Fig. 5.1. (*b*)

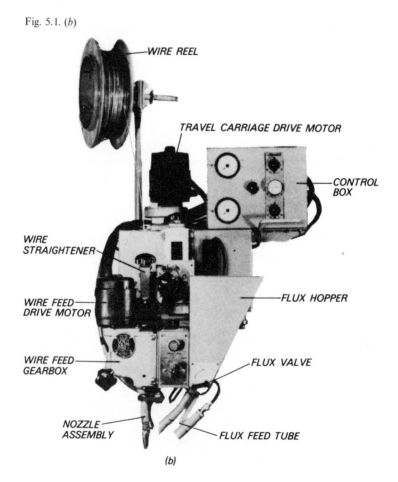

(b)

slide adjusters for horizontal and vertical movement and has angular adjustment also. The wire feed motor has an integral gearbox and wire-straightening rolls give smooth wire feed. The gear ratio for the metal arc process is much higher than for submerged arc and each wire diameter usually requires its own feed rolls, which are easily interchanged. For fine wires less than 3 mm diameter a fine-wire-straightening unit can be fitted.

Current is passed to the electrode wire through a contact tube and jaws

Fig. 5.2. Various mountings for automatic welding equipment.
(*a*) An automatic welding machine in which the head is mounted on a carriage which travels along a beam.
(*b*) An automatic welding head as in (*a*) designed for stationary mounting on a manipulator column or boom in order to be an integral part of a mechanized welding system.
(*c*) A tractor-mounted automatic welding head as in (*a*). The machine has a single welding head and is designed for welding butt joints and for making fillet welds in the flat or horizontal–vertical position.

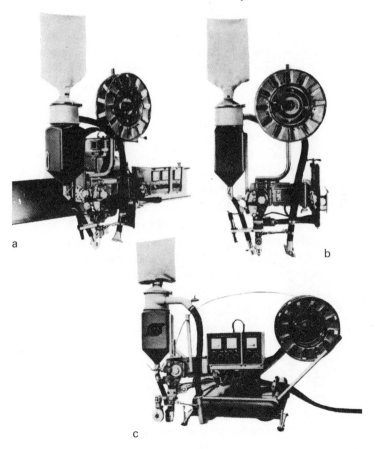

a

b

c

which fit the wire diameter being used, and the contact tube is used for the shielding gas when gas shielded welding is being performed and is water cooled. The coil arm holder has a brake hub with adjustable braking effect and carries 300 mm i.d. coils, and a flux hopper is connected to a flux funnel attached to the contact tube by a flexible hose.

A guide lamp which is attached to the contact tube provides a spot of light which indicates the position of the wire, thus enabling accurate positioning of the head along the joint, and a flux recovery unit collects unfused flux and returns it to the hopper.

A typical sequence of operations for a boom-mounted carriage carrying multiple welding head is: power on, carriage positioned, welding heads 1, 2, etc., down, electrode feed on, wire tips set, flux valve open, welding speed set, welding current set, welding voltage set, flux recovery on; press switch to commence welding.

Power unit

The power unit can be a motor- or engine-driven d.c. generator or transformer-rectifier with outputs in the 30–55 V range and with currents from 200 to 1600 A with the wire generally positive. In the case of multiple head units in which the leading electrode is d.c. and the trailing electrode is a.c. a transformer is also required. In general any power source designed for automatic welding is usually suitable when feeding a single head.

Wires are available in diameters of 1.6, 2.0, 2.4, 3.2, 4.0, 5.0, 6.0 mm on plastic reels or steel formers. Wrappings should be kept on the wire until ready for use and the reel should not be exposed to damp or dirty conditions.

A variety of wires are available including the following (they are usually copper coated). For mild steel types with varying manganese content, e.g. 0.5%, 1.0%, 1.5%, 2.0% manganese, a typical analysis being 0.1% C, 1.0% Mn, 0.25% Si, with S and P below 0.03%. For low alloy steels, 1,25% Cr, 0.5% Mo; 2.25% Cr, 1.0% Mo; 0.2% Mn, 0.5% Mo; and 1.5% Mn 0.5% Mo are examples, whilst for the stainless steel range there are: 20% Cr, 10% Ni for unstabilized steels; 20% Cr, 10% Ni, 0.03% C, for low-carbon 18/8 steels; 19% Cr, 12% Ni, 3% Mo for similar steels; 20% Cr, 10% Ni, nobium stabilized; 24% Cr, 13% Ni for steels of similar composition and for welding mild and low-alloy steels to stainless steel.

Many factors affect the quality of the deposited weld metal: electrode wire, slag basicity, welding variables (process), cleanliness, cooling rate, etc. For hardfacing, the alloy additions necessary to give the hard surface usually come from the welding wire and a neutral flux, and tubular wire with internal flux core is also used in conjunction with the external flux.

Hardness values as welded, using three layers on a mild steel base, are between 230 and 650 HV depending upon the wire and flux chosen.

Flux (see also Vol. 1, p. 57)

Fluxes are suitable for use with d.c. or a.c. They are graded according to their form, whether (1) fused or (2) agglomerated. Fused fluxes have solid glassy particles, low tendency to form dust, good recycling properties, good slag-flux compatibility, low combined water and little sensitivity to humid conditions. Agglomerated fluxes have irregular-sized grains with low bulk density, low weight consumption at high energy inputs with active deoxidizers and added alloying elements where required.

Fluxes are further classified as to whether they are acidic or basic, the basicity being the ratio of basic oxides to acidic oxides which they contain. In general the higher the basicity the greater the absorption of moisture and the more difficult it is to remove.*

The general types of flux include manganese silicate, calcium manganese aluminium sulphate, rutile, zirconia and bauxite, and the choice of flux affects the mechanical properties of the weld metal. Manufacturers supply full details of the chemical composition and mechanical properties of the deposited metal when using wires of varying compositions with various selected fluxes (i.e. UTS, % elongation Charpy impact value, and CTOD† figures at various temperatures).

Fluxes that have absorbed moisture should be dried in accordance with the makers' instructions, as the presence of moisture will affect the mechanical properties of the deposited metal.

Joint preparation

Joint edges should be carefully prepared and free from scale, paint, rust and oil, etc., and butt seams should fit tightly together. If the fit-up has gaps greater than 0.8 mm these should be sealed with a fast manual weld.

When welding curved circumferential seams there is a tendency for the molten metal and slag, which is very fluid, to run off the seam. This can be avoided or reduced by having the welding point 15–65 mm before top dead centre in the opposite direction to the rotation of the work and in some cases the speed of welding and current can be reduced. Preparation of joints is dependent upon the service to which the joint is to be put and the following preparations are given as examples only (Fig. 5.3).

* Basicity index (BI) $= \dfrac{CaO + CaF_2 + MgO + K_2O + Na_2O + \frac{1}{2}(MnO + FeO)}{SiO_2 + \frac{1}{2}(Al_2O_3 + TiO_2 + ZrO_2)}$

† Crack tip opening displacement.

Backing

As the cost of back-chipping and making a sealing run has escalated it becomes more and more necessary to be able to weld plates and cylinders of large size with a run on one side only. This may be the case, for example, if the fit-up of the sections is poor and a weld in the root of the section may not be able to bridge a wide root gap successfully. In these cases a backing can be used so that the weld is performed from one side only and with which a good profile of underbead is obtained even when fit-up and alignment are not good. The following are examples of differing types of backing strips available.

Fig. 5.3. Types of butt welds.

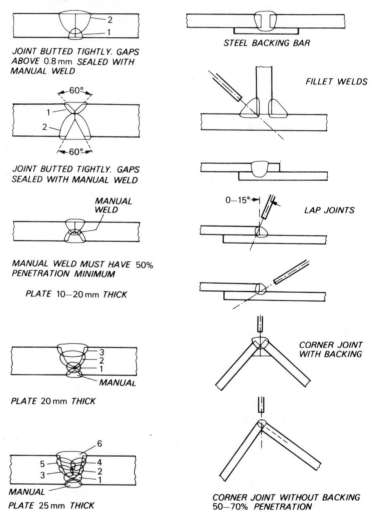

JOINT BUTTED TIGHTLY. GAPS
ABOVE 0.8 mm *SEALED WITH
MANUAL WELD*

JOINT BUTTED TIGHTLY. GAPS
SEALED WITH MANUAL WELD

MANUAL WELD

MANUAL WELD MUST HAVE 50%
PENETRATION MINIMUM

PLATE 10–20 mm THICK

PLATE 20 mm THICK

MANUAL

PLATE 25 mm THICK

STEEL BACKING BAR

FILLET WELDS

0–15° LAP JOINTS

CORNER JOINT
WITH BACKING

CORNER JOINT WITHOUT BACKING
50–70% PENETRATION

Ceramic tile backing strip. This is shown in section in Fig. 5.4*a* and is suitable for slag-forming processes such as submerged arc, or flux cored and MMA can be used for vertical and horizontal–vertical butts. A recess in the tile allows the slag to form below the underbead and is stripped off and discarded with the aluminium foil which holds it in position as the weld progresses.

Fibreglass tape backing strip is a closely woven flexible material of about four to six layers and fibreglass tape which gives good support to the underbead or root run of the weld and is usually used in conjunction with a copper or aluminium backing bar. It is non-hygroscopic and has low fume level. Sizes are from 30 mm wide heavy single layer, 35 mm wide four layer and 65 mm wide six layer.

Fibreglass tape, sintered sand backing plate. This is typical for submerged arc single or twin wire as in large structures, e.g. deck plates in shipyards, etc. The backing is of sintered sand (silica) about 600 mm long, 50 mm wide and 10 mm thick reinforced with steel wires. It has a fibreglass tape fitted to the upper surface to support the root of the weld and has adhesive outer edges to allow for attachment to the joint, which should be dry and which should have a 40–50° included angle preparation and a root gap of about

Fig. 5.4. (*a*) Ceramic tile backing strip.

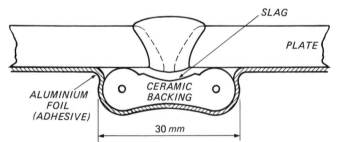

Fig. 5.4. (*b*)

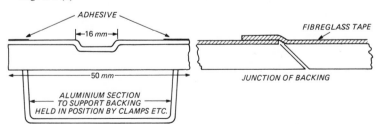

4 mm. The backing is slightly flexible to allow for errors of alignment and has 45° bevelled ends. Overlapping tape prevents burn-through at the backing junction. An aluminium section can be used as undersupport if required (Fig. 5.4b).

Electroslag welding (Fig. 5.5)

Developed in Russia, this process is used for butt welding steel sections usually above 60 mm in thickness although plates down to 10 mm thick can be welded. The sections to be welded are fixed in the vertical position and part of the joint line, where welding is to commence, is enclosed with water-cooled copper plates or dams which serve to position the molten weld metal and slag. The dams are pressed tightly against each side of the joint to prevent leakage. There may be from one to three electrode wires depending upon the thickness of the section and they are fed continuously from spools. The self-adjusting arc is struck on to a run-off plate beneath a coating of powder flux which is converted into a liquid in about half a minute. The current is then transferred, not as an arc but

Fig. 5.5. Principle of electroslag or vertical submerged melt welding.

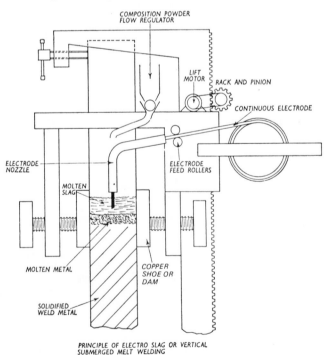

PRINCIPLE OF ELECTRO SLAG OR VERTICAL
SUBMERGED MELT WELDING

through the liquid slag, which gives the same order of voltage drop as would the arc. During welding some slag is lost in forming a skin between the molten metal and the copper dams, and a flow of powder, carefully metered to avoid disturbing the welding conditions, is fed in to make good wastage. The vertical traverse may be obtained by mounting the welding head on a carriage which is motor-driven and travels up a rack on a vertical column in alignment with the joint to be welded. The rate of travel is controlled so that the electrode nozzle and copper dams are in the correct position with regard to the molten pool, and since the electrode is at right angles to the pool, variations in fit-up are not troublesome. For thick sections the electrodes are given a weaving motion across the metal.

The welds produced are free from slag inclusions, porosity and cracks, and the process is rapid, preparation is reduced, and there is no deslagging. Composite wires containing deoxidizers and alloying elements can be used when required. A variation of the process uses a CO_2 shield instead of the flux powder, the CO_2 being introduced through pipes in the copper dams just above the molten metal level.

Electroslag welding with consumable guides or nozzles

Consumable guide welding is a development of the electroslag process for welding straight joints in thick plate in the vertical or near vertical position, in a range of 15–40 mm thick plate and with joints up to 2 m long. The set-up gap between plates is 25–30 mm, but when welding thicknesses less than 20 mm the joints can be reduced to 18–24 mm, the gap ensuring that the guide tube does not touch the plate edges. Water-cooled copper shoes act as dams and position the molten metal and also give it the required weld profile. As with electroslag welding the current passing through the molten slag generates enough heat to melt the electrode end, guide and edges of the joint, ensuring a good fusion weld. If a plain uncoated guide tube is used, flux is added to cover the electrode and guide end before welding commences. To start the process the arc is struck on the work. It continues burning under the slag with no visible arc or spatter. The slag should be viewed through dark glasses as in gas cutting because of its brightness when molten.

An a.c. or d.c. power source in the range 300–750 A is suitable, such as is used for automatic and MMA processes. Striking voltage is of the order of 70–80 V, with arc voltages of 30–50 V, higher with a.c. than d.c.

The advantages claimed for the process are: relatively simpler, cheaper, and more adaptable than other similar types, faster welding than MMA of thick plate, cheaper joint preparation, even heat input into the joint thus reducing distortion problems, no spatter losses, freedom from weld metal defects and low consumption of flux. Fig. 5.6 illustrates the layout of the machine.

Mechanized MIG welding with robots (robotics)

Fully automatic welding using the gas shielded metal arc or submerged arc processes in conjunction with columns and booms, positioners, rotators and other equipment such as jigs is extensively used for the making of welds either straight, circumferential or circular. The use

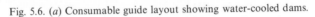

Fig. 5.6. (*a*) Consumable guide layout showing water-cooled dams.

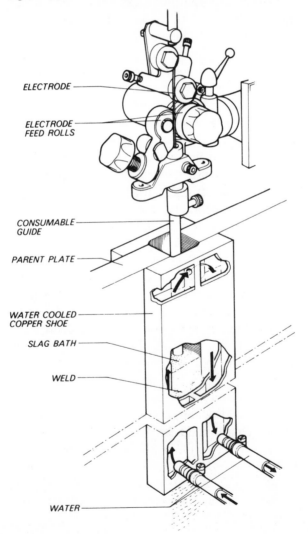

ELECTRODE

ELECTRODE FEED ROLLS

CONSUMABLE GUIDE

PARENT PLATE

WATER COOLED COPPER SHOE

SLAG BATH

WELD

WATER

Fig. 5.6. (*b*) Completed weld.

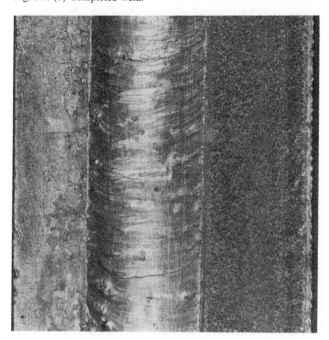

Fig. 5.6. (*c*) The consumable guide welding process using, in this case, twin wire/tube system. The dams have been removed to show the position of the guide tubes.

Fig. 5.7

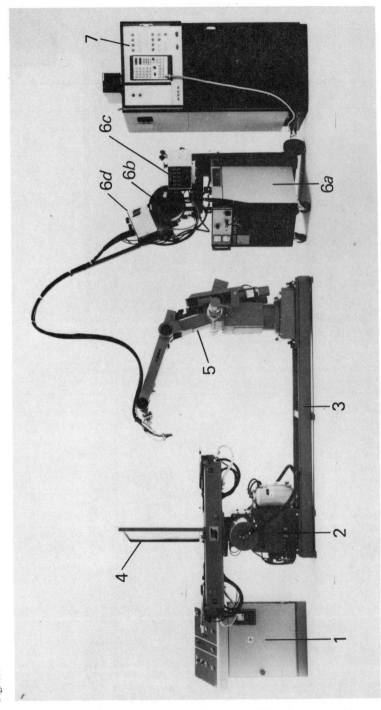

Key:
1. Control panel, by which operator determines when robot should start welding the next work piece.
2. Handling unit. Different types are available. On this one fixtures for the component are mounted on tilting turntables.
3. Handling unit on same foundation as robot giving stability and accuracy.
4. Screen.
5. Welding robot with rapid, accurate movements.

6a. Constant voltage power source for welding head. Wire feed unit can be positioned separately.
6b. Wire reel.
6c. Programming unit.
6d. Wire feed.
7. Control cabinet, controlling movements and where removable programming unit is located. Program is stored on tape cassette.

of robots in the car industry for spot welding the body shell, spray painting and general assembly is well known and it has resulted in increased speed of production and less overall cost with reduced monotony and fatigue for the operator.

Now consider the robot adapted to the fabrication of large numbers of similar components with welds in all positions performed by the gas shielded metal arc process.

The word robot comes from the Czech word *robota* which, in translation, means any class of work that involves monotony, repetition or drudgery. When operating the robot the welder becomes the supervisor for whom the robot works, performing this monotonous work at speed, with accuracy and not getting tired or needing breaks.

To serve as an example we can consider one particular modern robot illustrated in Fig. 5.7.

This has six degrees of freedom, (one being optional) based on the human hand, wrist and forearm and the following gives the approximate movement values:

> rotational 340°, radial arm 550 mm max., vertical arm 850 mm max., wrist bending 90°, wrist turning 180°.

The accuracy of these movements is 0.1–0.2 mm at 500 mm distance and it should be noticed that the robot can be equipped with heads for selection, grinding, polishing and spot welding, if required.

Fig. 5.8. Computer control, servo-powered positioner. Dual axes, fast and accurate handling, Self-braking worm gears.

Evidently there must be large production runs of components with repetitive welding to be performed for the cost of the station, consisting of positioners, control cabinets, robot with gas shielded metal arc welding, head and power and control unit, to be justified. Once the decision has been taken for the outlay to be justified the advantages that accrue are great.

The production station, which varies according to the size of the work pieces, can consist, for example, of two positioners, on which the work is jig held. These positioners (Fig. 5.8) and the data required in the welding operations are controlled by a microcomputer and are servo-steered.

The robot head has a gas shielded metal arc welding gun adapted to fit the head, the power source is of the thyristor-controlled constant voltage type and the wire is fed from a unit which controls speed of feed and compensates for variations of mains voltage and friction of the feed rollers.

To program the robot, the programming unit is taken from the control cabinet and the robot run through the complete welding sequence for the part to be fabricated. The welding gun is moved from point to point and each section is fed with the speed required and the welding parameters (current and voltage, etc.), and the accelerated movement from one welding point to the next is also programmed and mistakes are easily corrected. The computer memory has say 500 position capacity with additional instructions with a tape recorder increasing the memory. The storage of the program is on a digital tape cassette so that the switch from one program to another can be made without starting from scratch. As is usual pre- and post-flow of the shielding gases and crater filling are all part of the program. (*Note.* The TIG method of welding may also be performed by robots. See Chapter 3.)

Pressure welding

This is the joining together of metals in the plastic condition (not fusion) by the application of heat and pressure as typified by the blacksmith's weld. In general the process is confined to butt welding. The parts (tubes are a typical example) are placed on a jig which can apply pressure to force the parts together.

The faces to be welded are heated by oxy-acetylene flames, and when the temperature is high enough for easy plastic flow to take place, heating ceases and the tubes are pushed together causing an upset at the welded face. The welding temperature is about 1200°C for steel. It is considered that atoms diffuse across the interfaces and recrystallization takes place, the grains growing from one side to the other of the welded faces since they are in close contact due to the applied pressure. Any oxide is completely broken up at a temperature well below that of fusion welding but due to the heating time concerned, grain growth is often considerable. Steel, some alloy steels, copper, brass, and silver can be welded by this process.

Cold pressure welding

Cold pressure welding is a method of joining sections of metal together by the application of pressure alone using no heat or flux. Pressure is applied to the points to be welded at temperatures below the recrystallization temperature of the metals involved. This applied pressure brings the atoms on the interface to be welded into such close contact that they diffuse across the interface and a cold pressure weld is made.

It is a method for relatively ductile metals such as aluminium, copper cupro-nickel, gold, silver, platinum, lead, tin and lead–tin alloys, etc., and it is particularly suited to welds in circular wire section. In the type described of the multiple upset type, the surfaces to be welded are placed in contact and held in position by gripping dies and are fed together in small increments by a lever (Fig. 5.9). Each lever movement giving interface pressure displaces the original surfaces by plastic flow and after about four to six upsets, the last movement completes the weld and the flash is easily removed. In the machine illustrated, copper wire 1.1–3.5 mm diameter max. or aluminium wire to 4.75 max. can be butt welded. Fig. 5.10a and b illustrates the micrographs of welds made in copper to copper and copper to aluminium wire.

Fig. 5.9

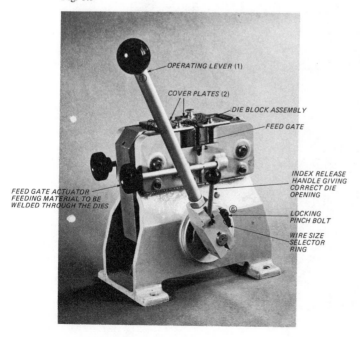

Fig. 5.10. (*a*) Cold pressure weld, aluminium to copper (wire 9.5 mm diameter). × 25.

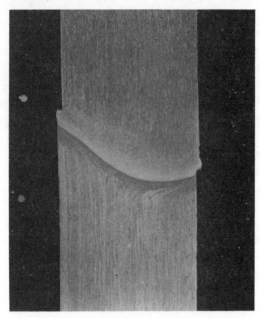

Fig. 5.10. (*b*) Cold pressure weld, copper to copper (wire 8 mm diameter). × 25.

Ultrasonic welding

Ultrasonic vibrations of several megacycles per second (the limit of audibility is 20 000–30 000 Hz) are applied to the region of the faces to be welded. These vibrations help to break up the grease and oxide film and heat the interface region. Deformation then occurs with the result that welding is possible with very greatly reduced pressure compared with an ordinary pressure weld. Very thin section and dissimilar metals can be welded and because of the reduced pressure there is reduced deformation (Fig. 5.11).

Fig. 5.11. Ultrasonic welding.

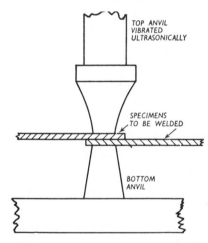

Friction welding

The principle of operation of this process is the changing of mechanical energy into heat energy. One component is gripped and rotated about its axis while the other component to be welded to it is gripped and does not rotate but can be moved axially to make contact with the rotating component. When contact is made between rotating and non-rotating parts heat is developed at the contact faces due to friction and the applied pressure ensures that the temperature rises to that required for welding. Rotation is then stopped and forging pressure applied, causing more extrusion at the joint area, forcing out surface oxides and impurities in the form of a flash (Fig. 5.12). The heat is concentrated and localized at the interface, grain structure is refined by hot work and there is little diffusion across the interface so that welding of dissimilar metals is possible.

In general at least one component must be circular in shape, the ideal situation being equal diameter tubes, and equal heating must take place over the whole contact area. If there is an angular relationship between the final parts the process is not yet suitable.

The parameters involved are: (1) the power required, (2) the peripheral speed of the rotating component, (3) the pressure applied and (4) the time of duration of the operation. By adjusting (1), (2) and (3), the time can be reduced to the lowest possible value consistent with a good weld.

Power required. When the interfaces are first brought into contact, maximum power is required, breaking up the surface film. The power required then falls and remains nearly constant while the joint is raised to welding temperature. The power required for a given machine can be chosen so that the peak power falls within the overload capacity of the driving motor. It is the contact areas which determine the capacity of a machine. The rotational speed can be as low as 1 metre per second peripheral and the pressure depends upon the materials being welded, for example for mild steel it can be of the order of 50 N/mm^2 for the first part of the cycle followed by 140 N/mm^2 for the forging operation. Non-ferrous metals require a somewhat greater difference between the two operations. The faster the rotation of the component and the greater the pressure, the shorter the weld cycle, but some materials suffer from hot cracking if the cycle is too short and the time is increased with lower pressure to increase the width of the HAZ. At the present time most steels can be welded including stainless, but excluding free cutting. Non-ferrous metals are also weldable and aluminium (99.7% Al) can be welded to steel.

Fig. 5.12. (*a*) Friction welding.

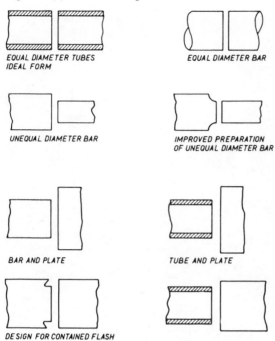

EQUAL DIAMETER TUBES
IDEAL FORM

EQUAL DIAMETER BAR

UNEQUAL DIAMETER BAR

IMPROVED PREPARATION
OF UNEQUAL DIAMETER BAR

BAR AND PLATE

TUBE AND PLATE

DESIGN FOR CONTAINED FLASH

Fig. 5.12. (*b*) Suitable forms of friction welding.

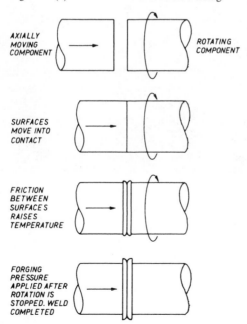

AXIALLY
MOVING
COMPONENT

ROTATING
COMPONENT

SURFACES
MOVE INTO
CONTACT

FRICTION
BETWEEN
SURFACES
RAISES
TEMPERATURE

FORGING
PRESSURE
APPLIED AFTER
ROTATION IS
STOPPED. WELD
COMPLETED

Fig. 5.13. Friction welder.

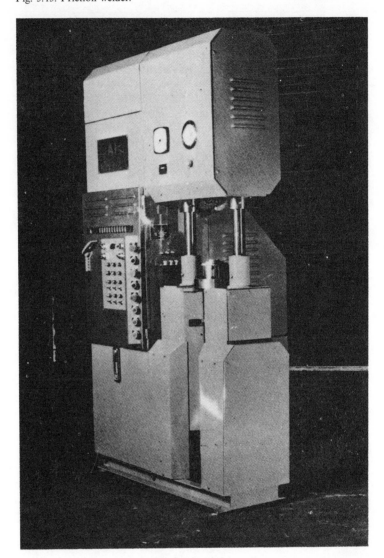

There are various control systems: (1) Time control, in which after a given set time period after contact of the faces, rotation is stopped and forge pressure is applied. There is no control of length with this method. (2) Burn-off to length: parts contact and heating and forging take place within a given pre-determined length through which the axially moving component moves. (3) Burn-off control: a pre-determined shortening of the component is measured off by the control system when minimum pressures are reached. Weld quality and amount of extrusion are thus controlled, but not the length. (4) Forging to length: the axially moving work holder moves up to a stop during the forging operation irrespective of the state of the weld and generally in this case extrusion tends to be excessive.

The extrusion of flash can be removed by a subsequent operation or, for example for tubes of equal diameter, a shearing unit can be built into the machine operating immediately after forging and while the component is hot, thus requiring much less power. Fig. 5.12(*a*) illustrates a joint designed to contain the flash.

In the process known as inertia welding the rotating component is held in a fixture attached to a flywheel which is accelerated to a given speed and then uncoupled from its drive. The parts are brought together under high thrust and the advantage claimed is that there is no possibility of the driving unit stalling before the flywheel energy is dissipated.

Friction welding machines resemble machine tools in appearance, as illustrated in Fig. 5.13.

Electron beam welding

If a filament of tungsten or tantalum is heated to high temperature in a vacuum either directly by means of an electric current or indirectly by means of an adjacent heater, a great number of electrons are given off from the filament, which slowly evaporates. This emission has been mentioned previously in the study of the tungsten arc welding process. The greater the filament current the higher the temperature and the greater the electron emission, and if a metal disc with a central hole is placed near the filament and charged to a high positive potential relative to the filament, so that the filament is the cathode and the disc the anode, the emitted electrons are attracted to the disc and because of their kinetic energy pass through the hole as a divergent beam. This can then be focused electrostatically, or magnetically, by means of coils situated adjacent to the beam and through which a current is passed. The beam is now convergent and can be spot focused. The basic arrangement, an electron 'gun', is similar to that used for television tubes and electron microscopes (Fig. 5.14).

If the beam is focused on to a metal surface the beam can have sufficient energy to raise the temperature to melting point, the heating effect depending upon the kinetic energy of the electrons. The kinetic energy of an electron is $\frac{1}{2}mV^2$, where m is the mass of an electron (9.1×10^{-28} g) and V its velocity. The electron mass is small, but increasing the emission from the filament by raising the filament current increases the number of electrons and hence the mass effect. Because the kinetic energy varies directly as the square of the velocity, accelerating the electrons up to velocities comparable with the velocity of light by using anode voltages (up to 200 kV) greatly increases the beam energy. The smaller the spot into which the beam is focused the greater the energy density but final spot size is often decided by working conditions, by aberration in the focusing system, etc., so that spot size may be of the order of 0.25–2.5 mm.

When the beam strikes a metal surface X-rays are generated, so that adequate precautions must be taken for screening personnel from the rays by using lead or other metal screens or making the metal walls of the gun chamber sufficiently thick. If the beam emerges into the atmosphere the energy is reduced by collision of the electrons with atmospheric molecules and focus is impaired. Because of this it has been the practice to perform many welding operations in a vacuum, either in the gun chamber in which case each time the component is loaded the chamber must be evacuated to high vacuum conditions, thus increasing the time and cost of the operation,

Fig. 5.14. Electron beam welding.

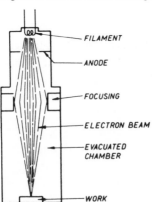

FILAMENT

ANODE

FOCUSING

ELECTRON BEAM

EVACUATED CHAMBER

WORK

or in a separate steel component chamber fixed to the gun chamber. This can be made of a size to suit the component being welded and is evacuated to a relatively low vacuum after each loading. In either case welds suffer no contamination because of vacuum conditions. Viewing of the spot for set-up, focusing and welding is done by various optical arrangements.

Welding in non-vacuum conditions requires much greater power than for the preceding method because of the effects of the atmosphere on the beam and the greater distance from gun to work, and a shielding gas may be required around the weld area. Research work is proceeding in this field involving guns of higher power consumption. Difficulties may also be encountered in focusing the beam if there is a variation in the gun-to-work distance, as on a weld on a component of irregular shape.

Welds made with this process on thicker sections are narrow with deep penetration with minimum thermal disturbance and at present welds are performed in titanium, niobium, tungsten, tantalum, beryllium, nickel alloys (e.g. nimonic), inconel, aluminium alloys and magnesium, mostly in the aero and space research industries. The advantages of the process are that being performed in a vacuum there is no atmospheric contamination and the electrons do not affect the weld properties, accurate control over welding conditions is possible by control of electron emission and beam focus, and there is low thermal disturbance in areas adjacent to the weld (Fig. 5.15). Because of the vacuum conditions it is possible to weld the more reactive metals successfully. On the other hand the equipment is very costly, production of vacuum conditions is necessary in many cases and there must be protection against radiation hazards.

Laser beam welding

Radio waves, visible light, ultra-violet and infra-red radiations are electro-magnetic radiations which have two component fields, one electric and the other magnetic. If either one of these components is suppressed the resulting radiations are said to be polarized and the direction in which the resultant electric or magnetic forces act is the plane of polarization. Light from a source such as a tungsten filament electric light consists of several frequencies involving various shades of colour. These waves are not in phase and are of various amplitudes and planes of polarization, and the light is said to be non-coherent. Light of a single wave-length or frequency is termed monochromatic. The wave-length of light is measured in metres or micro-metres, termed microns (μm), visible light being in the range 0.4–0.7 μm. The frequency is related to the wave-length by the expression $V = n\lambda$, where n is the frequency in Hz, λ the wave-length in metres and V

the velocity of electro-magnetic radiation, 3×10^8 m/s, so that the frequency range of visible light is in the range 430–750×10^{12} Hz.

Atoms of matter can absorb and give out energy and the energy of any atomic system is thereby raised or lowered about a mean or 'ground' level. Energy can only be absorbed by atoms in definite small amounts (quanta) termed photons, and the relationship between the energy level and the frequency of the photon is $E = h\nu$, where E is the energy level, h is Planck's constant and ν the photon frequency, so that the energy level depends upon the frequency of the photon. An atom can return to a lower energy level by emitting a photon and this takes place in an exceedingly short space of time from when the photon was absorbed, so that if a photon of the correct frequency strikes an atom at a higher energy level, the photon which is released is the same in phase and direction as the incident photon.

The principle of the laser (Light Amplification by Stimulated Emission of Radiation) is the use of this stimulated energy to produce a beam of coherent light, that is one which is monochromatic, and the radiation has the same plane of polarization and is in phase. At the present time lasers

Fig. 5.15. Electron beam welding machine with indexing table, tooled for welding distributor shafts to plates at a production rate of 450 per hour. The gun is fitted with optical viewing system. Power 7.5 kW at 60 kV, 125 mA. Vacuum sealing is achieved by seals fitted in the tooling support plate and at the bottom of the work chamber. As the six individual tooling stations reach the welding station they are elevated to the weld position and then rotated by an electronically controlled d.c. motor.

operate with wave-lengths in the visible and infra-red region of the spectrum. When the beam is focused into a small spot and there is sufficient energy, welding, cutting and piercing operations can be performed on metals.

The ruby laser has a cylindrical rod of ruby crystal (Al_2O_3) in which there is a trace of chromium as an impurity. An electronic flash gun, usually containing neon, is used to provide the radiation for stimulation of the atoms. This type of gun can emit intense flashes of light of one or two milliseconds duration and the gun is placed so that the radiations impinge on the crystal. The chromium atoms are stimulated to higher energy levels, returning to lower levels with the emission of photons. The stimulation continues until an 'inversion' point is reached when there are more chromium atoms at the higher levels than at the lower levels, and photons impinging on atoms at the higher energy level cause them to emit photons. The effect builds up until large numbers of photons are travelling along the axis of the crystal, being reflected by the ends of the crystal back along the axis, until they reach an intensity when a coherent pulse of light, the laser beam (of wave-length about $0.63\,\mu$m), emerges from the semi-transparent rod end. The emergent pulses may have high energy for a short time period, in which case vaporization may occur when the beam falls on a metal surface, or the beam may have lower energy for a longer time period, in which case melting may occur, while a beam of intermediate power and duration may produce intermediate conditions of melting and vaporization, so that control of the time and energy of the beam and focusing of the spot exercise control over the working conditions.

Developments of the ruby laser include the use of calcium tungstate and glass as the 'host' material with chromium, neodymium, etc., as impurities, a particular example being yttrium–aluminium–garnet with neodymium (YAG), used for operations on small components.

Gas lasers operate on the same basic principles as the solid state type. The gas is contained in a long tube of quartz or pyrex with end windows, and specially designed mirrors are arranged to reflect the beam back along the tube axis. Neon with a trace of helium was first used and an electromagnetic radiation at chosen frequency is applied from a radio-frequency generator to electrodes around the tube. The helium atoms are stimulated and their energy level raised. Collision with neon atoms causes energy to be transferred to the neon atoms until inversion occurs, the radiation stimulating release of photons from atoms faster than by normal emission, and a coherent beam is emitted. The CO_2 laser uses CO_2 with some nitrogen and/or helium added, in a tube some metres long, the wavelength of the beam ($10.6\,\mu$m) being longer than that of the solid state lasers, and either continuous wave or pulsed, the power increasing as the length of the tube increases.

A pulsed or continuous wave laser beam can produce enough energy to heat, melt and vaporize a metal surface and refractory metals such as niobium, tantalum and tungsten can be welded in thin sections, and welds made between dissimilar metals, the operations being performed within an inert gas shield when required. Because of the vaporization which can occur, holes can be pierced even in diamonds.

2 kW CO_2 lasers can be used to weld up to 3 mm thick material and are an alternative to the electron beam for thin gauge material. The width of the weld may be increased at speeds below 12 mm/s due to interaction between the beam and an ionized plasma which occurs near the work. At speeds of 20 mm/s and over, laser and electron beam welds are practically indistinguishable from one another.

Lasers of 10–20 KW and above are now being developed and, applied to welding and cutting (with oxygen) of thicker sections, will greatly widen the scope of the process.

Stud welding

This is a rapid, reliable and economical method of fixing studs and fasteners of a variety of shapes and diameters to parent plate. The studs may be of circular or rectangular cross-section, plain or threaded internally or externally and vary from heavy support pins to clips used in component assembly.

There are two main methods of stud welding: (1) arc (drawn arc), (2) capacitor discharge, and the process selected for a given operation depends upon the size, shape and material of the stud and the composition and thickness of the parent plate, the arc method generally being used for heavier studs and plate, and capacitor discharge for lighter gauge work.

Arc (drawn arc) process

This is used in both engineering and construction work and the equipment consists of a d.c. power source, controller and a hand-operated or bench-mounted tool or gun.

A typical unit consists of a forced-draught-cooled power source with a 380/440 V, 3-phase, 50 Hz transformer, the secondary of which is connected to a full-wave, bridge-connected silicon rectifier (p. 93), with tappings giving ranges of 120–180 A, 170–320 A, and 265–1500 A approximately (equipment is also available up to 1800 A), at about 65 V. The two lower ranges can also be used for MMA welding and two or more units can be connected in parallel if required, the loading being shared automatically.

The power controller contains a main current contactor and timer relays and a pilot arc device, energized from the d.c. source. The solid-state timer has a multi-position switch or switches for the selection of operating time and current for the diameter of stud in use. Work is usually connected to the positive pole with the stud negative except for aluminium, when the polarity is reversed.

The hand gun (Fig. 5.16) has the operating solenoid and return spring within the gun body, which also carries the operating switch and adjustable legs to accommodate varying lengths of studs, an interchangeable chuck for varying stud diameters and a foot adaptor to maintain concentricity between the stud and ferrule are arc shield, this latter being held by ferrule grips. Studs are fluxed on the contact end, which is slightly pointed (Fig. 5.17), and are supplied with ferrules.

To operate the equipment, the welding current and time for the diameter of stud in use are selected, the stud is loaded into the appropriate chuck, the legs adjusted for length and the stud positioned on the plate. A centre punch can be used for locating the stud point and the plate should be free of contamination.

Sequence of operations. The gun switch is pressed, a low current flows between pointed stud end and workpiece and immediately the stud is raised, drawing an arc and ionizing the gap. The main current contactor now closes and full welding current flows in an arc, creating a molten state

Fig. 5.16

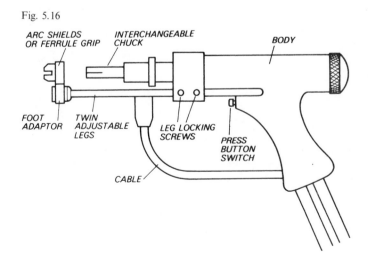

in plate and stud end. The solenoid is de-energized and the stud is pushed under controlled spring pressure into the molten pool in the plate. Finally the main current contactor opens, the current is switched off, and the operation cycle is complete, having taken only a few hundredths of a second (Fig. 5.17).

This method, which is usually employed, whereby the welding current is kept flowing until the return cycle is completed, is termed 'hot plunge'. If the current is cut off just before the stud enters the molten pool it is termed 'cold plunge'. The metal displaced during the movement of the stud into the plate is moulded into fillet form by the ceramic ferrule or arc shield held against the workpiece by the foot. This also protects the operator, retains heat and helps to reduce oxidation.

Studs from 3.3 mm to 20 mm and above in diameter can be used on parent plate thicker than 1.6 mm, and the types include split, U shaped, J bent anchors, etc. in circular and rectangular cross-section for the engineering and construction industries, and can be in mild steel (low carbon), austenitic stainless steel, aluminium and aluminium alloys (3–4% Mg). The rate of welding varies with the type of work, jigging, location, etc., but can be of the order of 8 per minute for the larger diameters and 20 per minute for smaller diameters.

Fig. 5.17. Arc stud welding, cycle of operations.

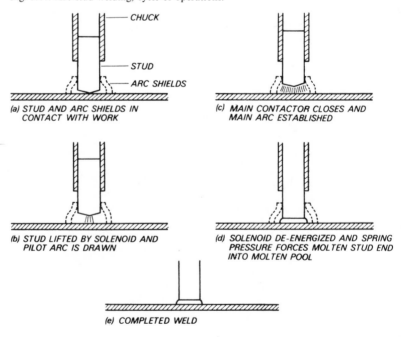

Capacitor discharge stud welding

In this process a small projection on the end of the stud makes contact with the workpiece and the energy from a bank of charged capacitors (see Volume 1) is discharged across the contact, melting the stud projection, ionizing the zone and producing a molten end on the stud and a shallow molten pool in the parent plate. At this time the stud is pushed into the workpiece under controlled spring pressure, completing the weld (Fig. 5.18).

If C is the capacitance of the capacitor, Q the charge and V the potential difference across the capacitor, then $C = Q/V$ or *capacitance = charge/potential*. The energy of a charged capacitor is $\frac{1}{2}QV$, and since $Q = CV$, the energy is $\frac{1}{2}CV^2$, so that the energy available when a capacitor is discharged is dependent upon: (1) the capacitance – the greater the capacitance the greater the energy; (2) the square of the potential difference (voltage) across the capacitance. The greater this voltage, the greater the energy. Thus the energy required for a given welding operation is obtained by selection of the voltage across the capacitor, i.e. that to which it is charged, and the total capacitance in the circuit.

In the contact method of capacitor discharge welding the small projection on the stud end is placed in contact with the workpiece as explained above. In the hold-off method, useful for thin gauge plate to avoid reverse marking, a holding coil in the hand gun is energized when the welding power is selected in the 'hold-off' position on the capacitor switch. When the stud is pushed into the gun-chuck both stud and chuck move into the hold-off position giving a pre-set clearance between the projection on the stud end and the workpiece. When the gun switch is pressed the hold-off

Fig. 5.18. Capacitor discharge stud welding, sequence of operations.

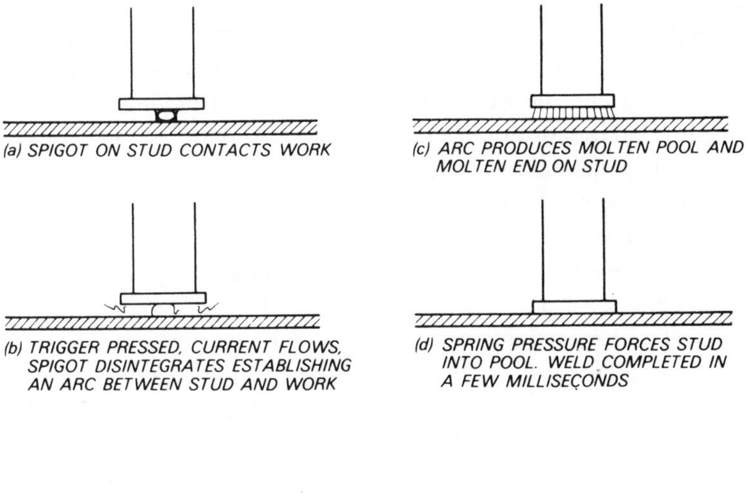

(a) SPIGOT ON STUD CONTACTS WORK

(b) TRIGGER PRESSED, CURRENT FLOWS, SPIGOT DISINTEGRATES ESTABLISHING AN ARC BETWEEN STUD AND WORK

(c) ARC PRODUCES MOLTEN POOL AND MOLTEN END ON STUD

(d) SPRING PRESSURE FORCES STUD INTO POOL. WELD COMPLETED IN A FEW MILLISECONDS

coil is de-energized and spring pressure pushes the stud into contact with the workpiece, the discharge takes place and the weld is made.

This process minimizes the depth of penetration into the parent metal surface and is used for welding smaller diameter ferrous and non-ferrous studs and fasteners to light gauge material down to 0.45 mm thickness in low-carbon and austenitic stainless steel, and 0.7 mm in aluminium and its alloys and brass, the studs being from 2.8 to 6.5 mm diameter. Studs are usually supplied with a standard flange on the end to be welded (Fig. 5.18) but this can be reduced to stud diameter if required and centre punch marks should not be used for location. The studs can be welded on to the reverse side of finished or coated sheets with little or no marking on the finished side. The studs are not fluxed and no arc shield or ferrule is required.

Typical equipment with a weld time cycle of 3–7 milliseconds consists of a control unit and a hand- or bench-mounted tool or gun.

The control unit houses the banks of capacitors of 100 000–200 000 μF capacitance depending upon the size of the unit, the capacitance required for a given operation being selected by a switch on the front panel. The control circuits comprise a charging stage embodying a mains transformer and a bridge-connected full-wave silicon rectifier and the solid-state circuits for charging the capacitors to the voltage predetermined by the voltage sensing module. Interlocking prevents the energy being discharged by operating the gun switch until the capacitors have reached the power pre-set by voltage and capacitance controls.

The printed circuit voltage sensing module controls the voltage to which the capacitors are charged and switches them out of circuit when they are charged to the selected voltage, and is controlled by a voltage dial on the panel. A panel switch is also provided to discharge the capacitors if required.

Solid-state switching controls the discharge of the capacitors between stud and work, so there are no moving contactors. The gun, similar in appearance to the arc stud welding gun, contain the adjustable spring pressure unit which enables variation to be made in the speed of return of the stud into the workpiece, and a chuck for holding the stud. Legs are provided for positioning or there can be a nosecap gas shroud for use when welding aluminium or its alloys using an argon gas shield. The weld cannot be performed until legs or shroud are firmly in contact with the workpiece.

Welding rates attainable are 12 per minute at 6.35 mm diameter to 28 per minute at 3.2 mm diameter in mild steel. Similar rates apply to stainless steel and brass but are lower for aluminium because of the necessity of argon purging. Partial scorch marks may indicate cold laps (insufficient energy) with the possibility of the stud seating too high on the work. The weld should show even scorch marks all round the stud, indicating a sound weld. Excessive spatter indicates the use of too much energy.

Automatic single- and multiple-head machines, pneumatically operated and with gravity feed for the fasteners, are currently available.

Explosive welding

This process is very successfully applied to the welding of tubes to tube plates in heat exchangers, feedwater heaters, boiler tubes to clad tube plates, etc.; and also for welding plugs into leaking tubes to seal the leaks.

The welds made are sound and allow higher operating pressures and temperatures than with fusion welding, and the tubes may be of steel, stainless steel or copper; aluminium brass and bronze tubes in naval brass tube plates are also successfully welded.

The tube and tube plate may be parallel to each other (parallel geometry) with a small distance between them and the tube plate can be counterbored as in Fig. 5.19a. The explosive (e.g. trimonite) must have a low detonation velocity, below the velocity of sound in the material, and there is no limit to the joint area so that the method can be used for cladding surfaces. For the tube plate welds several charges can be fired simultaneously, the explosive being in cartridge form.

In the oblique geometry method now considered (YIMpact patent) the two surfaces are inclined at an angle to each other, Fig. 5.19b, the tube plate being machined or swaged as shown. As distance between tube and plate continuously increases because of the obliquity, there is a limit to the surface area which can be welded because the distance between tube and plate becomes too great.

The detonation value of the explosive can be above the velocity of sound

Fig. 5.19

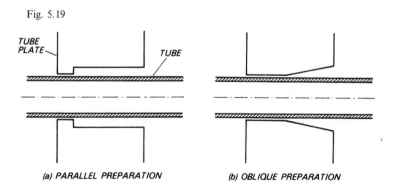

(a) PARALLEL PREPARATION   (b) OBLIQUE PREPARATION

in the material and the explosive (e.g. PETN) can be of pre-fabricated shape and is relatively cheap. The charge is fired electrically from a fuse head on the inner end of the charge and initiates the explosion, the detonation front then passing progressively through the charge.

The size of the charge depends upon the following variables: surface finish, angle of inclination of tube and plate, yield strength and melting point of the materials, the tube thickness and diameter; and the upper limit of the explosive is dependent upon the size of ligament of the tube plate between tubes, this usually being kept to a minimum in the interest of efficient heat exchange.

The tube plate is tapered towards the outer surface otherwise there would be a bulge in the tube on the inner side after welding and the tube would be difficult to remove. The charge must be fired from the inner end so that the weld will progress from contact point of tube and plate and thus the detonator wires must pass backwards along the charge to the outer end of the tube, the charge being witnn a polythene insert (Fig. 5.20*a*).

Upon initiation of the explosion, tube and tube plate collide at the inner end of the taper and, due to the release of energy, proceed along it, and ideally the jet of molten metal formed at the collision point is ejected at the tube mouth. In effect the welded surfaces assume a sinusoidal wave form, some of the molten metal, which is rather porous and brittle, being entrapped in the troughs and crests of the wave. This entrapment can be reduced to a minimum by having a good surface finish (e.g. of the order of 0.003 mm) and the angle between tube and tube plate from 10° to 15°. At the lower angle the waves are pronounced and of shorter wavelength while at the higher angle the wavelength is longer and the waves more undulating so that 15° is the usual angle (Fig. 5.20*b* and *c*).

Surface oxide and impurities between the surfaces increase the charge required compared with a smooth surface and the positioning of the charge is important. If it is too far in, the energy at the mouth of the tube is not sufficient to produce a weld in this region, while if the charge is not far enough in the tube, welding is not commenced until some distance along the taper. To position the charge correctly and quickly a polythene insert has been developed to contain the charge and is positioned by a brass plug.

At present tubes of any thickness and of diameters 16–57 mm are weldable, with ·plate thickness greater than 32 mm and 9.5 mm plate ligaments.

The plug for explosive plugging of leaking tubes is of tubular form, the end of which is swaged to give the necessary taper, the polythene insert protruding beyond the open end of the plug to allow for extraction in case of misfire, thus increasing the safety factor (Fig. 5.21).

Fig. 5.20

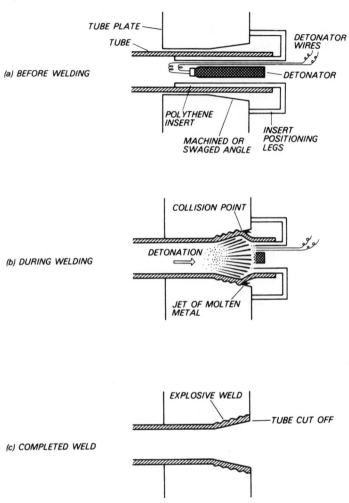

(a) BEFORE WELDING

TUBE PLATE

TUBE

DETONATOR WIRES

DETONATOR

POLYTHENE INSERT

MACHINED OR SWAGED ANGLE

INSERT POSITIONING LEGS

(b) DURING WELDING

COLLISION POINT

DETONATION

JET OF MOLTEN METAL

(c) COMPLETED WELD

EXPLOSIVE WELD

TUBE CUT OFF

Fig. 5.21. A swaged explosive plug.

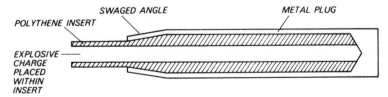

SWAGED ANGLE

METAL PLUG

POLYTHENE INSERT

EXPLOSIVE CHARGE PLACED WITHIN INSERT

Since all the configuration is confined to the plug, the tube plate hole is cleaned by grinding, the plug with explosive charge is inserted and the detonator wires connected. The plugs are available in diameters with 1.5 mm increments.

This method has proved very satisfactory in reducing the time and cost in repair work of this nature.

Personnel can be trained by the manufacturers or trained personnel are available under contract.

Gravity welding

This is a method for economically welding long fillets in the flat position using gravity to feed the electrode in and to traverse it along the plate. The equipment is usually used in pairs, welding two fillets at a time, one on each side of a plate, giving symmetrical welds and reducing stress and distortion. The electrode holder is mounted on a ball-bearing carriage and slides smoothly down a guide bar, the angle of which to the weld can be adjusted to give faster or slower traverse and thus vary the length of deposit of the electrode and the leg length of the weld.

The special copper alloy electrode socket is changed for varying electrode diameters and screws into the electrode holder, the electrode being pushed into a slightly larger hole in the socket and held there by the weight of the carriage. Turning the electrode holder varies the angle of electrode inclination. The base upon which the guide and support bar is mounted has two small ball bearings fitted so that it is easy to move along the base plate when resetting. If the horizontal plate is wider than about 280 mm a counterweight can be used with the base, otherwise the base can be attached to the vertical plate by means of two magnets or a jig can be used in place of the base to which the segment is fixed. A flexible cable connects the electrode holder to a disconnector switch carried on the support arm. This enables the current to be switched on and off so that electrodes can be changed without danger of shock. A simple mechanism at the bottom of the guide bar switches the arc off when the carriage reaches the bottom of its travel (Fig. 5.22).

At the present time electrodes up to 700 mm long are available in diameters of 3.5, 4.0, 4.5, 5.0 and 5.5 mm using currents of 220–315 A with rutile, rutile-basic and acid coatings suitable for various grades of steel.

Gravity welding is usually used for fillets with leg lengths of 5–8 mm, the lengths being varied by altering the length of deposit per electrode. An a.c. power source is used for each unit with an OCV of 60 and arc volts about 40 V with currents up to 300 A. Sources are available for supplying up to 6 units (three pairs), manageable by one welding operator and so arranged that when the current setting for one unit is chosen, the remaining units are supplied at this value. In general gravity welding is particularly suitable for welding, for example, long parallel stiffeners on large unit panels, enabling one operator to carry out three or four times the deposit length as when welding manually.

Thermit welding

Thermit (or alumino-thermic) is the name given to a mixture of finely divided iron oxide and powdered aluminium. If this mixture is placed in a fireclay crucible and ignited by means of a special powder, the action, once started, continues throughout the mass of the mixture, giving out great heat. The aluminium is a strong reducing agent, and combines with the oxygen from the iron oxide, the iron oxide being reduced to iron.

The intense heat that results, because of the chemical action, not only melts the iron, but raises it to a temperature of about 3000°C. The aluminium oxide floats to the top of the molten metal as a slag. The crucible is then tapped and the superheated metal runs around the parts to be welded, which are contained in a mould. The high temperature of the iron results in excellent fusion taking place with the parts to be welded. Additions may be made to the mixture in the form of good steel scrap, or a small percentage of manganese or other alloying elements, thereby producing a good quality thermit steel. The thermit mixture may consist of about 5 parts of aluminium to 8 parts of iron oxide, and the weight of thermit used will depend on the size of the parts to be welded. The ignition

Fig. 5.22. Gravity welding.

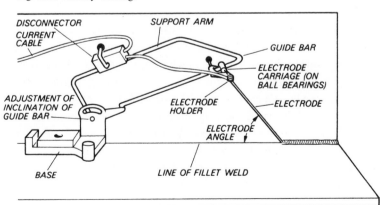

powder usually consists of powdered magnesium or a mixture of aluminium and barium peroxide.

Preparation

The ends which are to be welded are thoroughly cleaned of scale and rust and prepared so that there is a gap between them for the molten metal to penetrate well into the joint. Wax is then moulded into this gap, and also moulded into a collar round the fracture. This is important, as it gives the necessary reinforcement to the weld section. The moulding box is now placed around the joint and a mould of fireclay and sand made, a riser, pouring gate and pre-heating gate being included. The ends to be welded are now heated through the pre-heating gate by means of a flame and the wax is first melted from between the ends of the joint. The heating is continued until the ends to be welded are at red heat. This prevents the thermit steel being chilled, as it would be if it came into contact with cold metal. The pre-heating gate is now sealed off with sand and the thermit process started by igniting the powder. The thermit reaction takes up to about 1 minute, depending upon the size of the charge and the additions that have been made in the form of steel scrap and alloying elements etc. When the action is completed the steel is poured from the crucible through the pouring gate, and it flows around the red-hot ends to be welded, excellent fusion resulting. The riser allows extra metal to be drawn by the welded section when contraction occurs on cooling, that is it acts as a reservoir. The weld should be left in the mould as long as possible (up to 12 hours), since this anneals the steel and improves the weld.

Thermit steel is pure and contains few inclusions. It has a tensile strength of about 460 N/mm^2. The process is especially useful in welding together parts of large section, such as locomotive frames, ships' stern posts and rudders, etc. It is also being used in place of flash butt welding for the welding together of rail sections into long lengths.

At the present time British Rail practice is to weld 18 m long rails into lengths of 91 to 366 m at various depots by flash butt welding. These lengths are then welded into continuous very long lengths *in situ* by means of the thermit process and conductor rails are welded in the same way. Normal running rails have a cross-sectional area of 7184 mm^2 and two techniques are employed, one requiring a 7 minute pre-heat before pouring the molten thermit steel into the moulds and the other requiring only $1\frac{1}{2}$ minutes pre-heat, the latter being the technique usually employed. Excess metal can be removed by pneumatic hammer and hot chisel or by portable trimming machine.

Underwater welding

The three main methods of underwater welding at present are (1) wet, (2) localized dry chamber, (3) dry habitat (surroundings).

Wet welding

Wet welding is performed by the diver-welder normally using surface diving or saturation diving techniques using the MMA process with electrodes specially coated with insulating varnishes to keep them dry.

Current is supplied by a generator of some 400 A capacity at 60–100 OCV, directly to the welding torch head. Interchangeable collets of 4 mm and 4.8 mm ($\frac{5}{32}''$ and $\frac{3}{16}''$), hold electrodes of these diameters and are tightened by the twist-grip control, which is also used to eject the stub when electrodes are changed. Watertight glands and washers prevent seepage of water into the body of the torch, which is tough rubber covered thus reducing the danger of electric shock.

Arc stability is less than that in air due to the large volumes of gas and steam which are evolved, making vision difficult and as a result touch welding is generally used. The presence of water in the immediate vicinity of the weld except at the molten pool under the arc stream results in the very rapid quenching of the weld metal, which produces a hard, narrow heat affected zone and can give rise to severe hydrogen cracking. At present this method is used for non-critical welds, critical welding being carried out under dry conditions.

Localized dry chamber (hyperbaric)

These chambers, in which welding is performed at pressure above one atmosphere, are generally made of steel and are constructed with antichambers around the section to be welded. Water is then displaced from the chamber by gas pressure. The gas in the chamber used by the diver-welder is typically a mixture of helium and oxygen (heliox). This mixture eliminates the harmful effect of nitrogen, the helium being the carrier for the oxygen – a partial pressure of oxygen above 2 bar can be poisonous. The oxygen content is adjusted until the diver-welder does not use a breathing mask, but at the moment of welding he puts on a breathing–welding mask which has a gas supply separate from that of the chamber and umbilicals connected to the mask to enable the breathing gases to be exhausted outside the chamber.

TIG, MMA and flux cored methods of welding are used down to a depth of 300 m of water but TIG, using high purity argon as the shielding gas, is the slowest method as it is a low deposition rate process. It can be used for root runs and hot passes, but MMA or the flux cored method is used for filling and capping. Stringer bead techniques may be used to impart toughness, and temper beads to minimize HAZ hardness. Generally the high-strength, basic-coated steel electrodes which may contain iron powder are used. They are pre-heated in a drying oven immediately before use and pre-heat of about 100°C can be applied locally to the joint.

After welding, the joints are subject to X-ray or ultra-sonic testing according to the welding codes used.

Coffer dam

A coffer dam is a watertight case of steel piling erected around a given point to keep out the water. By pumping the dam dry, welding can be performed at atmospheric pressure up to about 18 m deep and welding can proceed irrespective of the state of the tides if in tidal water.

In this case the welding is performed exactly as if it were on shore and the welders are in audio and visual contact with the welding process engineer. Much of the damage to offshore platforms is in the splash zone so there is quite a saving in cost using this method as opposed to hyperbaric welding. The figure shows a typical underwater torch, which can also be used with small modifications for underwater cutting, Fig. 5.23.

Fig. 5.23. Underwater welding (and cutting) torch.

INTERCHANGEABLE COLLETS
FOR VARIOUS SIZES OF
ELECTRODES

ELECTRIC POWER
CABLE

HAND GRIP

ELECTRODE HELD IN COLLET
AND IS FREED BY TWISTING
HAND GRIP. (WELDING OR
CUTTING)

OXYGEN CONTROL LEVER
(USED WHEN CUTTING)

OXYGEN SUPPLY
(WHEN USED FOR CUTTING)

UNDERWATER WELDING AND CUTTING TORCH

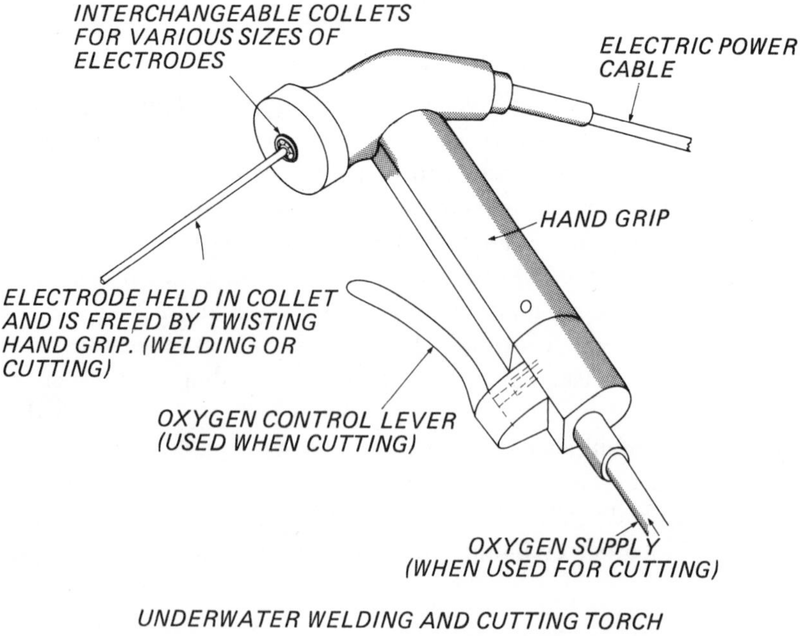

6

Oxy-acetylene welding

Principles and equipment
(See also pp. 309–11 for general safety precautions)

For oxy-acetylene welding, the oxygen is supplied from steel cylinders and the acetylene either from cylinders or from an acetylene generator which can be of the medium-pressure or low-pressure type.

With cylinder gas, the pressure is reduced to 0.06 N/mm² or under, according to the work, by means of a pressure-reducing valve and the acetylene is passed to the blowpipe where it is mixed with oxygen in approximately equal proportions, and finally passed into the nozzle or tip to be burnt.

The medium-pressure acetylene generator delivers gas to any desired pressure up to a maximum of 0.06 N/mm² (0.6 bar) in the same way as cylinder gas.

The low-pressure generator produces gas at a pressure of only a few millimetres water column, necessitating the use of 'injector' blowpipes where the high-pressure oxygen injects or sucks the acetylene into the blowpipe. In some cases, as for example if the supply pipes are small for the volume of gas to be carried or if it is desirable to use the equal pressure type of blowpipe, a booster is fitted to the low-pressure generator and the gas pressure increased to 0.06 N/mm².

Without Home Office approval, the maximum pressure at which acetylene may be used in England is 0.06 N/mm². With approval, the pressure may be increased to 0.15 N/mm², but this is rare and applies only in special cases.

Oxygen
The oxygen for both high- and low-pressure systems is supplied in solid drawn steel cylinders at a pressure of 17.2 N/mm² (172 bar).* The

---

* *Note.* 1 bar = 0.1 N/mm² = approximately 15 lbf/in².

cylinders are rated according to the amount of gas they contain, 8500 litres (8.5 m³) being the usual size but very large cylinders containing 800 m³ (2400 cu. ft.) are available.

The volume of oxygen contained in the cylinder is approximately proportional to the pressure; hence for every 10 litres of oxygen used, the pressure drops about 0.02 N/mm². This enables us to tell how much oxygen remains in a cylinder. The oxygen cylinder is provided with a valve threaded right hand and is painted black. On to this valve, which contains a screw-type tap, the pressure regulator and pressure gauge are screwed. The regulator adjusts the pressure to that required at the blowpipe. Since grease and oil can catch fire spontaneously when in contact with pure oxygen under pressure, they must never be used on any account upon any part of the apparatus. Leakages of oxygen can be detected by the application of a soap solution, when the leak is indicated by the soap bubbles. Never test for leakages with a naked flame.

Liquid oxygen

Liquid oxygen, nitrogen, argon and LPG are available in bulk supply from tankers to vacuum-insulated evaporators (VIE) in which the liquid is stored at temperatures of -160 to $-200\,°C$ and are very convenient for larger industrial users.

There is no interruption in the supply of gas nor drop in pressure during filling.

The inner vessel, of austenitic stainless steel welded construction, has dished ends and is fitted with safety valve and bursting disc and is available in various sizes with nominal capacities from 844 to 33 900 m³. Nominal capacity is the gaseous equivalent of the amount of liquid that the vessel will hold at atmospheric pressure. The outer vessel is of carbon steel and fitted with pressure release valve. The inner vessel vacuum and pearlite insulated from the outer vessel, thus reducing the thermal conductivity to a minimum. The inner vessel A (Fig. 6.1) contains the liquid with gas above, and gas is withdrawn from the vessel through the gaseous withdrawal line B and rises to ambient temperature in the superheater-vaporizer C, from which it passes to the supply pipeline. If the pressure in the supply falls below the required level the pressure control valve D opens and liquid flows under gravity to the pressure-raising vaporizer E, where heat is absorbed from the atmosphere, and the liquid vaporizes and passes through the gas withdrawal point H raising the pressure to the required pre-set level which can be up to 1.7 N/mm² (250 lbf/in²), and the valve D then shuts.

In larger units, to allow for heavy gas demand, when the pressure falls on the remote side of the restrictive plate F, liquid flows from the vessel via the withdrawal line G and passes to the superheater-vaporizer where it changes to the gaseous form and is heated to ambient temperature, finally passing to the pipeline.

During periods when the VIE is not in use the valve D remains shut. Heat from the outside atmosphere gradually flows through the insulation between the vessels so that more liquid is vaporized and the pressure of the gas rises. This rate of heat leakage is slow, however, and it usually takes about seven days for the pressure to rise sufficiently to lift the safety valve, so that under normal working conditions there is almost zero loss.

Acetylene*

In the high-pressure system the acetylene is stored in steel cylinders similar to the oxygen cylinders. Acetylene gas, however, is unstable when compressed to high pressures, and because of this it is contained in the bottles dissolved in a chemical called acetone; hence the name 'dissolved acetylene'.

The acetone is contained in a porous spongy mass of a substance such as charcoal, asbestos, kapok or other such material. Acetone can absorb 25 times its own volume of acetylene at normal temperature and pressure and for every increase of one atmosphere of pressure (0.1 N/mm^2 or 1 bar) it can absorb an equal amount.

Fig. 6.1. Vacuum insulated evaporator.

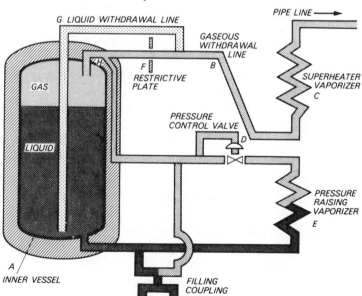

* The impurities in crude acetylene are salts of ammonia, hydrides of phosphorus, sulphur and nitrogen as well as water vapour and particles of lime.
  They can be removed by salts of ferric iron.

The pressure of the acetylene is usually about 1.5 N/mm^2 or 15 bar, a typical capacity being 5700 litres (5.7 m^3). The gas leaves the cylinder through a valve after passing through a filter pad. The valve had a screw tap fitted and is screwed left-hand. The cylinder is painted maroon and a regulator (also screwed left-hand) is necessary to reduce the pressure to 0.013–0.05 N/mm^2 or 0.13–0.5 bar as required at the blowpipe.

The amount of dissolved acetylene in a cylinder cannot be determined with any accuracy from the pressure gauge reading since it is in the dissolved condition. The most accurate way to determine the quantity of gas in a cylinder is to weigh it, and subtract this weight from the weight of the full cylinder, which is usually stamped on the label attached to the cylinder. The volume of gas remaining in the cylinder is calculated by remembering that 1 litre of acetylene weighs 1.1 g.

As long as the volume of acetylene drawn from the cylinder is not greater than $\frac{1}{5}$ of its capacity per hour, there is no appreciable amount of acetone contained in gas; hence this rate of supply should not be exceeded. For example, a 5700 litre cylinder can supply up to 1200 litres of gas per hour. The advantages of dissolved acetylene are that no licence is required for storage of the cylinders, there is no fluctuation of pressure in use, and the gas is always dry, clean and chemically pure, resulting in a reliable welding flame, and it avoids the need to charge and maintain a generator and to dispose of the sludge. There is no discernible difference in efficiency between generated and dissolved acetylene when both are used under normal conditions. Acetylene is highly inflammable and no naked lights should be held near a leaking cylinder, valve or tube. Leaks can be detected by smell or by soap bubbles. If any part of the acetylene apparatus catches fire, immediately shut the acetylene valve on the cylinder. The cylinder should be stored and used in an upright position.

The following is a summary of the main safety precautions to be taken when storing and using cylinders of compressed gas.

Storage
(1) Store in a well ventilated, fire-proof room with flame-proof electrical fittings. Do not smoke, wear greasy clothing or have exposed flames in the storage room.
(2) Protect cylinders from snow and ice and from the direct rays of the sun, if stored outside.
(3) Store away from sources of heat and greasy and oily sources. (Heat increases the pressure of the gas and may weaken the cylinder wall. Oil and grease may ignite spontaneously in the presence of pure oxygen.)

(4) Store acetylene cylinders in an upright position and do not store oxygen and combustible gases such as acetylene and propane together.

(5) Keep full and empty cylinders apart from each other.

(6) Avoid dropping and bumping cylinders violently together.

Use

(7) Keep cylinders away from electrical apparatus or live wires where there may be danger of arcing taking place.

(8) Protect them from the sparks and flames of welding and cutting operations.

(9) Always use pressure-reducing regulators to obtain a supply of gas from cylinders.

(10) Make sure that cylinder outlet valves are clear of oil, water and foreign matter, otherwise leakage may occur when the pressure-reducing regulators are fitted.

(11) Do not use lifting magnets on the cylinders. Rope slings may be used on single cylinders taking due precautions against the cylinder slipping from the sling. Otherwise use a cradle with chain suspension.

(12) Deload the diaphragm of the regulator by unscrewing before fitting to a full cylinder, and open the cylinder valve slowly to avoid sudden application of high pressure on to the regulator.

(13) Do not overtighten the valve when shutting off the gas supply; just tighten enough to prevent any leakage.

(14) Always shut off the gas supply when not in use for even a short time, and always shut off when moving cylinders.

(15) Never test for leaks with a naked flame; use soapy water.

(16) Make sure that oxygen cylinders with round bases are fastened when standing vertically, to prevent damage by falling.

(17) Thaw out frozen spindle valves with hot water NOT with a flame.

(18) Use no copper or copper alloy fitting with more than 70% copper because of the explosive compounds which can be formed when in contact with acetylene.

(19) Do not use oil or grease or other lubricant on valves or other apparatus, and do not use any jointing compound.

(20) Blow out the cylinder outlet by quickly opening and closing the valve before fitting the regulator.

(21) Should an acetylene cylinder become heated due to any cause, immediately take it outdoors, immerse in water or spray it with water, open the valve and keep as cool as possible until the cylinder is empty. Then contact the suppliers.

—

(22) Do not force a regulator on to a damaged outlet thread. Report damage to cylinders to the suppliers.

Note. Also refer to Form 1704, 'Safety measures required in the use of acetylene gas and in oxy-acetylene processes in factories' (HMSO) and also 'Safety in the use of compressed gas cylinders', a booklet published by the British Oxygen Gases Ltd.

The cylinder outlet union is screwed left hand for combustible gases and right hand for non-combustible gases, the thread being $\frac{5}{8}$ in (16 mm) BSP except for CO_2 which is 21.8 mm, 14 TPI male outlet.

Acetylene (dissolved) is contained in a maroon coloured cylinder with the name ACETYLENE. Oxygen (commercial) is contained in a black coloured cylinder. (See Chapter 2 for full description of colour codes used on cylinders (BS349, BS 381 C).) Consult BS 179, Gas welding filters (GWF) for the correct shade of filter for eye protection and pp. 309–10.

Medium-pressure acetylene generators

The medium-pressure generator uses small granulated carbide. 50 kg of good calcium carbide will produce about 14 000 litres of acetylene, an average practical value being 250 litres of acetylene per kilogram of carbide. The most popular sizes are 20/50 mm, 15/25 mm and 10/15 mm. The generator is self-contained (i.e. there is no separate gas holder) and consists of a water tank surmounted by a carbide hopper, in the top of which is a diaphragm. The carbide feed valve is controlled by the diaphragm, which is actuated by the pressure of the gas generated in the tank. When the pressure falls, carbide flows into the tank; as the pressure builds up the flow ceases. The gas pressure at which the generator will work is adjusted by means of a spring fitted to the opposite side of the diaphragm, ensuring close control of pressure with generation strictly in accordance with the demand. The carbide hopper is either made of glass or is fitted with windows so that the quantity of carbide remaining in the hopper can be ascertained at a glance.

Owing to the relatively large volume of water into which the small grains of carbide fall, there is no possibility of overheating and the carbide is completely slaked. The sludge, which collects at the bottom of the tank, and is emptied each time the generator is charged, consists of a thin milky fluid.

The impurities in crude acetylene consist chiefly of ammonia, hydrides of phosphorus, sulphur and nitrogen, and there are also present water vapour and particles of lime.

These impurities must be removed before the gas is suitable for welding use; the gas is filtered and washed and chemically purified by passing it through salts of ferric iron.

The normal method of testing acetylene to ascertain whether it is being efficiently purified is to hold a silver nitrate test paper (a piece of filter paper soaked in a solution of silver nitrate) in the stream of gas for about 10 seconds. If the acetylene is being properly purified, there will be no trace of stain on the silver nitrate paper.

The reducing valve or pressure regulator

In order to reduce the pressure of either oxygen or dissolved acetylene from the high pressure of the storage cylinder to that required at the blowpipe, a regulator or reducing valve is necessary. Good regulators are essential to ensure the even flow of gas to the blowpipe. A reference to Fig. 6.2 will make the principle of operation of the regulator clear. The gas enters the regulator at the base and the cylinder pressure is indicated on the first gauge. The gas then enters the body of the regulator R through the aperture A, which is controlled by the valve V. The pressure inside the regulator rises until it is sufficient to overcome the pressure of the spring S, which loads the diaphragm D. The diaphragm is therefore pushed back and

Fig. 6.2. Reducing valve or pressure regulator.

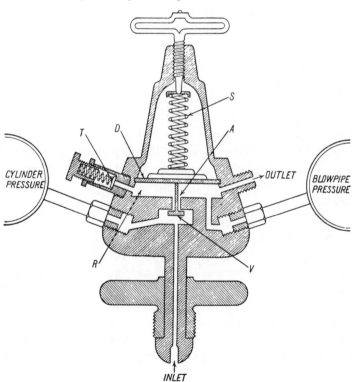

the valve V, to which it is attached, closes the aperture A and prevents any more gas from entering the regulator.

The outlet side is also fitted with a pressure gauge (although in some cases this may be dispensed with) which indicates the working pressure on the blowpipe. Upon gas being drawn off from the outlet side the pressure inside the regulator body falls, the diaphragm is pushed back by the spring, and the valve opens, letting more gas in from the cylinder. The pressure in the body R therefore depends on the pressure of the springs and this can be adjusted by means of a screw as shown.

Regulator bodies are made from brass forgings and single stage regulators are fitted with one safety valve set to relieve pressures of 16–20 bar, and should it be rendered inoperative, as for example by misuse, it ruptures at pressures of 70–80 bar and vents to the atmosphere through a vent in the bonnet. Single-stage regulators are suitable for general welding with maximum outlet pressure of 2.1 bar and for scrap cutting and heavy-duty cutting, thermic lancing and boring with outlet pressures 8.3–14 bar.

Figure 6.3*a*, *b* and *c* shows a two-stage regulator. This reduces the pressure in two stages and gives a much more stable output pressure than the single-stage regulator.

It really consists of two single stages in series within one body forging. The first stage, which is pre-set, reduces the pressure from that of the cylinder to 13–16 bar, and gas at that pressure passes into the second stage, from which it emerges at a pressure set by the pressure-adjusting control screw attached to the diaphragm. High-pressure regulators are designed for inlet pressures of 200 bar and tested to four times working pressure. Two-stage regulators have two safety valves, the first relieves at pressures of 60–75 bar and the second at 16–20 bar, so that if there is any excess pressure there will be no explosion. If a safety valve is blowing, the main valve is not seating and the regulator should immediately be taken out of service and sent for overhaul. Needle-type control valves can be fitted to regulator outlets.

The correct blowpipe pressure is obtained by adjusting the pressure of the spring with the control knob and noting the pressure in bar on the blowpipe pressure gauge. On changing a tip the regulator is set with its finger on the tip number on the scale, and final accurate adjustment of the flame is made with the blowpipe regulating valves. This is a simple and convenient method. The regulators are supplied with a table indicating the suitable pressures for various nozzles, which are stamped with their consumption of gas in litres per minute or suitable numbers. With practice the welder soon recognizes the correct pressures for various tips without reference to the table.

Regulators can be obtained for a wide range of gases; oxygen, acetylene, argon, nitrogen, propane, hydrogen, CO_2, etc., with outlet pressures to suit.

To enable two, three or more cylinders of gas to be connected together, as may be required when heavy cutting work is to be done and the oxygen consumption is very great, special adaptors are available, and these feed the bottles into one gauge. In this way a much steadier supply of oxygen is obtained.

Two operators may also be fed from one cylinder of oxygen or acetylene by using a branched gauge with two regulators. The type of branched gauge which has only one regulator feeding two outlet pipes is not recommended,

Fig. 6.3. (*a*) Two-stage regulator.

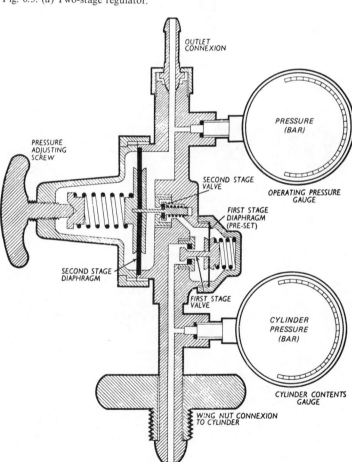

Fig. 6.3. (*b*) Multi-stage regulator.

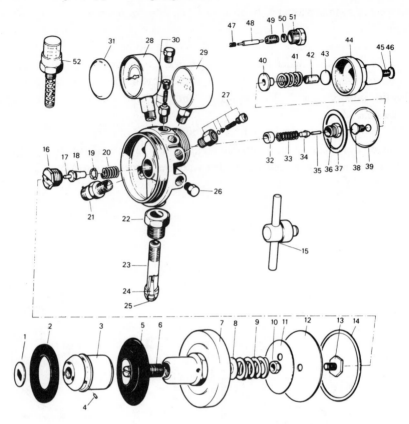

Key:

1. Disc monogram.
2. Ring cover.
3. Knob.
4. Set screw.
5. Name plate
6. Screw P. A.
7. Bonnet.
8. Spring centre.
9. Spring.
10. Nut.
11. Packing plate.
12. Diaphragm.
13. Diaphragm carrier.
14. Washer.
15. Screw P. A.
16. Nozzle.
17. Valve pin.
18. Valve.

19. Washer.
20. Spring.
21. Outlet adaptor LH.
    Outlet adaptor RH.
22. Inlet nut LH.
    Inlet nut RH.
23. Inlet nipple.
24. Filter.
25. Retaining ring.
26. Plug.
27. Safety valve LP.
28. Gauge.
29. Gauge.
30. Relief valve HP.
31. Gauge glass.
32. Sleeve.
33. Spring.
34. Valve.

35. Plunger.
26. Nozzle and seat.
37. Sealing ring.
38. Diaphragm carrier.
39. Diaphragm.
40. Disc.
41. Spring.
42. Damper plug.
43. Pivot.
44. Bonnet.
45. Screw.
46. Disc anti-tamper.
47. Spring.
48. Valve.
49. Seat retainer.
50. Seat.
51. Valve holder.
52. Indicator assembly.

since any alteration of the blowpipe pressure by one operator will affect the flame of the other operator.

Owing to the rapid expansion of the oxygen in cases where large quantities are being used, the regulator may become blocked with particles of ice, causing stoppage. This happens most frequently in cold weather, and

Fig. 6.3. (*c*)

can be prevented by use of an electric regulator heater. The heater screws into the cylinder and the regulator screws into the heater. The heater is plugged into a source of electric supply, the connexion being by flexible cable.

Hoses

Hoses are usually of a seamless flexible tube reinforced with plies of closely woven fabric impregnated with rubber and covered overall with a tough, flexible, abrasion-resistant sheath giving a light-weight hose. They are coloured blue for oxygen, red for fuel gases, black for non-combustible gases and orange for LPG, Mapp and natural gas. Available lengths are from 5 to 20 m, with bore diameters 4.5 mm for maximum working pressure of 7 bar, 8 mm for a maximum of 12 bar and 10 mm for a maximum working pressure of 15 bar. Nipple- and nut-type connexions and couplers are available for 4.5 mm ($\frac{3}{8}$ in.), 8 and 10 mm hoses with 6.4 mm ($\frac{1}{4}$ in. BSP) and 10 mm $\frac{3}{8}$ in. BSP) nuts. A hose check valve is used to prevent feeding back of gases from higher or lower pressures and reduces the danger of a flashback due to a blocked nozzle, leaking valve, etc. It is connected in the hose at the blowpipe end or to the economizer or regulator, and consists of a self-aligning spring-loaded valve which seals off the line in the event of backflow. BS 924 J and 796 J apply to hoses.

The welding blowpipe or torch

There are two types of blowpipes: (1) high pressure, (2) low pressure, and each type consists of a variety of designs depending on the duty for which the pipe is required. Special designs are available for rightward and leftward methods of welding (the angle of the head is different in these designs), thin gauge or thick plate, etc., in addition to blowpipes designed for general purposes.

The high-pressure blowpipe is simply a mixing device to supply approximately equal volumes of oxygen and acetylene to the nozzle, and is fitted with regulating valves to vary the pressure of the gases as required (Fig. 6.4a, b). A selection of shanks is supplied with each blowpipe, having orifices of varying sizes, each stamped with a number or with the consumption in litres per hour (l/h). Various sizes of pipes are available, from a small light variety, suitable for thin gauge sheet, to a heavy duty pipe. A high-pressure pipe cannot be used on a low-pressure system.

The low-pressure blowpipe has an injector nozzle inside its body through which the high-pressure oxygen streams (Fig. 6.5). This oxygen draws the low-pressure acetylene into the mixing chamber and gives it the necessary velocity to preserve a steady flame, and the injector also helps to prevent

Fig. 6.4. (*a*) Principle of the high-pressure blowpipe.

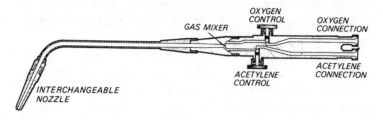

Fig. 6.4. (*b*) Blowpipe.

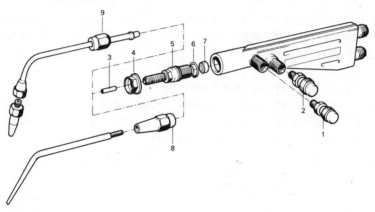

Key:
1. Spindle assembly RH.
2. Spindle assembly LH.
3. Insert.
4. Nut–locking.
5. Mixer.

6. 'O' ring.
7. Mixer spool.
8. Adaptor nut.
9. Neck assembly.

Fig. 6.5. Principle of the low-pressure blowpipe.

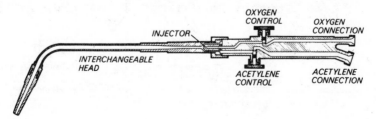

backfiring. The velocity of a 1/1 mixture of oxygen/acetylene may be 200 m per minute, while the maximum gas velocity occurs for a 30% acetylene mixture and may be up to 460 m per min. (These figures are approximate only.)

It is usual for the whole head to be interchangeable in this type of pipe, the head containing both nozzle and injector. This is necessary, since there is a corresponding injector size for each nozzle. Regulating valves, as on the high-pressure pipes, enable the gas to be adjusted as required. The low-pressure pipe is more expensive than the high-pressure pipe; and it can be used on a high-pressure system if required, but it is now used on a very small scale.

A very useful type of combined welding blowpipe and metal-cutting torch is shown in Fig. 6.6. The shank is arranged so that a full range of nozzles, or a cutting head, can be fitted. The design is cheaper than for a corresponding separate set for welding and cutting, and the cutter is sufficient for most work.

The oxy-acetylene flame

The chemical actions which occur in the flame have already been discussed in Volume 1, and we will now consider the control and regulation of the flame to a condition suitable for welding.

Fig. 6.6. Combined welding and cutting pipes. Will weld sections from 1.6 mm to 32 mm thick, and cut steel up to 150 mm thick with acetylene and 75 mm with propane.

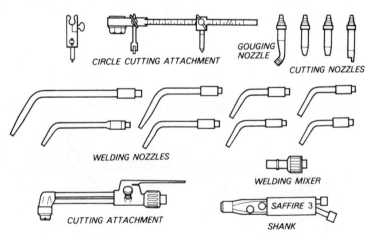

Adjustment of the flame.* To adjust the flame to the neutral condition the acetylene is first turned on and lit. The flame is yellow and smoky. The acetylene pressure is then increased by means of the valve on the pipe until the smokiness has just disappeared and the flame is quite bright. The condition gives approximately the correct amount of acetylene for the particular jet in use. The oxygen is then turned on as quickly as possible, and as its pressure is increased the flame ceases to be luminous. It will now be noticed that around the inner blue luminous cone, which has appeared on the tip of the jet, there is a feathery white plume which denotes excess acetylene (Fig. 6.7a). As more oxygen is supplied this plume decreases in size until there is a clear-cut blue cone with no white fringe (Fig. 6.7b). This is the neutral flame used for most welding operations. If the oxygen supply is further increased, the inner blue cone becomes smaller and thinner and the outer envelope becomes streaky; the flame is now oxidizing (Fig. 6.7c). Since the oxidizing flame is more difficult to distinguish than the carbonizing or carburizing (excess acetylene) flame, it is always better to start with excess acetylene and increase the oxygen supply until the neutral condition is reached, than to try to obtain the neutral flame from the oxidizing condition.

Some welders prefer to regulate the oxygen pressure at the regulator itself. The acetylene is lighted as before, and with the oxygen valve on the blowpipe turned full on, the pressure is adjusted correctly at the regulator until the flame is neutral. In this way the welder is certain that the regulator is supplying the correct pressure to the blowpipe for the particular nozzle being used.

Fig. 6.7

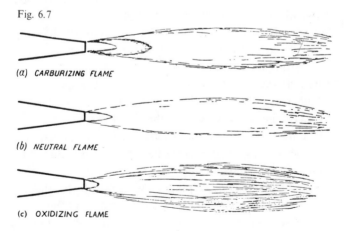

(a) CARBURIZING FLAME

(b) NEUTRAL FLAME

(c) OXIDIZING FLAME

* The pressure on oxygen and acetylene gauges is approximately that given in the table on p. 262.

Selection of correct nozzle. As the thickness of the work to be welded increases, the flame will have to supply more heat, and this is made possible by increasing the size of the nozzle. The nozzle selected may cover one or two thicknesses of plate; for example, a nozzle suitable for welding 6.4 mm plate will weld both 4.8 mm and 8 mm plate by suitable regulation of the pressure valves. This is because the blowpipe continues to mix the gases in the correct proportion over a range of pressures. If, however, one is tempted to weld a thickness of plate with a nozzle which is too large, by cutting down the supply of gas at the valves instead of changing the nozzle for one smaller, it will be noticed that explosions occur at the nozzle when welding, these making the operation impossible. These explosions indicate too low a pressure for the nozzle being used.

If, on the other hand, one attempts to weld too great a thickness of metal with a certain nozzle, it will be noticed that as one attempts to increase the pressure of the gases beyond a certain point to obtain a sufficiently powerful flame, the flame leaves the end of the nozzle. This indicates too high a pressure and results in a *hard* noisy flame. It is always better to work with a soft flame which is obtained by using the correct nozzle and pressure. Thus, although there is considerable elasticity as to the thickness weldable with a given nozzle, care should be taken not to overtax it.

Use and care of blowpipe

Oil or grease should upon no account be used on any part of the blowpipe, but a non-oily graphite may be used and is useful for preventing wear and any small leaks.

A backfire is the appearance of the flame in the neck or body of the blowpipe and which rapidly extinguishes itself.

A flashback is the appearance of the flame beyond the blowpipe body into the hose and even the gauge, with subsequent explosion. It can be prevented by fitting a flashback arrestor.

Backfiring may occur at the pipe through several causes:

(1) Insufficient pressure for the nozzle being used. This can be cured by increasing the pressure on the gauge.

(2) Metal particles adhering to the nozzle. The nozzle can be freed of particles by rubbing it on a leather or wooden surface. (The gases should be first shut off and then relit.)

(3) The welder touching the plate or weld metal with the nozzle. In this case the gases should be shut off and then relit.

(4) Overheating of the blowpipe. A can of water should be kept near so as to cool the nozzle from time to time, especially when using a large flame. Oxygen should be allowed to pass slowly through the

nozzle, when immersed in the water, to prevent the water entering the inside of the blowpipe.

(5) Should the flame backfire into the mixing chamber with a squealing sound, and a thin plume of black smoke be emitted from the nozzle, serious damage will be done to the blowpipe unless the valves are immediately turned off. This fault may be caused by particles håving lodged inside the pipe, or even under the regulating valves. The pipe should be thoroughly inspected for defects before being relit.

In the event of a backfire, therefore, immediately shut off the acetylene cylinder valve, and then the oxygen, before investigating the cause.

Blowpipe nozzles should be cleaned by using a soft copper or brass pin. They should be taken off the shank and cleaned from the inside, as this prevents enlarging the hole. A clean nozzle is essential, since a dirty one gives an uneven-shaped flame with which good welds are impossible to make. Special sets of reamers can be obtained for this work.

Flashback arrestor

Note. Flashback arrestors should be fitted to all welding equipment. A flashback occurs when the flame moves from the blow torch into the supply system against the flow of the gases. It is potentially dangerous and its effects can range from sooty deposits in the blowpipe and hoses to a fire in the gauge or cylinder often accompanied by squealing or popping noises. It is generally due to incorrect operating practice such as overheated blowpipe and wrong pressures.

The automatic flashback arrestor is made for acetylene, propane, hydrogen, and oxygen and is generally connected to the regulator outlet (Fig. 6.3c) and prevents flame movement into gauges and cylinder, causing regulator damage and even cylinder fires.

A dense sintered stainless steel plate filter of up to 100 micron mesh (1) in Fig. 6.8a arrests the flame and is designed to quench the flame from even the most violent explosion of the gas mixture. The large surface area of the filter does not offer much resistance to the forward flow of the gas. The pressure-sensitive cut-off valve (2) automatically cuts off the supply of gas to the blow torch preventing sustained flashback and it will remain closed until it is set manually, a spring plus the gas pressure holding it closed. There should never be any attempt to reset the valve until the cause of the flashback has been thoroughly investigated and put right and it cannot be reset until the pressure is taken off the system, so close the cylinder valve to cut off the supply at source. To reset, on the model shown, the reset pin (attached to the gauge by a chain to prevent loss) is inserted into the gas inlet orifice, centred against the valve and pushed hard home resetting the valve mechanism. On some models there is a visible lever which is actuated when the arrestor cut-off valve operates. This lever actuates the mechanism but should never be reset until the cause of the flashback is determined and put to rights.

After several actuations, carbon deposits may interfere with the correct functioning of the arrestor so it should be exchanged. Arrestors are generally used up to oxygen pressures of 10 bar, propane of 5 bar and acetylene of 1.5 bar.

Cheaper models are made to fit into the hose line at blowpipe or regulator end (Fig. 6.8*b*).

Technique of welding

Before attempting any actual welding operations, the beginner should acquire a sense of fusion and a knowledge of blowpipe control. This can be obtained by running lines of fusion on thin-gauge steel plate.

The flame is regulated to the neutral condition and strips of 1.6 or 2 mm steel plate are placed on firebricks on the welding bench. Holding the

Fig. 6.8. (*a*) Section through flashback arrestor.

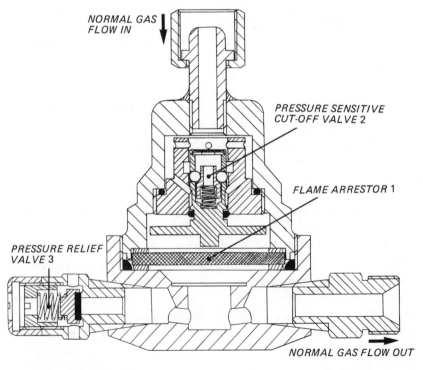

NORMAL GAS FLOW IN

PRESSURE SENSITIVE CUT-OFF VALVE 2

FLAME ARRESTOR 1

PRESSURE RELIEF VALVE 3

NORMAL GAS FLOW OUT

blowpipe at approximately 60° to the plate, with the inner blue cone near the metal surface, and beginning a little from the right-hand edge of the sheet, the metal is brought to the melting point and a puddle formed with a rotational movement of the blowpipe. Before the sheet has time to melt through into a hole, the pipe is moved steadily forward, still keeping the steady rotating motion, and the line of fusion is made in a straight line. This exercise should be continued on various thicknesses of thin-gauge plate until an even line is obtained, and the underside shows a regular continuous bead, indicating good penetration, the student thus acquiring a sense of fusion and of blowpipe control.

In the following pages various methods of welding techniques are considered, and it would be well to state at this point what constitutes a good weld, and what features are present in a bad weld.

Fig. 6.9*a* indicates the main features of a good fusion weld, with the following features:

    (*a*) Good fusion over the whole side surface of the V.

    (*b*) Penetration of the weld metal to the underside of the parent plate.

    (*c*) Slight reinforcement of the weld above the parent plate.

    (*d*) No entrapped slag, oxide, or blowholes.

Fig. 6.8. (*b*)

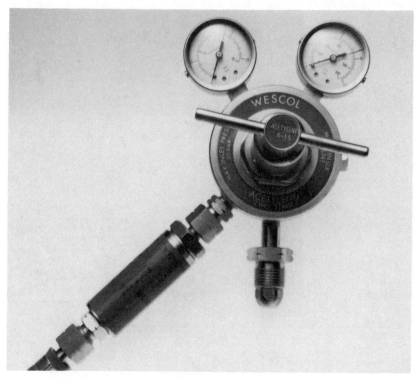

Fig. 6.9*b* indicates the following faults in a weld:

(*a*) Bad flame manipulation and too large a flame has caused molten metal to flow on to unfused plate, giving no fusion (i.e. adhesion).

(*b*) Wrong position of work, incorrect temperature of molten metal, and bad flame manipulation has caused slag and oxide to be entrapped and channels may be formed on each line of fusion, causing undercutting.

(*c*) The blowpipe flame may have been too small, or the speed of welding too rapid, and this with lack of skill in manipulation has caused bad penetration.

(N.B. Reinforcement on the face of a weld *will not* make up for lack of penetration.)

Methods of welding

The following British Standards apply to this section: BS 1845, *Filler alloys for brazing*; BS 1724, *Bronze welding by gas*; BS 1453, *Filler rods and wires for gas welding; group A, steel filler rods and wires; group B cast iron filler rods; group C, copper and copper alloy filler rods and wires; group D, magnesium alloy filler rods and wires*; BS 1821, *Oxyacetylene welding of pipe lines, Class 1*; BS 2640, *Oxy-acetylene welding of pipe lines, Class 2*.

Fig. 6.9

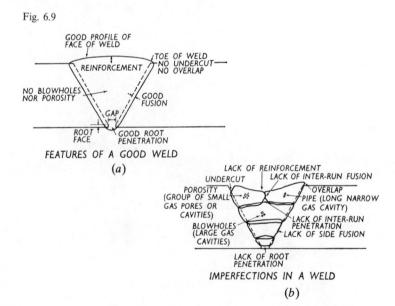

FEATURES OF A GOOD WELD

(*a*)

IMPERFECTIONS IN A WELD

(*b*)

Leftward or forward welding

This method is used nowadays for welding steel plate under 6.5 mm thick and for welding non-ferrous metals. The welding rod precedes the blowpipe along the seam, and the weld travels from right to left when the pipe is held in the right hand. The inner cone of the flame, which is adjusted to the neutral condition, is held near the metal, the blowpipe making an angle of 60 to 70° with the plate, while the filler rod is held at an angle of 30 to 40°. This gives an angle of approximately 90° between the rod and the blowpipe. The flame is given a rotational, circular, or side-to-side motion, to obtain even fusion on each side of the weld. The flame is first played on the joint until a molten pool is obtained and the weld then proceeds, the rod being fed into the molten pool and not melted off by the flame itself. If the flame is used to melt the rod itself into the pool, it becomes easy to melt off too much and thus reduce the temperature of the molten pool in the parent metal to such an extent that good fusion cannot be obtained. Fig. 6.10 will make this clear.

The first exercise in welding with the filler rod is done with the technique just described and consists of running lines of weld on 1.6 or 2 mm plate, using the filler rod. Butt welds of thin plate up to 2.4 mm can be made by flanging the edges and melting the edges down. When a uniform weld is obtained, with good penetration, the exercises can be repeated on plate up to 3.2 mm thick, and butt welds on this thickness attempted. Above 3.2 mm thick, the plates are bevelled, chamfered, or V'd to an angle of 80 to 90° (Fig. 6.11). the large area of this V means that a large quantity of weld metal is required to fill it. If, however, the V is reduced to less than 80°, it is found that as the V becomes narrower the blowpipe flame tends to push the molten metal from the pool, forward along onto the unmelted sides of the

Fig. 6.10. Leftward welding.

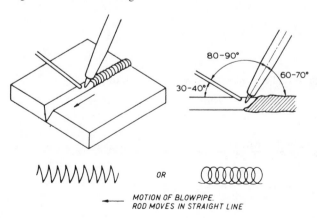

80–90°

60–70°

30–40°

OR

MOTION OF BLOWPIPE.
ROD MOVES IN STRAIGHT LINE

V, resulting in poor fusion or adhesion. This gives an unsound weld, and the narrower the V the greater this effect.

As the plate to be welded increases in thickness, a larger nozzle is required on the blowpipe, and the control of the molten pool becomes more difficult; the volume of metal required to fill the V becomes increasingly greater, and the size of nozzle which can be used does not increase in proportion to the thickness of the plate, and thus welding speed decreases. Also with thicker plates the side-to-side motion of the blowpipe over a wide V makes it difficult to obtain even fusion on the sides and penetration to the bottom, while the large volume of molten metal present causes considerable expansion. As a result it is necessary to weld thicker plate with two or more layers if this method is used. From these considerations it can be seen that above 6.4 mm plate the leftward method suffers from several drawbacks. It is essential, however, that the beginner should become efficient in this method before proceeding to the other methods, since for general work, including the non-ferrous metals (see later), it is the most used.

The preparation of various thicknesses of plate for butt joints by the leftward method is given in the table accompanying Fig. 6.11.

Edge preparation (Letters refer to Fig. 6.11)	Thickness of plate (mm)	Nozzle size (mm)	Oxygen and acetylene pressure (bar)	Oxygen and acetylene gas consumption (l/h)
(*a*)	0.9–1.6	0.9–1	0.14	28
		1.2–2	0.14	57
(*b*)	2.4–3.0	2–3	0.14	86
		2.6–5	0.14	140
(*c*)	3.0–4.0	3.2–7	0.14	200
		4.0–10	0.21	280

Fig. 6.11. Leftward welding: edge preparations.

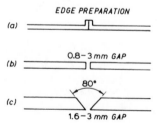

Rightward welding

This method was introduced some years ago to compete with electric arc welding in the welding of plate over 4.8 mm thick, since the leftward method has the disadvantage just mentioned on welding thick plate. This method has definite advantages over the leftward method on thick plate, but the student should be quite aware of its limitations and use it only where it has a definite advantage.

In this method the weld progresses along the seam from left to right, the rod following the blowpipe. The rod is given a rotational or circular motion, while the blowpipe moves in practically a straight line, as illustrated in Fig. 6.12. The angle between blowpipe and rod is greater than that used in the leftward method.

When using this method good fusion can be obtained without a V up to 8 mm plate. Above 8 mm the plates are prepared with a 60° V, and since the blowpipe has no side motion the heat is all concentrated in the narrow V, giving good fusion. The blowpipe is pointing backwards towards the part that has been welded and thus there is no likelihood of the molten metal being pushed over any of the unheated surface, giving poor fusion.

A larger blowpipe nozzle is required for a given size plate than in leftward welding, because the molten pool is controlled by the pipe and rod but the pipe has no side to side motion. This larger flame gives greater welding speed, and less filler rod is used in the narrower V. The metal is under good control and plates up to 16 mm thick can be welded in one pass. Because the blowpipe does not move except in a straight line, the molten metal is agitated very little and excess oxidation is prevented. The flame playing on the metal just deposited helps to anneal it, while the smaller volume of molten metal in the V reduces the amount of expansion. In addition, a better view is obtained of the molten pool, resulting in better penetration.

Fig. 6.12. Rightward welding.

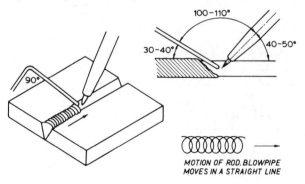

MOTION OF ROD. BLOWPIPE
MOVES IN A STRAIGHT LINE

It is essential, however, in order to ensure good welds by this method, that blowpipe and rod should be held at the correct angle, the correct size nozzle and filler rod used, and the edges prepared properly (Fig. 6.13). The rod diameter is about half the thickness of the plate being welded up to 8 mm plate, and half the thickness + 0.8 mm when welding V'd plate. The blowpipe nozzle is increased in size from one using about 300 litres per hour, with the leftward method, to one using about 350 litres per hour, when welding 3.2 mm plate. If too large diameter filler rods are used, they melt too slowly causing poor penetration, and poor fusion. Small rods melt too quickly and reinforcement of the weld is difficult. Rightward welding has no advantage on plates below 6.4 mm thick and is rarely used below this thickness, the leftward method being preferred.

The advantages of the rightward method on thicker plate are:

(1) Less cost per metre run due to less filler rod being used and increased speed.

(2) Less expansion and contraction.

(3) Annealing action of the flame on the weld metal.

(4) Better view of the molten pool, giving better control of the weld.

See later for 'all-position rightward welding'.

Edge preparation (Letters refer to Fig. 6.13)	Thickness of plate (mm)	Nozzle size (mm)	Oxygen and acetylene pressure (bar)	Oxygen and acetylene gas consumption (l7h)
(a)	4.8–8.2	5.5–13	0.28	370
		6.5–18	0.28	520
		8.2–25	0.42	710
(b)	8.2–15	10–35	0.63	1000
		13–45	0.35 (heavy duty mixer)	1300

Fig. 6.13. Rightward welding: edge preparations.

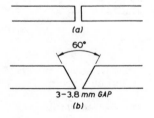

(a)

60°

3–3.8 mm GAP

(b)

Vertical welding

The preparation of the plate for welding greatly affects the cost of the weld, since it takes time to prepare the edges, and the preparation given affects the amount of filler rod and gas used. Square edges need no preparation and require a minimum of filler rod. In leftward welding square edges are limited to 3.2 mm thickness and less. In vertical welding, up to 4.8 mm plate can be welded with no V'ing with the single-operator method while up to 16 mm plate can be welded with no V'ing with the two-operator method, the welders working simultaneously on the weld from each side of the plate. The single-operator method is the most economical up to 4.8 mm plate.

The single-operator method (up to 4.8 mm plate) requires more skill in the control of the molten metal than in downhand welding. Welding is performed either from the bottom upwards, and the rod precedes the flame as in the leftward method, or from the top downwards, in which case the metal is held in place by the blowpipe flame. This may be regarded as the rightward method of vertical welding, since the flame precedes the rod down the seam. In the upward method the aim of the welder is to use the weld metal which has just solidified as a 'step' on which to place the molten pool. A hole is first blown right through the seam, and this hole is maintained as the weld proceeds up the seam, thus ensuring correct penetration and giving an even back bead.

In the vertical welding of thin plate where the edges are close together, as for example in a cracked automobile chassis, little filler rod is needed and the molten pool can be worked upwards using the metal from the sides of the weld. Little blowpipe movement is necessary when the edges are close together, the rod being fed into the molten pool as required. When the edges are farther apart, the blowpipe can be given the usual semicircular movement to ensure even fusion of the sides.

From Fig. 6.14 it will be noted that as the thickness of the plate increases, the angle of the blowpipe becomes much steeper.

When welding downwards much practice is required (together with the correct size flame and rod), in order to prevent the molten metal from falling downwards. This method is excellent practice to obtain perfect control of the molten pool.

Double-operator vertical welding

The flames of each welder are adjusted to the neutral condition, both flames being of equal size. To ensure even supply of gas to each pipe the blowpipes can be supplied from the same gas supply. It is possible to use much smaller jets with this method, the combined consumption of which is less than that of a single blowpipe on the same thickness plate. Blowpipes and rods are held at the same angles by each operator, and it is well that a third person should check this when practice runs are being done. To avoid fatigue a sitting position is desirable, while, as for all types of vertical welding, the pipes and tubes should be as light as possible. Angles of blowpipes and rods are shown in Fig. 6.15.

This method has the advantage that plate up to 16 mm thick can be welded without preparation, reducing the gas consumption and filler rod used, and cutting out the time required for preparation. When two operators are welding 12.5 mm plate, the gas used by both is less than 50%

Fig. 6.14. Single-operator vertical welding.

MOVEMENT OF ROD AND TORCH

DIRECTION OF WELDING

30°

1.6 mm 30°
3.2 mm 60°
4.8 mm 80°

3.2 mm
1.6 mm

3.2 mm
3.2 mm

80°
4.8 mm

3.2 mm

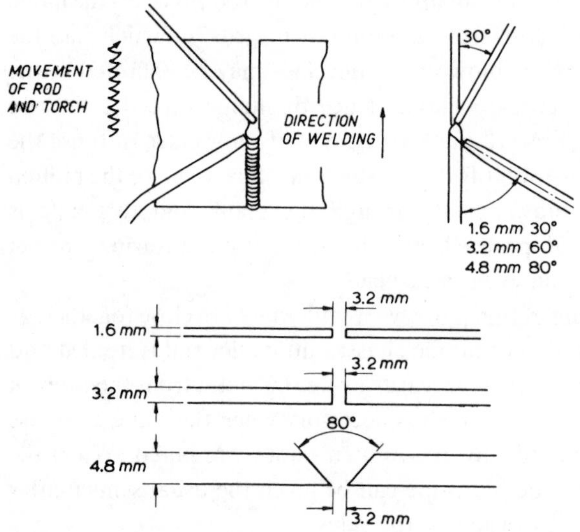

Fig. 6.15. Double-operator vertical welding.

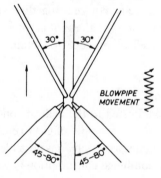

30°   30°

BLOWPIPE MOVEMENT

45-80°   45-80°

of the total consumption of the blowpipe when welding the same thickness by the downhand rightward method. Owing to the increased speed of welding and the reduced volume of molten metal, there is a reduction in the heating effect, which reduces the effects of expansion and contraction.

Overhead welding

Overhead welding is usually performed by holding the blowpipe at a very steep angle to the plate being welded. The molten pool is entirely controlled by the flame and by surface tension, and holding the flame almost at right angles to the plate enables the pool to be kept in position.

Difficulty is most frequently found in obtaining the correct amount of penetration. This is due to the fact that as sufficient heat to obtain the required penetration is applied, the molten pool becomes more fluid and tends to become uncontrollable. With correct size of flame and rod, however, and practice, this difficulty can be entirely overcome and sound welds made. Care should also be taken that there is no undercutting along the edges of the weld.

A comfortable position and light blowpipe and tubes are essential if the weld is to be made to any fair length, as fatigue of the operator rapidly occurs in this position and precludes the making of a good weld.

The positions of blowpipe and rod for the leftward techniques are shown in Fig. 6.16a, while Fig. 6.16b shows their relative positions when the rightward method is used. The rightward method is generally favoured, but

Fig. 6.16. (*a*) Leftward overhead welding. The flame is used to position the molten metal. (*b*) Rightward overhead welding. Blowpipe has little motion. Rod moves criss-cross from side to side.

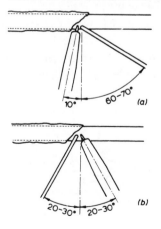

it must again be stressed that considerable practice is required for a welder to become skilled in overhead welding.

Lindewelding

This method was devised by the Linde Co. of the United States for the welding of pipelines (gas and oil), and for this type of work it is very suitable. Its operation is based on the following facts:

(1) When steel is heated in the presence of carbon, the carbon will reduce any iron oxide present, by combining with the oxygen and leaving pure iron. The heated surface of the steel then readily absorbs any carbon present.

(2) The absorption of carbon by the steel lowers the melting point of the steel (e.g. pure iron melts at 1500°C, while cast iron with $3\frac{1}{2}\%$ carbon melts at 1130°C).

In Lindewelding the carbon for the above action is supplied by using a carburizing flame. This deoxidizes any iron oxide present and then the carbon is absorbed by the surface layers, lowering their melting point. By using a special rod, a good sound weld can be made in this way at increased speed. The method is almost exclusively used on pipe work and is performed in the downhand position only.

Block welding

Block welding is a method especially applicable to steel pipes and thick walled tubes in which the weld is carried out to the full depth of the joint in steps. This can more easily be understood by reference to the figure. The first run is laid giving good penetration for as great a length as is convenient, say *AB* (Fig. 6.17).

The second run is now started a little way short of *A* and finished short of *B*, as *CD*. The third run is laid in a similar manner from *E* to *F*, if required, and so on for the full depth of the weld required. We thus have a series of ledges or platforms at the beginning and end of the weld. The welding is now continued with the first run from *B* with full penetration. The second run starts at *D* and has the ledge *BD* deposited before it gets to where the

Fig. 6.17

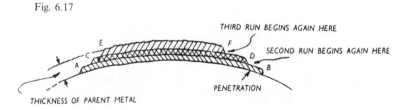

first run started. Similarly with the third run, which starts at *F*. Upon completing the weld in the case of a pipe, the first run finishes at *A*, the second one at *C* and the third one at *E*, giving a good anchorage on to the previous run.*

Horizontal–vertical fillet welding

In making fillet welds (Fig. 6.18), care must be taken that, in addition to the precautions taken regarding fusion and penetration, the vertical plate is not undercut as in Fig. 6.19*b*, and the weld is not of a weakened section. A lap joint may be regarded as a fillet. No difficulty will be experienced with undercutting, since there is no vertical leg, but care should be taken not to melt the edge of the lapped plate.

The blowpipe and rod must be held at the correct angles. Holding the

Fig. 6.18. Types of fillet joints.

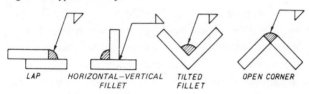

LAP   HORIZONTAL–VERTICAL FILLET   TILTED FILLET   OPEN CORNER

Fig. 6.19. (*a*)

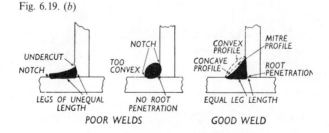

(*a*) SINGLE V PREPARATION   (*b*) DOUBLE V PREPARATION

Fig. 6.19. (*b*)

UNDERCUT NOTCH   NOTCH   TOO CONVEX   CONVEX PROFILE   MITRE PROFILE   CONCAVE PROFILE   ROOT PENETRATION

LEGS OF UNEQUAL LENGTH   NO ROOT PENETRATION   EQUAL LEG LENGTH

POOR WELDS   GOOD WELD

---

* Refer also to BS 1821 and 2640, *Class I and class II oxy-acetylene welding of steel pipelines and assemblies for carrying fluids.*

flame too high produces undercutting, and the nozzle of the cone should be held rather more towards the lower plate, since there is a greater mass of metal to be heated in this than in the vertical plate (Fig. 6.19*a* and *b*).

Figure 6.20 *a* and *b* show the angles of the blowpipe and rod, the latter being held at a steeper angle than the blowpipe. Fillet welding requires a larger size nozzle than when butt welding the same section plate, owing to the greater amount of metal adjacent to the weld. Because of this, multi-jet blowpipes can be used to great advantage for fillet welding. The single (Fig. 6.19*a*) preparation is used for joints which are subjected to severe loading, while the double V preparation (Fig. 6.19*b*) is used for thick section plate when the welding can be done from both sides. The type of preparation therefore depends entirely on the service conditions, the unprepared joint being quite suitable for most normal work.

All-position rightward welding

This method can be used for vertical, overhead and horizontal–vertical positions, the blowpipe preceding the rod as for downhand welding. For vertical welding the blowpipe is held 10° below the horizontal line (welding upwards) while the rod is held alongside the pipe at 45–60° to the vertical plate. Overhead welding is done similarly. The advantages are similar to those of the downhand position but considerable practice is required to become proficient. See Fig. 6.21.

For the preparation and welding of steel pipes the student should refer to BS 1821, *Class I steel pipelines* and BS 2640, *Class II steel pipelines*.

Fig. 6.20. (*a*) Fillet weld.

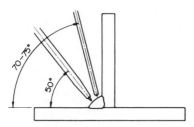

Fig. 6.20. (*b*) Fillet weld.

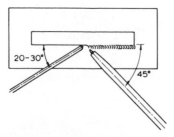

Fig. 6.21. All-position rightward welding.

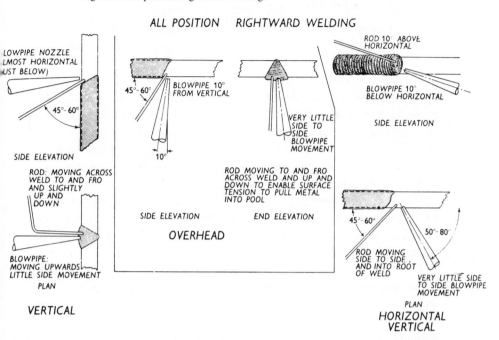

Steel filler rods for oxy-acetylene welding

Rod type	Composition %					Application
	C	Si	Mn	Ni	Cr	
Mild steel (copper coated)	0.1	—	0.6	0.25	—	General utility low-carbon-steel rod for low and mild carbon steel and wrought iron. UTS 386 N/mm² hardness 120 Brinell, melting point 1490°C.
Medium carbon steel (copper coated)	0.25–0.3	0.3–0.5	1.3–1.6	0.25	0.25	For medium carbon steels, high strength with toughness, UTS 552 N/mm², hardness 150 Brinell, melting point 1400°C.
Pipe-welding steel	0.1–0.2 Plus Al, Ti, Zr as deoxidizers up to 0.15% max.	0.1–0.35	1.0–1.6	—	—	Low carbon steel for high-strength welds in steel pipes. UTS 492 N/mm², hardness 145 Brinell, melting point 1450°.

Note. Phosphorus and sulphur 0.04% max. for all types.

Cast iron welding

Cast iron, because of its brittleness, presents a different problem in welding from steel. We may consider three types – grey, white and malleable.

The grey cast iron is softer and tougher than the white, which is hard and brittle. The good mechanical properties of grey cast iron are due to the presence of particles of free carbon or graphite, which separate out during slow cooling. When the cooling is rapid, it is impossible for the cementite (iron carbide) to decompose into ferrite and graphite; hence the structure consists of masses of cementite embedded in pearlite, this giving the white variety of cast iron with its hardness and brittleness.

The other constituents of cast iron are silicon, sulphur, manganese and phosphorus. Silicon is very important, because it helps to increase the formation of graphite, and this helps to soften the cast iron. Manganese makes the casting harder and stronger. It has a great affinity for sulphur and, by combining with it, prevents the formation of iron sulphide, which makes the metal hard and brittle. Phosphorus reduces the melting point and increases the fluidity. If present in a greater proportion than 1% it tends to increase the brittleness. Sulphur tends to prevent the formation of graphite and should generally not exceed 0.1%. It may be added to enable the outer layers to have a hard surface (chill casting), while the body of the casting is still kept in the grey state.

The aim of the welder should be always, therefore, to form grey cast iron in the weld (Fig. 6.22).

Fig. 6.22. Oxy-acetylene fusion weld in cast iron. × 100.

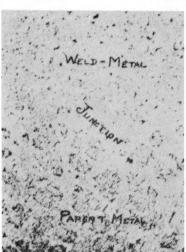

Preparation. The edges are V'd out to 90° on one side only up to 10 mm thickness and from both sides for greater thickness. Pre-heating is essential, because not only does it prevent cracking due to expansion and contraction, but by enabling the weld to cool down slowly, it causes grey cast iron to be formed instead of the hard white unmachinable deposit which would result if the weld cooled off rapidly. Pre-heating may be done by blowpipe, forge or furnace, according to the size of the casting.

Blowpipe, flame, flux and rod. The neutral flame is used, care being taken that there is not the slightest trace of excess oxygen which would cause a weak weld through oxidation, and that the metal is not overheated. The inner cone should be about 3–4 mm away from the molten metal. If it touches the molten metal, hard spots will result. The flux (of alkaline borates) should be of good quality to dissolve the oxide and prevent oxidation, and it will cause a coating of slag to form on the surface of the weld, preventing atmospheric oxidation. Ferro-silicon and super-silicon rods, containing a high percentage of silicon, are the most suitable rods to use.

Technique. The welding operation should be performed on the dull red-hot casting with the rod and blowpipe at the angles shown in Fig. 6.23 and done in the leftward manner. The rod is dipped into the flux at intervals, using only enough flux to remove the oxide. It will be noticed that, as the flux is added, the metal flows more easily and looks brighter, and this gives a good indication as to when flux is required. The excessive use of flux causes blowholes and a weak weld. The rod should be pressed down well into the weld, removing a good percentage of the slag and oxide. Very little motion of the blowpipe is required and the rod should not be stirred round continuously, as this means the formation of more slag with more danger of entrapping it in the weld. Cast-iron welds made by beginners often have brittle parts along the edges, due to their not getting under the oxide and floating it to the top.

Fig. 6.23. Cast iron welding.

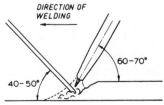

After treatment. The slag and oxide on the surface of the finished weld can be removed by scraping and brushing with a wire brush, but the weld should not be hammered. The casting is then allowed to cool off very slowly, either in the furnace or fire, or if it has been pre-heated with the flame, it can be put in a heap of lime, ashes or coke, to cool. Rapid cooling will result in a hard weld with possibly cracking or distortion.

Malleable cast iron welding

This is unsuitable for welding with cast-iron rods because of its structure. If attempted it invariably results in hard, brittle welds having no strength. The best method of welding is with the use of bronze rods, and is described below.

Braze welding and bronze welding

Braze welding is the general term applied to the process in which the metal filler wire or strip has a lower melting point than that of the parent metal, using a technique similar to that used in fusion welding, except that the parent metal is not melted nor is capillary action involved as in brazing. When the filler metal is made of a copper-rich alloy the process is referred to as bronze welding.

By the use of bronze filler wires such as those given in the accompanying table, a sound weld can be made in steel, wrought iron, cast iron and copper and between dissimilar metals such as steel and copper, at a lower temperature than that usually employed in fusion welding. The bronze filler metal, which has a lower melting point than the parent metal, is melted at the joint so as to flow and 'wet' the surfaces of the joint to form a sound

Fig. 6.24. When specimen (*a*) is subjected to a tensile pull as shown the faces *F* of the joint are partly in shear. (*c*) 'Shear V' preparation.

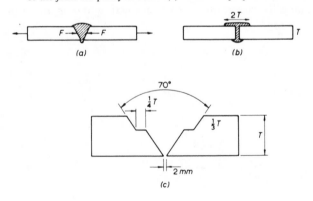

bond without fusion of the parent metal. There is less thermal disturbance than with fusion welding due to the lower temperatures involved and the process is simple and relatively cheap to perform. Unlike capillary brazing (see pp. 303–6) the strength of the joint is not solely dependent upon the areas of the surfaces involved in the joint but rather upon the tensile strength of the filler metal. A bronze weld has relatively great strength in shear, and joints are often designed to make use of this (Fig. 6.24*a*, *b* and *c*).

The final strength of the welded joint depends upon the bond between filler metal and parent metal so that thorough cleanliness of the joints and immediate surroundings is essential to ensure that the molten filler metal should flow over the areas and 'wet' them completely without excessive penetration of the parent metal, and there should be freedom from porosity. Care, therefore, should be taken to avoid excessive overheating. Figs. 6.25 and 6.26 show various joint designs for plate and tube.

If bronze filler wires are used on alloys such as brasses and bronzes the melting points of wire and parent metal are so nearly equal that the result is a fusion weld.

General method of preparation for bronze welding. All impurities such as scale, oxide, grease, etc., should be removed, as these would prevent the bronze wetting the parent metal. The metal should be well cleaned on both upper and lower faces for at least 6 mm on each side of the joint, so that the bronze can overlap the sides of the joint, running through and under on the lower face.

Bronze welding is unlike brazing in that the heat must be kept as local as possible by using a small flame and welding quickly. The bronze must flow in front of the flame for a short distance only, wetting the surface, and by having sufficient control over the molten bronze, welding may be done in

Fig. 6.25. Typical joint designs for bronze welding (sheet and plate).

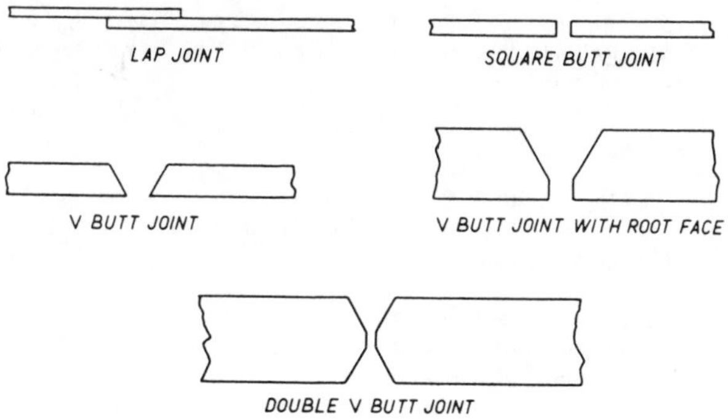

LAP JOINT   SQUARE BUTT JOINT

V BUTT JOINT   V BUTT JOINT WITH ROOT FACE

DOUBLE V BUTT JOINT

the overhead position. Too much heat prevents satisfactory wetting. We will now consider the bronze welding of special metals using a flux of mixture of alkaline fluoride and borax, but bronze filler rods are also supplied flux-coated.

Cast iron

Bevel the edges to a 90°V, round off the sharp edges of the V, and clean the casting well. Pre-heating may be dispensed with unless the casting is of complicated shape, and the welding may often be done without dismantling the work. If pre-heating is necessary it should be heated to

British Standards designations, compositions and recommended usage of filler alloys for bronze welding

Alloy designation		Composition % (by weight) Deleterious impurities, e.g. Al and Pb, are each restricted to 0.3% max.		Recommended for usage on	Approx. melting point (°C)	Applications
BS 1453	BS 1845 (Group CZ)					
C2	CZ6	Cu 57.00 Si 0.20 Zn balance Sn optional	to 63.00 to 0.50 to 0.50 max.	Copper Mild steel	875–895	A silicon-bronze used for copper sheet and tube mild steel and line production applications.
C4	CZ7	Cu 57.00 Si 0.15 Mn 0.05 Fe 0.10 Zn balance Sn optional	to 63.00 to 0.30 to 0.25 to 0.50 to 0.50 max.	Copper Cast iron Wrought iron	870–900	Similar to C2. (CZ6).
C5	CZ8	Cu 45.00 Si 0.15 Ni 8.00 Zn balance Sn optional Mn optional Fe optional	to 53.00 to 0.50 to 11.00 to 0.50 max. to 0.50 max. to 0.50 max.	Mild steel Cast iron Wrought iron	970–980	A nickel-bronze for bronze welding steel and malleable iron, building up worn surfaces and welding Cu–Zn–Ni alloys of similar composition.
C6	——	Cu 41.00 Si 0.20 Ni 14.00 Zn balance Sn optional Mn optional Fe optional	to 45.00 to 0.50 to 16.00 to 1.00 max. to 0.20 max. to 0.30 max.	Cast iron Wrought iron	—— ——	Similar to C5. (CZ8).

450°C, and on completion cooling should be as slow as possible as in the fusion welding of cast iron (Fig. 6.27).

Blowpipe, flame and rod. The blowpipe nozzle may be about two sizes smaller than for the same thickness steel plate, and the flame is adjusted so as to give a slight excess of oxygen. If a second deposit is to be run over the first, the flame is adjusted to a more oxidizing condition still for the subsequent runs, the inner cone being usually only about $\frac{3}{4}$ of its neutral length. The best flame condition can easily be found by trial. Suitable filler

Fig. 6.26. Typical joint designs for bronze welding (tube).

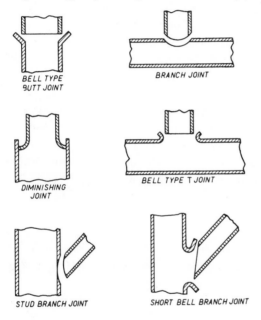

BELL TYPE BUTT JOINT

BRANCH JOINT

DIMINISHING JOINT

BELL TYPE T JOINT

STUD BRANCH JOINT

SHORT BELL BRANCH JOINT

Fig. 6.27. Bronze weld in cast iron.

alloys are given in the table, those containing nickel giving greater strength, the bronze flux being of the borax type.

Technique. The leftward method is used with the rod and blowpipe held as in Fig. 6.28, the inner cone being held well away from the molten metal. The rod is wiped on the edges of the cast iron and the bronze wets the surface. It is sometimes advisable to tilt the work so that the welding is done uphill, as shown in Fig. 6.28. This gives better control. Do not get the work too hot.

Vertical bronze welding of cast iron can be done by the two-operator method, and often results in saving of time, gas and rods and reduces the risk of cracking and distortion.

The edges are prepared with a double 90° V and thoroughly cleaned for 12 mm on each side of the edges. The blowpipes are held at the angle shown in Fig. 6.29a. Blowpipe and rod are given a side to side motion, as indicated in Fig. 6.29b as the weld proceeds upwards, so as to tin the surfaces.

Malleable cast iron

The bronze welding of malleable castings may be stated to be the only way to ensure any degree of success in welding them. Both types (blackheart and whiteheart) can be welded satisfactorily in this way, since the heat of the process does not materially alter the structure. The method is the same as for cast iron, using nickel bronze rods (C5) and a borax-type flux.

Steel

In cases where excessive distortion must be avoided, or where thin sections are to be joined to thick ones, the bronze welding of steel is often

Fig. 6.28. Bronze welding cast iron.

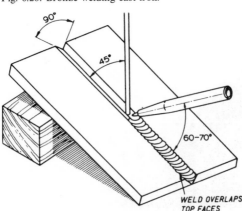

WELD OVERLAPS
TOP FACES

used, the technique being similar to that for cast iron, except, of course, that no pre-heating is necessary.

Galvanized iron

This can be easily bronze welded, and will result in a strong corrosion-resisting joint, with no damage to the zinc coating. If fusion welding is used, the heat of the process would of course burn the zinc (or galvanizing) off the joint and the joint would then not resist corrosion.

Preparation, flame and rod. For galvanized sheet welding, the edges of the joint are tack welded or held in a jig and smeared with a silver–copper flux. Thicker plates and galvanized pipes are bevelled 60–80° and tacked to position them. The smallest possible nozzle should be used (for the sheet thickness) and the flame adjusted to be slightly oxidizing. Suitable filler alloys are given in the table as for steel.

Technique. No side to side motion of the blowpipes is given, the flame being directed on to the rod, so as to avoid overheating the parent sheet. The rod is stroked on the edges of the joint so as to wet them. Excessive flux *must* be washed off with hot water.

Fig. 6.29. (*a*) Two-operator vertical bronze welding of cast iron. (*b*) Showing motion of pipe and rod.

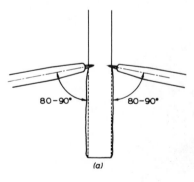

(a)

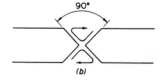

(b)

Copper

Tough pitch copper can be readily bronze welded owing to the much lower temperature of the process compared with fusion welding (Fig. 6.30).

Preparation, flame and rod. Preparation is similar to that for cast iron. Copper tubes can be bell mouthed (Fig. 6.31). Special joints are available for multiple branches. The blowpipe nozzle should be small and will depend on the size of the work or the diameter of the pipe to be welded, and should be chosen so that the bronze flows freely but no overheating occurs. The flame should be slightly oxidizing, and if a second run is made, it should be adjusted slightly to be more oxidizing still (inner cone about $\frac{3}{4}$ of its normal neutral length). Suitable filler alloys are given in the table and are used with a bronze-base flux.

Fig. 6.30. Bronze weld in copper. $\times$ 250.

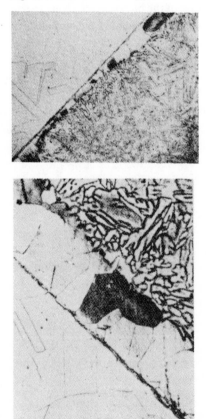

Technique. The method is similar to that for cast iron, and the final difference between the bronze-welded and brazed joint is that the former has the usual wavy appearance of the oxy-acetylene weld, while the latter has a smooth appearance, due to the larger area over which the heat was applied. The bronze joint is, of course, much stronger than the brazed one.

Brasses and zinc-containing bronzes

Since the filler rod now melts at approximately the same temperature as the parent metal, this may now be called fusion welding. When these alloys are heated to melting point, the zinc is oxidized, with copious evolution of fumes of zinc oxide, and it this continued, the weld would be full of bubble holes and weak (Fig. 6.32). This can be prevented by using an oxidizing flame, so as to form a layer of zinc oxide over the molten metal, and thus prevent further formation of zinc oxide and vaporization.

Preparation. The edges of the faces of the joint are cleaned and prepared as usual sheets above 3.2 mm thickness being V'd to 90°. Flux can be applied by making it into a paste, or by dipping the rod into it in the usual manner, or a flux-coated rod can be used.

Flame and rod. Suitable filler alloys are given in the table, while a brass rod is used for brass welding, the colour of the weld then being similar to that of the parent metal. Owing to the greater heat conductivity, a larger size jet is required than for the same thickness of steel plate. The flame is adjusted to be oxidizing, as for bronze welding cast iron, and the exact flame condition is best found by trial as follows. A small test piece of the brass or bronze to be welded is heated with a neutral flame and gives off copious fumes of zinc oxide when molten. The acetylene is now cut down until no more fumes are given off. If any blowholes are seen in the metal on solidifying, the acetylene should be reduced slightly further. The inner cone will now be about half its

Fig. 6.31

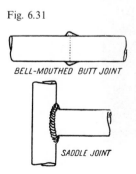

BELL-MOUTHED BUTT JOINT

SADDLE JOINT

Fig. 6.32. (*a*) Unsatisfactory brass weld made with neutral flame,
Unetched × 2.5. (*b*) Brass weld made with insufficient excess of oxygen.
Unetched. (*c*) Correct brass weld made with adequate excess of oxygen.
Unetched. × 2.5.

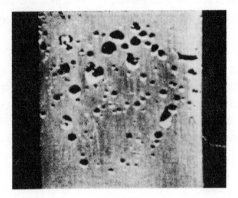

a

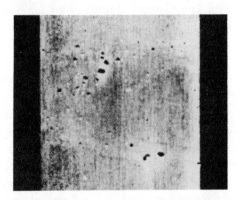

b

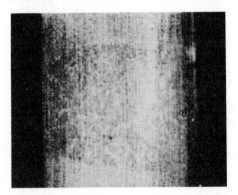

c

normal neutral length. Too much oxygen should be avoided, as it will form a thick layer of zinc oxide over the metal and make the filler rod less fluid.

The weld is formed in the 'as cast' condition, and hammering improves its strength. Where 60/40 brass rods have been used the weld should be hammered while hot, while if 70/30 rods have been used the weld should be hammered cold and finally annealed from dull red heat.

Tin bronze

Tin bronze cannot be welded using an oxidizing flame. Special rods and fluxes are available, however, with which good welds can be made. Urgent repairs may be safely carried out using a *neutral* flame and a silicon–bronze rod with borax-type flux.

Gilding metal

For the weld to be satisfactory on completion, the weld metal must have the same colour as the parent metal. Special rods of various compositions are available, so that the colours will 'match'.

Aluminium bronze

Aluminium bronze can be welded using a filler rod of approximately 90% copper, 10% aluminium (C13, BS 2901 Pt 3), melting point 1040°C, a rod also suitable for welding copper, manganese bronze and alloy steels where resistance to shock, fatigue and sea-water corrosion is required. The aluminium bronze flux (melting point 940°C) can be mixed with water to form a paste if required.

Preparation. The edges of the joint should be thoroughly cleaned by filing or wire brushing to remove the oxide film which is difficult to dissolve.

Up to 4.8 mm thickness no preparation is required – just a butt joint with gap. Above 4.8 mm the usual V preparation is required and a double V above 16 mm thick. Sheets should not be clamped for welding, as this tends to cause the weld to crack, they must be allowed to contract freely on cooling, and it is advisable to weld a seam continuously and not make starts in various places.

Flame and rod. A neutral flame is usually used – any excess of acetylene tends to produce hydrogen with porosity of the weld, while excess oxygen causes oxidation. Flame size should be carefully chosen according to the thickness of the plate – too small a flame causes the weld metal to solidify too quickly while there is a danger of burning through with too large a flame.

The filler rod should be a little thicker (0.8 mm) than the sheet to be welded to avoid overheating, and it should be added quickly to give complete penetration without deep fusion.

Technique. The leftward method is used with a steep blowpipe angle (80°) to start the weld, this being reduced to 60–70° as welding progresses. The parent metal should be well pre-heated prior to starting welding and during welding a large area should be kept hot to avoid cracks. The rod should be used with a scraping motion to clean the molten pool and remove any entrapped gas.

In welding the single-constituent (or α phase) aluminium bronze, i.e. 5–7% Al, 93% Cu, the weld metal should be deposited in a single run or at most two runs to avoid intergranular cracks. Since the metal is hot short in the range 500–700 °C it should cool quickly through the range, and should not be peened. Cold peening is sometimes an advantage. The two-constituent (α and β) or duplex aluminium bronzes contain 10% aluminium. They have a wide application, are not as prone to porosity, and are easier to weld than the 7% Al type, and also their hot short range is smaller.

After treatment. Stresses can be relieved by heat treating at low temperature, and any required heat treatment can be carried out as required after welding.

Copper welding

Tough pitch copper (that containing copper oxide) is difficult to weld, and so much depends on the operator's skill that it is advisable to specify deoxidized copper for all work in which welding is to be used as the method of jointing. Welds made on tough pitch copper often crack along the edge of the weld if they are bent (Fig. 6.33a), showing that the weld is unsound due to the presence of oxide, often along the lines of fusion. A *good* copper weld (Fig. 6.33b), on the other hand, can be bent through 180° without cracking and can be hammered and twisted without breaking. This type of copper weld is strong and sound, free from corrosion effects, and is eminently satisfactory as a method of jointing.

Preparation. The surfaces are thoroughly cleaned and the edges are prepared according to the thickness, as shown in Fig. 6.34. In flanging thin sheet the height of the flange is about twice the plate thickness and the flanges are bent at right angles. Copper has a high coefficient of expansion, and it is necessary therefore to set the plates diverging at the rate of 3–4 mm per 100 mm run, because they come together so much on being welded. Since copper is weak at high temperatures, the weld should be well supported if possible and an asbestos sheet between the weld and the backing strip of steel prevents loss of heat.

Tacking to preserve alignment is not advised owing to the weakness of the copper tacks when hot. When welding long seams, tapered spacing clamps or jigs should be used to ensure correct spacing of the joint, care

Fig. 6.33. (a) Poor copper weld. Crack developed when bent. (b) Oxy-acetylene weld in deoxidized copper. = 100.

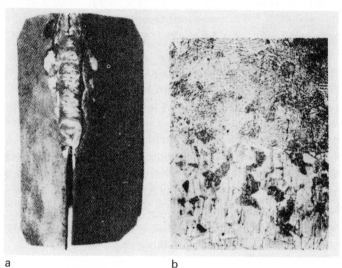

a   b

Fig. 6.34. Preparation of copper plates for welding.

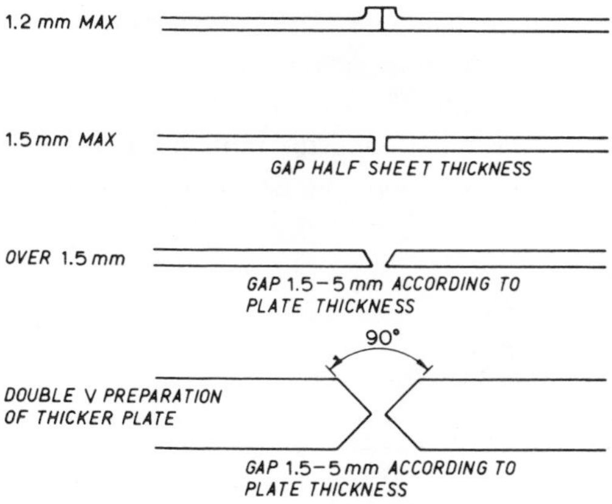

1.2 mm MAX

1.5 mm MAX

GAP HALF SHEET THICKNESS

OVER 1.5 mm

GAP 1.5 – 5 mm ACCORDING TO
PLATE THICKNESS

90°

DOUBLE V PREPARATION
OF THICKER PLATE

GAP 1.5 – 5 mm ACCORDING TO
PLATE THICKNESS

being taken that these do not put sufficient pressure on the edges to indent them when hot. A very satisfactory method of procedure is to place a clamp *C* at the centre of the seam and commence welding at a point say about one-third along the seam.

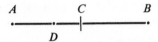

Welding is performed from *D* to *B* and then from *D* to *A*.

Because of the high conductivity of copper it is essential to pre-heat the surface, so as to avoid the heat being taken from the weld too rapidly. If the surface is large or the metal thick, two blowpipes must be used, one being used for pre-heating. When welding pipes they may be flanged or plain butt welded, while T joints can be made as saddles.

Blowpipe, flame and rod. A larger nozzle than for the same thickness of steel should be used and the flame adjusted to be neutral or very slightly carbonizing. Too much oxygen will cause the formation of copper oxide and the weld will be brittle. Too much acetylene will cause steam to form, giving a porous weld, therefore close the acetylene valve until the white feathery plume has almost disappeared. The welding rod should be of the deoxidized type, and many alloy rods, containing deoxidizers and other elements such as silver to increase the fluidity, are now available and give excellent results.

The weld may be made without flux, or a flux of the borax type used. Proprietary fluxes contained additional chemicals greatly help the welding operation and make it easier.

Technique. The blowpipe is held at a fairly steep angle, as shown in Fig. 6.35, to conserve the heat as much as possible. Great care must be taken to keep the tip of the inner cone 6–9 mm away from the molten metal, since the weld is then in an envelope of reducing gases, which prevent oxidation. The weld proceeds in the leftward manner, with a slight sideways motion of the blowpipe. Avoid agitating the molten metal, and do not remove the rod from the flame but keep it in the molten pool. Copper may

Fig. 6.35. Copper welding.

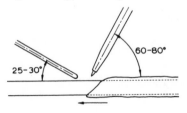

also be welded by the rightward method, which may be used when the filler rod is not particularly fluid. The technique is similar to that for rightward welding of mild steel, with the flame adjusted as for leftward welding of copper.

Welding can also be performed in the vertical position by either single- or double-operator method, the latter giving increased welding speed.

After treatment. Light peening, performed while the weld is still hot, increases the strength of the weld. The effect of cold hammering is to consolidate the metal, but whether or not it should be done depends on the type of weld and in general is not advised. Annealing, if required on small articles, can be carried out by heating to 600–650°C.

Aluminium welding

The welding of aluminium, either pure or alloyed, presents no difficulty (Fig. 6.36) provided the operator understands the problems which must be overcome and the technique employed.

The oxide of aluminium (alumina Al_2O_3), which is always present as a surface film and which is formed when aluminium is heated, has a very high melting point, much higher than that of aluminium, and if it is not removed it would become distributed throughout the weld, resulting in weakness and brittleness. A good flux, melting point 570°C, is necessary to dissolve this oxide and to prevent its formation.

Fig. 6.36. Oxy-acetylene weld in aluminium. = 45.

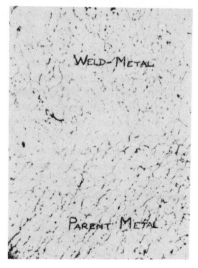

Aluminium and its alloys

Preparation. The work should be cleaned of grease and brushed with a wire brush. Sheets below 1 mm thickness can be turned up at right angles (as for mild steel) and the weld made without a filler rod. Over 3.2 mm thick the edges should have a 90° V and over 6 mm thick a double 90° V. Tubes may be bevelled if thick or simply butted with a gap between them. It is advisable always to support the work with backing strips of asbestos or other material, to prevent collapse when welding. Aluminium, when near its melting point, is extremely weak, and much trouble can be avoided by seeing that no collapsing can occur during the welding operation.

Blowpipe, flame, flux and rod. The flame is adjusted to have a very slight excess of acetylene and then adjusted to neutral, and the rod of pure aluminium or 5% silicon–aluminium alloy should be a little thicker than the section to be welded. A good aluminium flux must be used and should be applied to the rod as a varnish coat, by heating the end of the rod, dipping it in the flux and letting the tuft, which adheres to the rod, run over the surface for about 150 mm of its length. This ensures an even supply. Too much flux is detrimental to the weld.

Technique. The angles of the blowpipe and rod are shown in Fig. 6.37 (a slightly larger angle between the blowpipe and rod than for mild steel), and the welding proceeds in the leftward manner, keeping the inner cone well above the molten pool. The work may be tacked at about 150 mm intervals to preserve alignment, or else due allowance made for the joint coming in as welded, like mild steel. As the weld progresses and the metal becomes

Fig. 6.37. (*a*) Aluminium flat welding. Rightward technique may also be employed using approximately the same angles of blowpipe and rod. (*b*) Aluminium welding by the double-operator method.

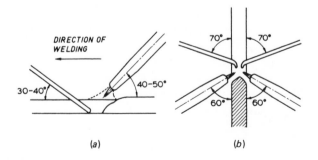

(a)   (b)

hotter, the rate of welding increases, and it is usual to reduce the angle between blowpipe and weld somewhat to prevent melting a hole in the weld. Learners are afraid of applying sufficient heat to the joint as a rule, because they find it difficult to tell exactly when the metal is molten, since it does not change colour and is not very fluid. When they do apply enough heat, owing to the above difficulty, the blowpipe is played on one spot for too long a period and a hole is the result.

If the rightward technique is used the blowpipe angle is 45° and the rod angle 30–40°. Distortion may be reduced when welding sheets, and the flame anneals the deposited metal.

The two-operator vertical method may be employed (as for cast iron) on sheets above 6 mm thickness, the angle of the blowpipes being 50–60° and the rods 70–80°. This method gives a great increase in welding speed (Fig. 6.37)

After treatment. All the corrosive flux must be removed first by washing and scrubbing in hot soapy water. This can be followed by dipping the article in a 5% nitric acid solution followed by a washing again in hot water.

Where it is not possible to get at the parts for scrubbing, such as in tanks, etc., the following method of removal is suitable. Great care, however, should be taken when using the hydrofluoric acid as it is dangerous, and rubber gloves should be worn, together with a face mask.

A solution is made up as follows in a heavy duty polythene container.

> Nitric acid – 100 g to 1 litre water.
> Hydrofluoric acid – 6 g to 1 litre water.

The nitric acid is added to the water first, followed by the hydrofluoric acid.

Articles immersed in this solution for about ten minutes will have all the flux removed from them, and will have a uniformly etched surface. They should then be rinsed in cold water followed by a hot rinse, the time of the latter not exceeding three minutes, otherwise staining may occur.

Hammering of the completed weld greatly improves and consolidates the structure of the weld metal, and increases its strength, since the deposited metal is originally in the 'as cast' condition and is coarse grained and weak. Annealing may also be performed if required.

Aluminium alloy castings and sheets

The process for the welding of castings is very similar to that for the welding of sheet aluminium. See Volume 1 for explanation of alloy coding letters.

Preparation. The work is prepared by V'ing if the section is thicker than 3 mm, and the joint is thoroughly cleaned of grease and impurities. Castings such as aluminium crank-cases are usually greasy and oily (if they have been in service), the oil saturating into any crack or break which may have occurred. If the work is not to be pre-heated, this oil *must* be removed. It may be washed first in petrol, then in a 10% caustic soda solution, and this followed by a 10% nitric acid or sulphuric acid solution. A final washing in hot water should result in a clean casting. In normal cases in which pre-heating is to be done, filing and a wire brush will produce a clean enough joint, since the pre-heating will burn off the remainder.

Aluminium alloys. Recommended filler rods

Alloys	Composition % (remainder aluminium)	Filler rod (old designation in brackets)
Casting alloys BS 1490		
LM2	0.7–2.5 Cu, 9–11.5 Si	4047A, 10–12% Si (NG2)
LM4	2–4 Cu, 4–6 Si	4047A or 4043A, 4.5–6% Si (NG2 or NG21)
LM5	3–6 Mg	5356, 4.6–5.5 Mg (NG6)
LM6	10–13 Si, 0.5 Mn	4043, 415–6.0 Si (NG21)
LM9	10–13 Si, 0.3–0.7 Mn	4043 or 4047 (NG21 or NG2)
LM18	4.5–6 Si	4043 (NG21)
LM20	10–13 Si, 0.4 Cu	4047 (NG2)
Wrought alloys BS 1470–1477		
1080A	99.8 Al	1080A (G1B)
1050A	99.5 Al	1080A (G1C)
3103 (N3)	1–1.5 Mn	3103 (NG3)
6063 (H9)	0.4–0.9 Mg, 0.3–0.7 Mn	4043 or 5056A (NG21 or NG6)
061 (H20)	0.15 C 0.4 Cu, 0.8–1.2 Mg, 0.4–0.8 Si, 0.2–0.8 Mn, 0.15–0.35 Cr	4043 or 5056A (NG21 or NG6)
6082 (H30)	0.5–1.2 Mg, 0.7–1.3 Si, 0.4–1.0 Mn	4043 or 5056A (NG21 or NG6)

If the casting is large or complicated, pre-heating should be done as for cast iron.* In any case it is advantageous to heat the work well with the blowpipe flame before commencing the weld.

---

* Large complicated castings can be pre-heated to about 400 °C, smaller castings to 300–350 °C and small castings to 250–300 °C. No visible change in the appearance of the aluminium occurs at these temperatures.

Blowpipe, flame, rod and flux. The blowpipe is adjusted as for pure aluminium and a similar flame used. The welding rod should preferably be of the same composition as the alloy being welded (see table on p. 290) but for general use a 5% silicon–aluminium rod is very satisfactory. This type of rod has strength, ductility, low shrinkage, and is reasonable fluid. A 10% silicon–aluminium rod is used for high silicon castings, while 5% copper–aluminium rods are used for the alloys containing copper, such as Y alloy, and are very useful in automobile and aircraft industries. The deposit from this type of rod is harder than from the other types.

When welding the Al–Mg alloys the oxide film consists of both aluminium and magnesium oxide making the fluxing more difficult so that as the magnesium proportion increases welding may become more difficult. Alloys containing more than $2\frac{1}{2}$% Mg, e.g. 5154A (N5) and 5183 (N8) are difficult to weld and require considerable experience as do the high strength alloys 6061 (H15) and 6082 (H30). The inert gas arc processes are to be preferred for welding these alloys.

Since there is also a loss of Mg in the welding process note that the filler rod recommended has a greater Mg content than the parent plate. The flux used is similar to that for pure aluminium and its removal must be carried out in the same way.

Technique. The welding is carried out as for aluminium sheet, and the cooling of the casting after welding must be gradual.*

After-treatment. After welding, the metal is in the 'as cast' condition and is weaker than the surrounding areas of parent metal, and the structure of the deposited metal may be improved by hammering. The area near the welded zone, however, is annealed during the welding process and failure thus often tends to occur in the area alongside the weld, and not in the weld itself. In the case of heat-treatable alloys, the welded zone can be given back much of its strength by first lightly hammering the weld itself and then heat-treating the whole of the work.

For this to be quite successful it is essential that the weld should be of the same composition as the parent metal. If oxidation has occurred, however, this will result in a weld metal whose structure will differ from that of the parent metal and the weld will not respond to heat treatment. Since many of this type of alloy are 'hot short', cracking may occur as a result of the welding process.

---

* When repairing cracked castings, any impurities which appear in the molten pool should be floated to the top, using excess flux if necessary.

If sheets are anodized, the welding disturbs the area and changes its appearance. Avoidance of overheating, localizing the heat as much as possible, and hammering, will reduce this disturbance to a minimum, but heat treatment will make the weld most inconspicuous.

Welding of nickel and nickel alloys

The alloys include Monel (nickel 69.4%, copper 29.1%, iron 1.2%, manganese 1.2%, carbon 0.12%) and Inconel (nickel 80%, chromium 20%) and modifications of these compositions to give variations in properties.

Oxy-acetylene welding is used only for welding nickel 200, monel alloys (90/10, 80/20, 70/30), Brightray alloys, Inconel 600, Incoloy DC and 800, and Nimonic 75. The welding of NiLO alloys is not usually performed.

Preparation. Sheets of thinner than 1 mm can be bent up through an angle of about 75°, as shown in Fig. 6.38, and the edges melted together. The ridge formed by welding can then be hammered flat. Sheets thicker than 1 mm are bevelled with the usual 90° V and butted together. For corner welds on sheet less than 1 mm the corners are flanged as shown, while for thicker plate the weld is treated as an open corner joint (see Fig. 6.38). Castings should be treated as for cast iron.

In tube welding, preparation should be an 80° V with no root gap.

Blowpipe, flame, flux and rod, High-purity acetylene as supplied from DA cylinders is required, and in general the blowpipe nozzle size is the same as that for the same thickness mild steel. For nickel 200 a size larger can be used. No flux is required for nickel 200 and for Incoloy DS the use of flux is

Fig. 6.38. Preparation of plates.

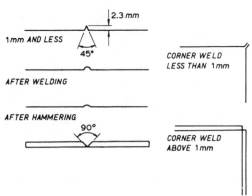

optional and special fluxes are available for the other alloys. Flux containing boron should not be used with alloys containing chromium as it tends to cause hot cracking in the weld. The joint is tacked in position without flux and the filler rod and allowed to dry before welding. Fused flux remaining should be removed by wire brushing and unfused flux with hot water. Flux remaining (after welding the nickel chromium alloys) can be removed by treatment with a solution of equal parts of nitric acid and water for 15–30 minutes followed by washing with water. Flux removal is important when high-temperature applications are involved, as the flux may react with the metal.

Technique. Leftward technique is used. Weaving and puddling of the molten pool should be avoided as agitation of the pool causes porosity due to the absorption of gases by the high-nickel alloys, and the filler rod should be kept within the protective envelope of the flame to prevent oxidation. Keep the flame tip above the pool but for Monel K 500 let it just touch the pool. Nickel 200 melts sluggishly, Inconel 600, Nimonic 75 and the Brightray alloys are more fluid while Monel 400 and Incoloy DS flow easily. A slightly carburizing soft flame (excess acetylene) should be used for the nickel and nickel-copper alloys, whilst for the chromium-containing alloys the flame should be a little more carburizing.

There are no pronounced ripples on the weld surface. They should be smooth without roughness, burning or signs or porosity.

Filler metals and fluxes

Material	Filler metal	Flux
Monel 400	Monel 40	Monel.
Monel K 500	Monel 64	Monel K 500.
Inconel 600		
Incoloy 800		Inconel.
Incoloy DS	NC 80/20	Boron-free.
Brightray alloys		

Note. Use of a flux is optional with Incoloy DS.

When welding tube with 80° feather edge preparation and no root gap, the first run is made with no filler rod, the edges being well fused together to give an even penetration bead, followed by filler runs. If the work is rotatable, welding in the two o'clock position gives good metal control. If the joint is fixed, vertical runs should be made first downwards followed by a run upwards.

Nickel clad steel*

This steel is produced by the hot rolling of pure nickel sheet on to steel plate, the two surfaces uniting to form a permanent bond. It gives a material having the advantage of nickel, but at much less cost than the solid nickel plate. It can be successfully welded by the oxy-acetylene process.

Preparation. The bevelled 90° butt joint is the best type. For fillet welds it is usual to remove (by grinding) the nickel cladding on the one side of the joint, as shown in Fig. 6.39, so as to ensure a good bond of steel to steel.

Technique. The weld is made on the steel side first, using a mild steel rod and the same technique as for mild steel. The nickel side is then welded, using a nickel rod and a slightly reducing flame as for welding monel and nickel. The penetration should be such that the nickel penetrates and welds itself into the steel weld.

Stainless steels

Stainless steels of the martensitic and austenitic class can be welded by the oxy-acetylene process.

Preparation. Thin gauges may be flanged at 90° as for mild steel sheet and the edges fused together. Thicker sections are prepared with the usual 90° V and the surfaces cleaned of all impurities. The coefficient of expansion of the 18/8 austenitic steels is about 50% greater than that of mild steel, and consequently the sheets should be set diverging much more than for mild

Fig. 6.39. Double fillet weld. Nickel clad steel.

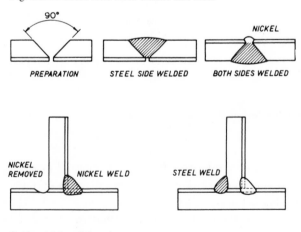

* Also stainless clad steel

steel to allow for them coming together during welding. An alternative method is to tack the weld at both ends first, and then for sheets thinner than 1 mm tack at 25 mm intervals, while for thicker sheets the tacks can be at 50 mm intervals. Cooling clamps are advised for the austenitic steels. The thermal conductivity is less than for mild steel, and thus the heat remains more localized. Unless care is taken, therefore, the beginner tends to penetrate the sheet when welding thin gauges.

Flame. The flame should be very slightly carburizing (excess acetylene), there being the smallest trace of a white plume around the inner cone. The flame should be checked from time to time to make sure that it is in this condition, since any excess oxygen is fatal to a good weld, producing porosity, while too great an excess of acetylene causes a brittle weld.

Rod and flux. The welding rod must be of the same composition as the steel being welded, the two types normally used being the 18/8 class, and one having a higher nickel and chromium content and suitable for welding steels up to a 25/12 chrome–nickel content. Since there is considerable variety in the amounts of the various elements added to stainless steels to improve their physical properties, such as molybdenum and tungsten in addition to titanium or niobium, it is essential that the analysis of the parent metal be known so that a suitable rod can be chosen. The steel makers and electrode makers will always co-operate in this matter. The rod should be of the same gauge or slightly thicker than the sheet being welded. No flux is necessary, but if difficulty is experienced with some steels in penetration, a special flux prepared for these steels should be used, and can be applied as a paste mixed with water.

Technique. Welding is performed in the leftward manner, and the flame is played over a larger area than usual because of the low thermal conductivity. This lessens the risk of melting a hole in the sheet. The tip of the inner cone is kept very close to the surface of the molten pool and the welding performed exactly as for the leftward welding of mild steel. No puddling should be done and the least possible amount of blowpipe motion used.

After treatment. Martensitic stainless steels should be heated to 700 to 800 °C and then left to air cool. The non-stabilized steels require heat treatment after welding if they are to encounter corrosive conditions, otherwise no heat treatment is necessary. The weld decay proof stabilized steels and those with a very low carbon content require no heat treatment after welding.

The original silver-like surface can be restored to stainless steels which have scaled or oxidized due to heat by immersing in a bath made up as follows: sulphuric acid 8%, hydrochloric acid 6%, nitric acid 1%, water 85%. The temperature should be 65°C and the steel should be left in for 10 to 15 minutes, and then immersed for 15 minutes or until clean in a bath of 10% nitric acid, 90% water, at 25°C.

Other baths made up with special proprietary chemicals with trade names will give a brighter surface. These chemicals are obtainable from the makers of the steel.

Stainless iron

The welding of stainless iron results in a brittle region adjacent to the weld. This brittleness cannot be completely removed by heat treatment, and thus the welding of stainless iron cannot be regarded as completely satisfactory. When, however, it has to be welded the welding should be done as for stainless steel, using a rod specially produced for this type of iron.

Hard surfacing and stelliting

Surfaces of intense hardness can be applied to steel, steel alloys, cast iron or monel metal, by means of the oxy-acetylene flame. The surfaces are hard and have intense resistance to wear and corrosion. By depositing a surface on a more ductile metal we have an excellent combination. Metals easily hardfaced are low- and medium-carbon steels, steels above 0.4% C, low-alloy steels, nickel, nickel–copper alloys, chrome-nickel and nobium bearing stainless steels (not titanium bearing nor free machining). Those metals faced with difficulty include cast iron, straight chromium stainless steels, tool and die steels and water-, air- and oil-hardening steels while copper, brass and other low melting point alloys are not suitable for hard surfacing. Hard surfaces may be deposited either on new parts so that they will have increased resistance to wear at reduced cost, or on old parts which may be worn, thus renewing their usefulness. In addition, non-ferrous surfaces, such as bronze or stellite, may be deposited and are described under their respective headings.

The surfaces deposited may be hard and/or wear and corrosion resisting. It is usual in the case of hard surfaces for the metal to be machinable as deposited, but to be capable of being hardened by quenching or by work hardening, e.g. 12/14% Mn (see table of rods available on p. 309).

We may divide the methods as follows:

(1) Building up worn parts with a deposit similar to that of the parent metal. This process is very widely used, being cheap and economical, and is used for building up, for example, gear wheels, shafts, key splines, etc.

The technique employed is similar to that for mild steel, using a neutral flame and the leftward method. The deposit should be laid a little at a time and the usual precautions taken for expansion and distortion. Large parts should be pre-heated and allowed to cool out very slowly, and there are no difficulties in the application.

(2) Building up surfaces with rods containing, for example, carbon, manganese, chromium or silicon to give surfaces which have the required degree of hardness or resistance to wear and corrosion. These surfaces differ in composition from that of the parent metal, and as a result the fusion method of depositing cannot be used, since the surface deposit would become alloyed with the base metal and its hardness or wear-resisting properties would be thus greatly reduced.

(3) Hard facing with tungsten carbide. The rods consist of a steel tube containing fused granules of tungsten carbide HV 1800 in a matrix of chromium iron HV 850. Various grades are available with granules of different mesh size. Large granules embedded in the steel surface do not round off as wear takes place, but chip because they are brittle and maintain good serrated cutting edges. Finer mesh granules give a more regular cutting edge so that the mesh of the granules is determined by the working conditions.

Technique. The flame is adjusted to have excess acetylene with the white plume from $2\frac{1}{2}$ to 3 times the length of the inner cone (Fig. 6.40a). As the heated surface absorbs carbon, its melting point is reduced and the surface sweats. The rod is melted on to this sweating surface and a sound bond is

Fig. 6.40. (a) Hard surfacing flame making the surface sweat.

EXCESS ACETYLENE FEATHER / INNER CONE

FEATHER = $2\frac{1}{2}$–3× (LENGTH OF INNER CONE)

made between deposit and parent metal with the minimum amount of alloying taking place.

This type of deposit, used for its wear-resisting properties, is usually very tough and is practically unmachinable. The surface is therefore usually ground to shape, but in many applications, such as in reinforcing tramway and rail crossings, it is convenient and suitable to hammer the deposit to shape while hot, and thus the deposit requires a minimum amount of grinding. The hammering, in addition, improves the structure.

In the case of the high-carbon deposits, they can be machined or ground to shape and afterwards heat treated to the requisite degree of hardness.

When using tungsten carbide rods the cone of the flame should be played on the rod and pool to allow gases to escape, preventing porosity. Weaving is necessary to give an even distribution of carbide granules in the matrix. Second runs should be applied with the same technique as the first, with no puddling of the first run since this would give dilution of the hard surface with the parent plate.

Another method frequently used to build up a hard surface is to deposit a surface of cast iron in the normal way, using a silicon cast iron rod and flux. Immediately the required depth of deposit has been built up, the part is quenched in oil or water depending on the hardness required.

This results in a hard deposit of white cast iron which can be ground to shape, and which possesses excellent wear-resisting properties. This method is suitable only for parts of relatively simple shape, that will not distort on quenching, such as camshafts, shackle pins, pump parts, etc.

The use of carbon and copper fences in building up and resurfacing results in the deposit being built much nearer to the required shape, reducing time in welding and finishing and also saving material.

When hard surfacing cast iron, the surface will not sweat. In this case the deposit is first laid as a fusion deposit, with a neutral flame, and a second layer is then 'sweated' on to this first layer. In this way the second layer is obtained practically free from any contamination of the base metal.

It will be seen, therefore, that the actual composition of the deposit will depend entirely upon the conditions under which it is required to operate. A table of alloy steel rods and uses is given at the end of this chapter.

Stellite

Stellite alloys of the cobalt–chrome type with carbon were first developed in 1900 in the US and were the first wear-resistant cobalt-based alloys to be used. Later tungsten and/or molybdenum were added. They are hard and have a great resistance to wear and corrosion and there are now about 20 stellite alloys available, covering a wide field of resistant surfaces. They can be divided into:

(1) Co–Cr–W–C,

(2) Co–Cr–W/Mo–Ni/Fe–C.

The latter alloys of group (2) have been developed for high-temperature, high-impact wear operations.

The carbon content of the alloys vary from 0.1% to 3.0% and over half of them have a carbon content above 1%. Those with less carbon have a combined carbon plus boron content greater than 1%. (The cobalt-based alloys are discussed in Chapter 1 on hard facing by MMA.

Stellite alloys available for use with the oxy-acetylene process with the hardness Rockwell on the C scale and VPH are:

> No. 1 53 R, 590 VPH; No. 4 47 R 480 VPH; No. 6 42 R, 420 VPH; No. 12 48 R, 500 VPH; No. 20 55 R, 620 VPH; No. 190 52 R, 580 VPH; No. 2006 44 R, 440 VPH; No. 2012 49 R, 520 VPH.

In general, wear resistance increases as the hardness increases, owing to the dispersal of the hard carbides in the somewhat softer matrix. Stellite alloys preserve their hardness at elevated temperatures dependent upon the W and/or the Mo content. Bare tube rods of granulated tungsten carbide or of a 60/40 percentage of tungsten carbide with either Stellite No. 6 or Hastealloy alloy B are steel tubes filled with the tungsten carbide or the 60/40 blend.

Tips of stellite can be brazed on to lathe and cutting tools of all types, giving an excellent cutting edge on a less brittle shank. Stellite, however, can be deposited directly on to surfaces, and in this form it is used for all types of duty, such as surfaces on shafts which have to stand up to great wear and corrosion, lathe centres, drill tips, etc.

Stelliting steel

Preparation. Scale, dirt and impurities are thoroughly removed and the parent metal is pre-heated. Pre-heating and slow cooling are essential to avoid cracking.

Small jets of water, playing on each side of the weld, can be used to limit the flow of heat and reduce distortion, when building up deposits on hardened parts such as camshafts.

Flame and rod. A flame with an excess of acetylene is used, the white plume being about $2\frac{1}{2}$–3 times the length of the inner blue cone. Too little acetylene will cause the stellite to foam and bubble, giving rise to blowholes, while too much acetylene will cause carbon to be deposited around the molten metal. The tip should be one size larger than that for the same thickness steel plate, but the pressure should be reduced, giving a softer flame. For small parts 5 mm diameter stellite rods can be used, while thicker rods are used for larger surfaces. Too much heat prevents a sound deposit being obtained, because some of the base metal may melt and mix with the molten stellite, thus modifying its structure.

Technique. The flame is directed on to the part to be surfaced, but the inner cone should not touch the work, both blowpipe and rod making an angle of 45–60° with the plate (Fig. 6.40*b* and *c*). When the steel begins to sweat the stelliting rod is brought into the flame and a drop of stellite melted on to the sweating surface of the base metal, and it will spread evenly and make an excellent bond with the base metal. The surfacing is continued in this way.

After treatment. The part should be allowed to cool out very slowly to prevent cracks developing.

Fig. 6.40. (*b*) and (*c*) Alternative methods of hard facing.

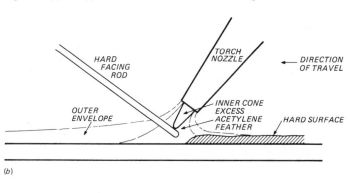

(b)

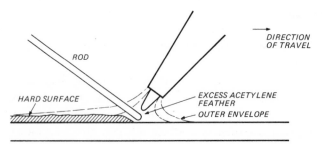

(c)

Stelliting cast iron

Preparation. Clean the casting thoroughly of oil and grease and pre-heat to a dull red heat.

Technique. Using the same type rod and flame as for stelliting steel, it is advisable first to lay down a thin layer and then build up a second layer on this. The reason for this is that more of the surface of the cast iron is melted than in the case of steel, and as a result the first layer of stellite will be diffused with impurities from the cast iron.

The flame is then played on the cast iron and the rod used to push away any scale. A drop of stellite is then flowed on to the surface, and the flame kept a little ahead of the molten pool so as to heat the cast iron to the right temperature before the stellite is run on. Cast iron flux may be used to flux the oxide and produce a better bond.

Heat treatment for depositing stellite

(*a*) Small components of mild steel and steel up to 0.4% carbon. Pre-heat with the torch the area to be faced; hard face the area and cool away from draughts.

(*b*) Large components of mild steel and steel up to 0.4% carbon; small components of high-carbon and low-alloy steels. Pre-heat to 250–350 °C. Hard face whilst keeping at this heat with auxiliary heating flame. Cover and bring to an even heat with flame and cool slowly in dry kieselguhr, mica, slaked lime, ashes or sand.

(*c*) Large components of high-carbon or low-alloy steel; cast iron components; bulky components of mild steel with large areas of stellite facing. Pre-heat to 400–500 °C. Hard face whilst at this temperature. Bring to a dull red even heat and cool slowly as for (*b*).

(*d*) Air-hardening steel (not stainless). When a large area of deposit is required these steels should be avoided. Otherwise pre-heat to 600–650 °C and deposit hard surface whilst at this heat. Then place in a furnace at 650 °C for 30 minutes and cool large components in the furnace and small ones as (*b*).

(*e*) 18/8 austenitic stainless steel non-hardening welding type. Pre-heat to 600–650 °C. Hard surface whilst at this temperature. Bring to an even temperature and cool out as (*b*).

(*f*) 12–14% manganese austenic steel. Use arc process.

Spray fuse process for depositing stellite

In this process a stellite powder (HV 425–750) or nickel powder (HV 375–750) is sprayed on to the part to be hardened and this layer is then fused on with an oxy-acetylene flame. In this way thin layers up to 2 mm thick can be deposited having little dilution with the parent metal. The alloys are self-fluxing, the deposit has a fine structure, and any inclusions are well distributed. Surfaces having sharp corners and sudden change of section should be avoided and the areas should be rough turned and then shot blasted to obtain a rough surface. The hardfacing powder, cobalt based, nickel based or tungsten carbide, is applied with a gas spray gun using aspirating gas pressure to mix the powder and to disperse it evenly through the flame on to the pre-heated surface to be treated. The particles are in the form of a semi-molten spray when they hit the surface. Jobs should be pre-heated and the gun held about 150 mm from the surface, and after deposition the deposit is porous like other cold-sprayed deposits. The next operation is to fuse the deposit into a sound, wear-resistant coating securely bonded to the parent metal. Because of stresses remaining in the deposit, fusing should be carried out immediately after spraying. It is done with an oxy-acetylene torch with a multi-jet nozzle, the part being first raised to about 350°C. One area is selected to begin fusing and the temperature raised to 700–800°C over a small area and part of this is then raised to 1100°C when glazing of the surface begins, indicating that fusing is taking place. The torch is moved over the area until it is all fused and the whole part finally brought up to an even temperature, cooling being carried out by covering it with heat-insulating substance or heat treatment being given, if required for the parent plate.

With this method of application localized heat is reduced, so that distortion is minimized and thin deposits can be applied with little dilution. Shrinkage amounting to 25% takes place during fusing so this must be allowed for, together with grinding tolerance if required, when calculating the thickness of the sprayed coat.

In the manual torch powder method, the gun and torch are combined in one unit and deposition and fusing are carried out in one operation. The incoming gas is used for mixing and carrying the powder to the flame. The usual method of deposit is by using a carburizing flame with a feather about $2\frac{1}{2}$–3 times the length of the inner cone. A dilution of 1–5% occurs and the thickness of deposit is governed by the rate of flow of powder and movement of the torch (Fig. 6.41a and b).

Typical applications are: dies, cast iron parts, cement dies, fan blades, feed screws, hammer mill hammers, knives, cams and other metal-to-metal parts, plough shares, pump parts, bearing surfaces of steel shafts, etc. See also hard facing by plasma spray, plasma transferred arc, Mechanized TIG, MIG and MMA methods.

Brazing (see also appendix 7)

Brazing may be defined (BS 499, Part 1, 1965) as 'a process of joining metals in which, during or after heating, molten filler metal is drawn by capillary attraction into the space between closely adjacent surfaces of the parts to be joined'. In general the melting point of the filler metal is above 500 °C but always below the melting temperature of the parent metal.

Since capillarity and hence surface tension are involved in the process it may be convenient to give a brief explanation of some of the principles involved.

The student should refer to BS 1845 which lists the chemical compositions and approximate melting ranges of filler metals grouped under the following headings: aluminium brazing alloys, silver brazing alloys, copper–phosphorus brazing alloys, copper brazing alloys, brazing brasses, nickel-base brazing alloys, palladium bearing brazing alloys, and gold bearing brazing alloys.

Fig. 6.41. (*a*) Manual torch operation.

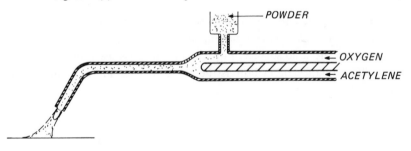

Fig. 6.41. (*b*) Manual power torch–stellite.

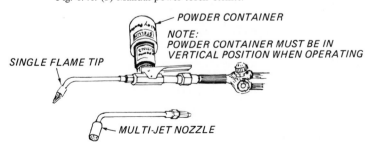

Surface tension

If drops of mercury rest on a level plate it will be noticed that the smaller the drop the more nearly spherical it is in shape, and if any drop is deformed it always returns to its original shape. If the only force acting on any drop were that due to its own weight, the mercury would spread out over the plate to bring its centre of gravity (the point at which the whole weight of the drop may be conceived to be concentrated) to the lowest point so that to keep the shape of the drop other forces must be present. As the drop gets smaller the force due to its own weight decreases and these other forces act so as to make the drop more spherical, that is to take up a shape which has the smallest surface area for a given volume. Other examples of these forces, termed surface tension, are the floating of a dry needle on the surface of water, soap bubbles and water dripping from a tap. In the first example the small dry needle must be laid carefully horizontally on the surface. If it is pushed slightly below the surface it will sink because of its greater density. Evidently the surface of the water exhibits a force (surface tension) which will sustain the weight of the needle. If a wire framework $ABCD$, with CD, length x, able to slide along BC and AD, holds a soap film, the film tends to contract, and to prevent this a force F must be applied at right angles to CD (Fig. 6.42). The surface tension is defined as the force per unit length S on a line drawn in the film surface, and since there are two surfaces to the film $F = 2Sx$.

Angle of contact(θ). The angle of contact between a liquid and a solid may be defined as the angle between the tangent to the liquid surface at the point of contact and the solid surface. For mercury on glass the contact angle is about 140°, while for other liquids the angle is acute and may approach zero (Fig. 6.43).

Fig. 6.42

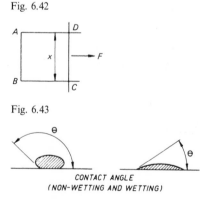

Fig. 6.43

CONTACT ANGLE
(NON-WETTING AND WETTING)

Wetting. If the contact angle approaches zero the liquid spreads and wets the surface and may do so in an upward direction. If the solid and liquid are such that the forces of attraction experienced by the molecules towards the interior of the liquid are less than the forces of attraction towards the solid, the area of contact will increase and the liquid spreads.

Capillarity. If a narrow bore (capillary) tube with open ends is placed vertically in a liquid which will wet the surface of the tube, the liquid rises in the tube and the narrower the bore of the tube the greater the rise. The wall thickness of the tube does not affect the rise and a similar rise takes place if the tube is replaced by two plates mounted vertically and held close together. If the tube or the plates are held out of the vertical the effect is similar and the vertical rise the same. If the liquid does not wet the tube (e.g. mercury) a depression occurs, and the shape of the liquid surface (the meniscus) is shown. The rise is due to the spreading or wetting action already considered – the liquid rises until the vertical upward force due to surface tension acting all round the contact surface with the tube is equal to the vertical downward force due to the weight of the column of liquid (Fig. 6.44).

This wetting action and capillary attraction are involved in the brazing process. The flux, which melts at a lower temperature than the brazing alloys, wets the surface to be brazed, removes the oxide film and gives clean surfaces which promote wetting by a reduction of the contact angle between the molten filler alloy and the parent plate at the joint. The molten filler alloy flows into the narrow space or joint between the surfaces by capillary attraction and the narrower the joint the further will be the capillary flow. Similarly solder flows into the narrow space between tube and union when 'capillary fittings' are used in copper pipe work.

Brazing can be performed on many metals including copper, steel and aluminium, and in all cases cleanliness and freedom from grease is essential. The filler alloy used for aluminium brazing has already been mentioned in the section on oxy-acetylene welding. In the case of the nickel alloys, time and temperature are important. For example copper alloys mix readily with nickel 200 or monel 400 and can pick up sufficient nickel to raise the

Fig. 6.44

CAPILLARY ELEVATION

CAPILLARY DEPRESSION

melting point and hinder the flow of the filler metal. Also chromium and aluminium form refractory oxides which make brazing difficult, so that the use of a flux is necessary. A wide range of brazing alloys are available having a variety of melting points. Fig. 6.45 shows a modern brazing torch using LPG. See appendix for table of general purpose silver brazing alloys and fluxes.

Aluminium brazing

The fusion welding of fillet and lap joints in thin aluminium sections presents considerable difficulty owing to the way in which the edges melt back. In addition the corrosive fluoride flux is very liable to be entrapped between contracting metal surfaces so it is advantageous to modify the design wherever possible so as to include butt joints instead of fillet and lap.

In many cases, however, corner joints are unavoidable and in these cases flame brazing overcomes the difficulty. It can be done more quickly and cheaply than fusion welding, less skill is required and the finished joint is neat and strong.

Fig. 6.45. LP gas brazing torch for brazing, hard and soft soldering. Propane gas recommended, cylinder pressure 7–8 bar. Standard torch pressure 4 bar. Use smaller nozzles for smaller flames. Temperatures up to 950 °C, piezo-electric ignition.

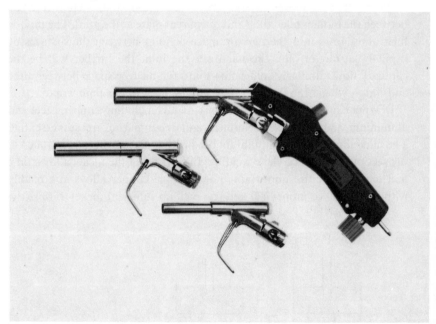

Aluminium brazing is suitable for pure aluminium and for alloys such as LM4, LM18 and for the aluminium–manganese and aluminium–manganese–magnesium alloys, as long as the magnesium content is not greater than 2%.

Preparation. It is always advisable to allow clearance at the joints since the weld metal is less fluid as it diffuses between adjacent surfaces. Clearance joints enable full penetration of both flux and filler rod to be obtained.

Surface oxide should be removed by wire brush or file and grease impurities by cleaning or degreasing. Burrs such as result from sawing or shearing and other irregularities should be removed so that the filler metal will run easily across the surface. Socket joints should have a 45° belling or chamfer at the mouth to allow a lead in for the flux and metal, and to prevent possibility of cracking on cooling, the sections of the surfaces to be jointed should be reduced to approximately the same thickness where possible.

Blowpipe, flame, flux and rod. The blowpipe and rod are held at the normal angle for leftward welding, and a nozzle giving a consumption of about 700 litres of both oxygen and acetylene each per hour is used. The flame is adjusted to the excess acetylene condition with the white acetylene plume approximately $1\frac{1}{2}$ to 2 times the length of the inner blue cone. The rod can be of 10–13% silicon–aluminium alloy melting in the range 565–595 °C (compared with 659 °C for pure aluminium). This is suitable for welding and brazing the high silicon aluminium alloys and for general aluminium brazing. Another type of rod containing 10–13% Si and 2–5% Cu is also used for pure aluminium and aluminium alloys except those with 5% silicon, or with more than 2% magnesium, but is not so suitable due to its copper content if corrosive conditions are to be encountered, but it has the advantage of being heat-treatable after brazing, giving greater mechanical strength.

BS 1845, Filler metals for aluminium brazing (Group AL)

Type	Major alloying elements only (%)				Melting range (°C)	
	Silicon	Copper	Iron	Aluminium	Solidus	Liquidus
AL1	10–13	2–5	0.6	Rem.	535	595
AL2	10–13	0–0.1	0.6	Rem.	565	595
AL3	7–8	0–0.1	—	Rem.	565	610
AL4	4.5–6.0	0–0.1	—	Rem.	565	630

In general, ordinary finely divided aluminium welding flux prepared commercially is quite suitable for brazing, but there is also available a brazing flux of similar composition which has a lower melting point.

Technique. The flame is held well away from the joint to be brazed and pre-heating is done with the outer envelope of the flame for about three minutes – this procedure ensures an even temperature, which is essential so that first the flux flows evenly into the joint followed by the filler metal which must displace *all* the flux, otherwise if islands of flux are entrapped corrosion of the joint will occur.

The rod is warmed, dipped in the flux and the tuft which adheres to the rod is touched down on the heated joint. When the correct temperature has been reached the flux will melt and flow over and between the surfaces smoothly and easily. The blowpipe is now lowered to the normal welding position, some rod is melted on to the joint and the blow-pipe moved forward along the seam running the filler rod into the joint. The blowpipe is then raised and brought back a little and lowered again, the above operation being repeated – the blowpipe thus describes an elliptical motion – each operator modifying the technique according to his individual style (Fig. 6.46).

Flux should only be added when the filler rod does not appear to be running freely. Too much flux is detrimental to the finished joint and great care should be taken that the filler rod flows freely into the joint, so as to displace all the flux.

After treatment. The corrosive flux should be removed by the treatment as given for aluminium welding.

Fig. 6.46. Aluminium brazing.

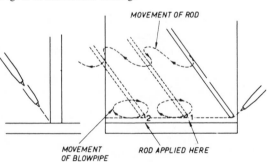

General precautions

The following general precautions should be taken in welding:
(1) Always use goggles of proved design when welding or cutting. The intense light of the flame is harmful to the eyes and in addition small particles of metal are continually flying about and may cause serious damage if they lodge in the eyes. Welding filters or glasses are graded according to BS 679 by numbers followed by the letters GW or GWF. The former are for welding operations without a flux and the latter with flux because there is an additional amount

Table giving some of the chief types of oxy-acetylene welding rods available for welding carbon and alloy steels.

Rod	Description and use
Low-carbon steel	A general purpose rod for mild steel. Easily filed and machined. Deposit can be case-hardened.
High-tensile steel	Gives a machinable deposit of greater tensile strength than the previous rod. Can be used instead of the above wherever greater strength is required.
High-carbon steel	Gives a deposit which is machinable as deposited, but which can be heat treated to give a hard abrasion-resisting surface. When used for welding broken parts, these should be of high carbon steel, and heat treatment given after welding.
High-nickel steel ($3\frac{1}{2}$–4%)	Produces a machinable deposit with good wear-resisting properties. Suitable for building up teeth in gear and chain wheels, splines and keyways in shafts, etc.
Wear-resisting steel (12–14% manganese)	Gives a dense tough unmachinable deposit which must be ground or forged into shape, and can be heat treated. Useful for building up worn sliding surfaces, cam profiles, teeth on excavators, tracks, etc.
Stellite	For hard surfacing and wear- and abrasion-resisting surfaces.
Chrome molybdenum steel (creep resisting)	High-tensile alloy steel deposit for pressure vessels and high-pressure steam pipes, etc. Rod should match the analysis of the parent metal.
Chrome-vanadium steel	A high-tensile alloy rod for very highly stressed parts.
Tool steel	Suitable for making cutting tools by tipping the ends of mild or low-carbon steel shanks.
Stainless steel	Decay-proof. Rod should match the analysis of the parent metal.

of glare. The grades range from 3/GW and 3/GWF to 6/GW and 6/GWF, the lightest shade having the lowest grade number. For aluminium welding and light oxy-acetylene cutting, 3/GW or 3/GWF is recommended, and for general welding of steel and heavier welding in copper, nickel and bronze, 5/GW or 5/GWF is recommended. A full list of recommendations is given in the BS.

(2) When welding galvanized articles the operator should be in a well-ventilated position and if welding is to be performed for any length of time a respirator should be used. (In cases of sickness caused by zinc fumes, as in welding galvanized articles or brass, milk should be drunk.)

(3) In heavy duty welding or cutting and in overhead welding, asbestos or leather gauntlet gloves, ankle and feet spats and protective clothing should be worn to prevent burns. When working inside closed vessels such as boilers or tanks, take every precaution to ensure a good supply of fresh air.

(4) In welding or cutting tanks which have contained inflammable liquids or gases, precautions must be taken to prevent danger of explosion. One method for tanks which have contained volatile liquids and gases is to pass steam through the tank for some hours according to its size. Any liquid remaining will be vaporized by the heat of the steam and the vapours removed by displacement.

Tanks should never be merely swilled out with water and then welded; many fatal explosions have occurred as a result of this method of preparation. Carbon dioxide in the compressed form can be used to displace the vapours and thus fill the tank, and is quite satisfactory but is not always available. Tanks which have contained heavier types of oil, such as fuel oil, tar, etc., present a more difficult problem since air and steam will not vaporize them. One method is to fill the tank with water, letting the water overflow for some time. The tank should then be closed and turned until the fracture is on top. The water level should be adjusted (by letting a little water out if necessary) until it is just below the fracture. Welding can then be done without fear as long as the level of the water does not drop much more than a fraction of an inch below the level of the fracture.

The welder should study the Department of Employment memorandum on *Safety measures for the use of oxy-acetylene equipment in factories* (Form 1704). Toxic gases and fine airborne particles can provide a hazard to a welder's health. The Threshold Limit Value (TLV) is a system by which concentrations of these are classified, and is explained in the Department of Employment Technical data notes No. 2, *Threshold limit values.*

Relevant publications. Safe working, maintenance and repair of gas cylinder and pipeline regulators used with compressed gases for welding, cutting and related processes. (BCGA CP1).

Safe working, maintenance and repair of hand held oxygen and fuel-gas blowpipes used for welding, cutting and related processes. (BCGA CP2).

Technical Information sheets; TIS 1 *Pressure gauges*, TIS2 *Hoses.*

All the above together with British Acetylene Association publications may be obtained from the British Compressed Gases Association, 93 Albert Embankment, London SE1 7TU.

In the 'Health and Safety at Work' series *Noise and the worker No. 25* and *Welding and flame-cutting using compressed gases No. 50* are obtainable from Her Majesty's Stationery Office.

7

Cutting processes

Gas cutting of iron and steel

Iron and steel can be cut by the oxy-hydrogen, oxy-propane, oxy-natural gas and oxy-acetylene cutting blowpipes with ease, speed and a cleanness of cut.

Principle of cutting operation

There are two operations in gas cutting. A heating flame is directed on the metal to be cut and raises it to bright red heat or ignition point. Then a stream of high-pressure oxygen is directed on to the hot metal. The iron is immediately oxidized to magnetic oxide of iron (Fe_3O_4) and, since the melting point of this oxide is well below that of the iron, it is melted immediately and blown away by the oxygen stream.

It will be noted that the metal is cut entirely by the exothermic chemical action and the iron or steel itself is not melted. Because of the rapid rate at which the oxide is produced, melted and blown away, the conduction of the metal is not sufficiently high to conduct the heat away too rapidly and prevent the edge of the cut from being kept at ignition point.

The heat to keep the cut going once it has started is provided partly by the heating jet, and partly by the heat of the chemical action.

The cutting torch or blowpipe (Fig. 7.1)

Cutting blowpipes may be either high or low pressure. The high pressure pipe, using cylinder acetylene or propane as the fuel gas,* can have the mixer in the head (Fig. 7.2), or in the shank, while the low pressure pipe with injector mixing can be used with natural gas at low pressure.

---

* Proprietary gases are available and contain mixtures of some of the following: methyl acetylene, propadiene, propylene, butane, butadiene, ethane, methane, diethyl ether, dimethyl ether, etc.

In the high pressure pipe, fuel gas and heating oxygen are mixed in the head (Fig. 7.2), and emerge from annular slots for propane or holes for acetylene (Fig. 7.4).

The cutting oxygen is controlled by a spring loaded lever, pressure on which releases the stream of cutting oxygen which emerges from the central

Fig. 7.1. Cutter.

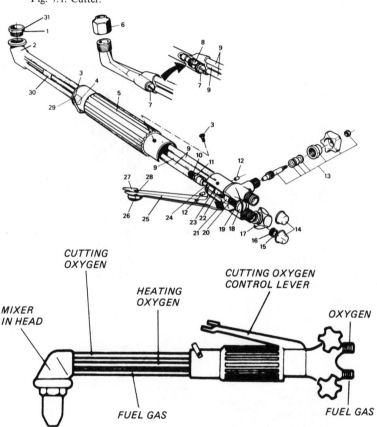

Key:

1. Nozzle nut.	10. Push rod.	22. Plunger.
2. Head 90°.	11. Lock nut.	23. O ring.
Head 75°.	12. Pivot pin.	24. Lever pin.
Head 180°.	13. Control valve fuel gas.	25. Lever.
3. Screw.	14. Red cap.	26. Button.
4. Bracket latch.	15. Spring clip.	Grub screw.
5. Handle.	16. Filter.	27. Lever latch.
6. Nozzle nut.	17. Control valve oxygen.	28. Spring washer.
7. Injector cap.	18. Rear cap.	29. Forward tube.
8. Injector assembly.	19. Cap washer.	30. Tube support.
9. Tube.	20. Valve spring disc.	31. Nut.
9a. Tube.	21. Rear valve.	

orifice the diameter of which increases as thickness of plate to be cut increases. An oxygen gauge giving a higher outlet pressure (up to 6.5 bar) than for welding is required, but the gauge for cylinder acetylene can be the same with pressures up to 0.28 bar. Propane pressure is usually up to 0.6 bar and if natural gas is used no regulator is required but a non-return valve should be fitted in the supply line to prevent flash back.

The size of blowpipe used depends upon whether it is for light duty or

Fig. 7.2

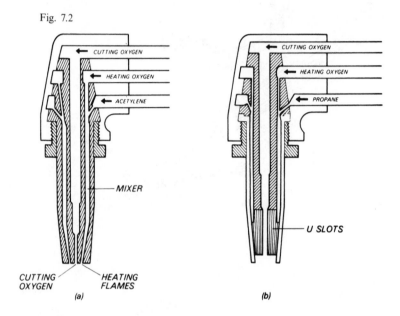

(a)   (b)

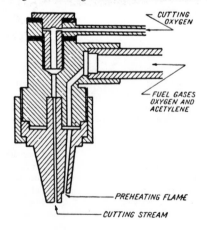

Fig. 7.3. Cutting head for thin steel sheet.

heavy continuous cutting and the volume of oxygen used is much greater than of fuel gas (measured in litres per hour, l/h).

Fig. 7.3 shows a stepped nozzle used for cutting steel sheet up to 4 mm thick and this type is also available with head mixing.

The size of torch varies with the thickness of work it is required to cut and whether it is for light duty, or heavy, continuous cutting.

Adjustment of flame

Oxy-hydrogen and Oxy-propane. The correctly adjusted pre-heating flame is a small non-luminous central cone with a pale blue envelope.

Oxy-natural gas. This is adjusted until the luminous inner cone assumes a clear, definite shape, that may be up to 8–10 mm in length for heavy cuts.

Oxy-acetylene. This flame is adjusted until there is a circular short blue luminous cone, if the nozzle is of the concentric ring type, or until there is a series of short, blue, luminous cones (similar to the neutral welding flame), if it is of the multi-hole type (Fig. 7.5*a* and *b*). The effect of too much oxygen is indicated in Fig. 7.5*c*.

Fig. 7.4

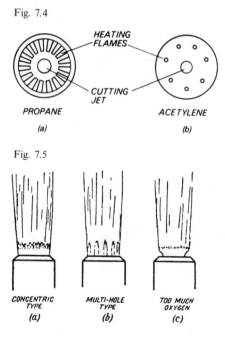

HEATING
FLAMES

CUTTING
JET

PROPANE

(a)

ACETYLENE

(b)

Fig. 7.5

CONCENTRIC
TYPE

(a)

MULTI-HOLE
TYPE

(b)

TOO MUCH
OXYGEN

(c)

It may be observed that when the cutting valve is released the flame may show a white feather, denoting excess acetylene. This is due to the slightly decreased pressure of the oxygen to the heating jet when the cutting oxygen is released. The flame should be adjusted in this case so that it is neutral when the cutting oxygen is released.

Care should be taken to see that the cutter nozzle is the correct size for the thickness to be cut and that the oxygen pressure is correct (the nozzle sizes and oxygen pressures vary according to the type of blowpipe used).

The nozzle should be cleaned regularly, since it becomes clogged with metallic particles during use. In the case of the concentric type of burner, the outer ring should be of even width all round, otherwise it will produce an irregular-shaped inner cone, detrimental to good cutting (Fig. 7.6).

Technique of cutting

The surface of the metal to be cut should be free of grease and oil, and the heating flame held above the edge of the metal to be cut, farthest from the operator, with the nozzle at right angles to the plate. The distance of the nozzle from the plate depends on the thickness of the metal to be cut, varying from 3 to 5 mm for metal up to 50 mm thick, up to 6 mm for metal 50 to 150 mm thick. Since the oxide must be removed quickly to ensure a good, clean cut, it is always preferable to begin on the edge of the metal.

The metal is brought to white heat and then the cutting valve is released, and the cut is then proceeded with at a steady rate. If the cutter is moved along too quickly, the edge loses its heat and the cut is halted. In this case, the cutter should be returned to the edge of the cut, the heating flame applied and the cut restarted in the usual manner. Round bars are best nicked with a chisel, as this makes the starting of the cut much easier. Rivet heads can be cut off flush by the use of a special type nozzle while if galvanized plates are to be cut for any length of time, a respirator is advisable, owing to the poisonous nature of the fumes.

Fig. 7.6

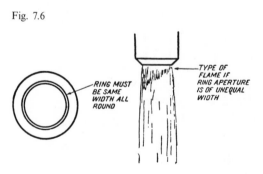

To cut a girder, for example, the cut may be commenced at *A* and taken to *B* (Fig. 7.7). Then commenced at *C* and taken to *B*, that is, the flange is cut first. Then the bottom flange is cut in a similar manner. The cut is then commenced at *B* and taken to *E* along the web, this completing the operation. By cutting the flanges first the strength of the girder is altered but little until the web itself is finally cut.

Rollers and point guides can be affixed to the cutter in order to ensure a steady rate of travel and to enable the operator to execute straight lines or circles, etc., with greater ease (Fig. 7.8*a*).

The position of the flame and the shape of the cut are illustrated in Fig. 7.8*b*.

To close down, first shut off the cutting stream, then the propane or acetylene and then the oxygen valve. Close the cylinder valves and release the pressure in the tubes by momentarily opening the cutter valves.

To cut holes in plates a slightly higher oxygen pressure may be used. The spot where the hole is required is heated as usual and the cutting valve

Fig. 7.7

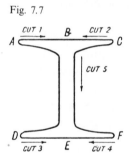

Fig. 7.8

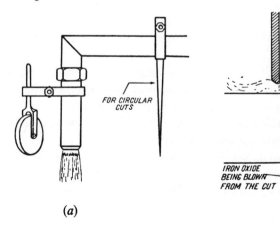

(*a*) (*b*)

released gently, at the same time withdrawing the cutter from the plate. The extra oxygen pressure assists in blowing away the oxide, and withdrawing the nozzle from the plate helps to prevent oxide from being blown on to the nozzle and clogging it. The cutting valve is then closed and the lower surface now exposed is heated again, and this is then blown away, these operations continuing until the hole is blown through. The edges of the hole are easily trimmed up afterwards with the cutter.

When propane or natural gas is used instead of acetylene, the flame temperature is lower; consequently it takes longer to raise the metal to ignition point and to start the cut, and it is not suitable for hand cutting over 100 mm thick nor for cast iron cutting. Because of this lower temperature, the speed of cutting is also slower. The advantages, however, are in its ease of adjustment and control; it is cheap to instal and operate and, most important of all, since the metal is not raised to such a high temperature as with the oxy-acetylene flame, the rate of cooling of the metal is slower and hence the edges of the cut are not so hard. This is especially so in the case of low-carbon and alloy steels. For this reason, oxy-propane or natural gas is often used in works where cutting machines are operated (see later).

The oxy-hydrogen flame can also be used (the hydrogen being supplied in cylinders, as the oxygen). It is similar in operation to oxy-propane and has about the same flame temperature. It is advantageous where cutting has to be done in confined spaces when the ventilation is bad, since the products of combustion are not so harmful as in the case of oxy-acetylene and oxy-propane, but is not so convenient and quick to operate.

The effect of gas cutting on steel

It would be expected that the cut edge would present great hardness, owing to its being raised to a high temperature and then subjected to rapid cooling, due to the rapid rate at which heat is dissipated from the cut edges. Many factors, however, influence the hardness of the edge.

Steels of below 0.3% carbon can be easily cut, but the cut edges will definitely harden, although the hardness rarely extends more than 3 mm inwards and the increase is only 30 to 50 points Brinell.

Steels of 0.3% carbon and above and also alloy steels are best preheated before cutting, as this reduces the liability to crack.

Nickel, molybdenum, manganese and chrome steels (below 5%) can be cut in this way. Steels having a high tungsten, cobalt or chrome content, however, cannot be cut satisfactorily. Manganese steel, which is machined with difficulty owing to the work hardening, can be cut without any bad effects at all.

The oxy-acetylene flame produces greater hardening effect than the oxy-propane flame, as before mentioned, owing to its higher temperature. Excessive cutting speeds also cause increased hardness, since the heat is thereby confined to a narrower zone near the cut and cooling is thus more rapid. Similarly, a thick plate will harden more than a thin one, owing to its more rapid rate of cooling from the increased mass of metal being capable of absorbing the heat more quickly. The hardening effect for low carbon steels, however, can be removed either by preheating or heat treatment after the cut. The hardening effect in mild steel is very small. On thicknesses of plate over 12 mm it is advisable to grind off the top edge of the cut, as this tends to be very hard and becomes liable to crack on bending.

The structure of the edges of the cut and the nearby areas will naturally depend on the rate of cooling. Should the cutting speed be high and the cooling be very rapid in carbon steels, a hard martensitic zone may occur, while with a slower rate of cutting and reduced rate of cooling the structure will be softer. A band of pearlite is usually found, however, very near to the cut edges and because of this, the hardness zone, containing increased carbon, is naturally very narrow. When the cut edge is welded on directly, without preparation, all this concentration of carbon is removed.

Thus, we may say that, for steels of less than 0.3% carbon, if the edges of the cut are smooth and free from slag and loose oxide, the weld can be made directly on to the gas-cut edge without preparation.

Oxygen or thermic lance

This is a method of boring or cutting holes in concrete, brick, granite, etc., by means of the heat generated by chemical reactions.

The lance consists of a tube about 3 m long and 6.5–9.5 mm diameter which is packed with steel wires. One end of the tube is threaded or snap connected and is connected by means of a flexible hose to an oxygen supply. To operate the lance the free end is heated and oxygen passed down the tube. Rapid oxidation of the wires begins at the heated end with great evolution of heat. Magnesium and aluminium are often added to the packing to increase the heat output. The operator can be protected by a shield and protective clothing should be worn. The exothermic reaction melts concrete and other hard materials to a fluid slag and cast iron is satisfactorily bored. Standard gas pipe can be used for thick steel sections.

As an example, a hole of 30 mm diameter and 300 mm deep can be bored in concrete using 1.9 m of lance and 1.0–1.3 m³ of oxygen at a pressure of 7 bar (100 lb per in²) in 120 seconds.

Cutting machines

Profiles cut by hand methods are apt to be very irregular and, where accuracy of the cut edge is required, cutting machines are used. The heating flame is similar to that used in the hand cutter and is usually oxy-propane or oxy-acetylene (either dissolved or generated), while the thickness of the cut depends on the nozzle and gas pressures.

The mechanical devices of the machine vary greatly, depending upon the type of cut for which they are required. In many types a tracing head on the upper table moves over the drawing or template of the shape to be cut. Underneath the table or on the opposite side of the machine (depending on the type) the cutting head describes the same motion, being worked through an intermediate mechanism. The steel being cut is placed on supports below the table (Fig. 7.9a and b).

Simpler machines for easier types of cuts, such as straight line and circles, bevels, etc., are also made.

A typical machine is capable of cutting from 1.5 m to 350 mm thick, 3 m in a straight line and up to 1.5 m diameter circle. The machine incorporates a magnetic tracing roller, which follows round a steel or iron template the exact shape of the cut required, while the cutting head cuts the replica of this shape below the table.

Fig. 7.9. (a) Cutting machine fitted with four cutting heads.

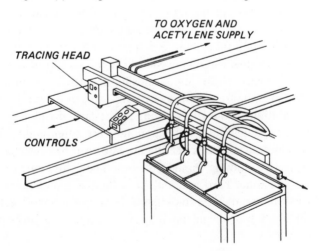

Stack cutting

Thin plates which are required in quantities can be cut by clamping them tightly in the form of a stack and, due to the accuracy of the modern machines, this gives excellent results and the edges are left smooth and even. Best results are obtained with a stack 75–100 mm thick, while G clamps can be used for the simpler types of stack cutting.

Cast iron cutting

Cast iron cutting is made difficult by the fact that the graphite and silicon present are not easily oxidized. Reasonably clean cuts can now be made, however, using a blowpipe capable of working at high pressure of oxygen and acetylene. Cast iron cannot be cut with hydrogen.

Since great heat is evolved in the cutting process, it is advisable to wear protective clothing, face mask and gloves.

The oxygen pressure varies from 7 N/mm² for 35 mm thick cast iron to 11 N/mm² for 350 to 400 mm thick, while the acetylene pressure is increased accordingly.

The flame is adjusted to have a large excess of acetylene, the length of the white plume being from 50–100 mm long (e.g. 75 mm long for 35–50 mm thick plate). The speed of cutting is low, being about 2.5 m per hour for 75–125 mm thick metal.

Fig. 7.9. (*b*)

Technique. The nozzle is held at 45° to the plate with inner cone 5–6 mm from the plate, and the edge where the cut is to be started is heated to red heat over an area about 12–18 mm diameter. The oxygen is then released and this area burnt out. The blowpipe is given a zig-zag movement, and the cut must not be too narrow or the slag and metal removed will clog the cut. About 12 to 18 mm is the normal width. After the cut is commenced the blowpipe may be raised to an angle of 70–80°, which will produce a lag in the cut, as shown in Fig. 7.10.

Owing to the fact that high pressures are used in order to supply sufficient heat for oxidation, large volumes of gas are required, and this is often obtained by connecting several bottles together.

Flame gouging by the oxy-acetylene process

Flame gouging is an important extension of the principle of oxy-acetylene cutting by which grooves with very smooth contours can be cut easily in steel plates without the plate being penetrated.

Principle of operation

This is the same as that in the oxy-acetylene process, except that a special type of nozzle is used in the standard cutting blowpipe. A pre-heating flame heats the metal to red heat (ignition temperature), the cutting oxygen is switched on, oxidation occurs and the cut continues as previously explained.

Equipment

The cutting torch may have a straight 75° or 180° angle head together with a range of special gouging nozzles, as shown in Fig. 7.11. The nozzle sizes are designated by numbers and they are bent at an angle which is best for the gouging process. Regulators and other equipment are as for cutting.

Fig. 7.10

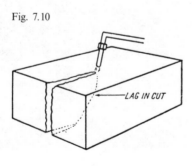
LAG IN CUT

Operation

There are two main techniques:

(1) progressive,

(2) spot.

In the former the groove is cut continuously along the plate – it may be started at the plate edge or anywhere in the plate area. It can be used for removing the underbeads of welds prior to depositing a sealing run, or it may be used for preparing the edges of plates. Spot gouging, however, is used for removing small localized areas such as defective spots in welds.

To start the groove at the edge of a plate for continuous or progressive gouging the nozzle is held at an angle of about 20° so that the pre-heating flames are just on the plate and when this area gets red hot the cutting oxygen stream is released and at the same time the nozzle is brought at a shallower angle to the plate as shown, depending upon depth of gouge required. The nozzle is held so that the nozzle end clears the bottom of the cut and the pre-heating flames are about 1.5 mm above the plate.

The same method is adopted for a groove that does not start at the plate edge. The starting point is pre-heated with the nozzle making a fairly steep angle with the plate at 20–40°. When the pre-heated spot is red hot the cutting oxygen stream is released and the angle of the nozzle reduced to 5–10° depending upon the depth of gouge required (Fig. 7.12).

To gouge a single spot, it is pre-heated as usual where required, but when red hot and the cutting stream of oxygen is turned on, the angle of the nozzle is *increased* (instead of, as previously, decreased) so as to make the gouge deep.

The depth of groove cut depends upon nozzle size, speed of progress and angle between nozzle and plate (i.e. angle at which the cutting stream of oxygen hits the plate). The sharper this angle, the deeper the groove. If the

Fig. 7.11. Flame gouging nozzle.

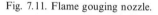

cutting oxygen pressure is too low ripples are left on the base of the groove. If the pressure is too high the cut at the surface proceeds too far in advance of the molten pool, and eventually the cut is lost and must be restarted.

Use of flame gouging

Certain specifications, such as those for fabrication of butt welded tanks, etc., stipulate that the underbead (or back bead) should be removed and a sealing run laid in place. This can easily and efficiently be done by flame gouging, as also can the removal of weld defects, tack welds, lugs, cleats, and also the removal of local areas of cracking in armour plate, and flashing left after upset welding (Fig. 7.13).

Oxy-arc cutting process

In this process the electric arc takes the place of the heating flame of the oxy-acetylene cutter. The covered electrodes of mild steel are in 4 sizes and are of tubular construction, the one selected depending upon whether it is required for cutting, piercing, gouging, etc. They are about 5 to 6 mm outside diameter with a fine hole about 1.5 mm diameter or more through which the cutting oxygen stream passes, down the centre. The gun type holder secures the electrode by means of a split collet and has a trigger controlling the oxygen supply. A d.c. or a.c. supply of 100–300 A is suitable.

The arc is struck with the oxygen off, the oxygen valve is released immediately and cutting begins, the electrode being held at an angle of 60° to the line of cut, except at the finish when it is raised to 90°, and is

Fig. 7.12. Method and uses of flame gouging.

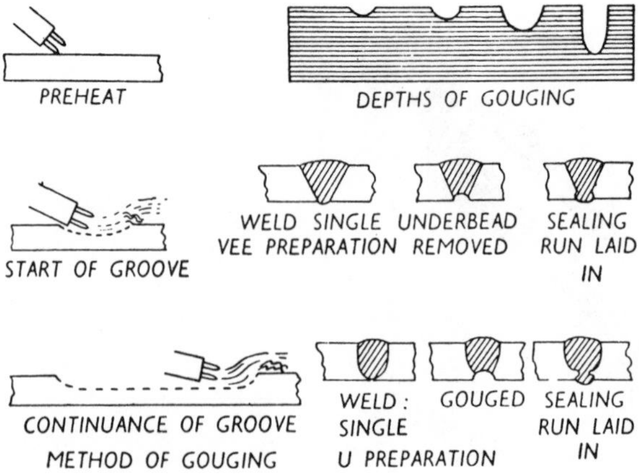

PREHEAT   DEPTHS OF GOUGING

START OF GROOVE

WELD SINGLE UNDERBEAD SEALING
VEE PREPARATION REMOVED RUN LAID
                                IN

CONTINUANCE OF GROOVE
METHOD OF GOUGING

WELD: GOUGED SEALING
SINGLE   RUN LAID
U PREPARATION   IN

consumed in the process. The oxygen pressure varies with the thickness of steel and with the size of electrode being used, about 4 bar (60 lb per in^2) for mild and low alloy steel plate of 8 to 10 mm thick and about 4.5–5.5 bar (70–80 lb per in.^{2}) for 23 to 25 mm thick. In addition to mild and low-alloy steel, and stainless steel copper, bronze, brass, monel and cupro-nickels can be roughly cut by this process.

Arc-air cutting and gouging process

In this process a carbon arc is used with a d.c. supply from a welding generator, or rectifier, together with a compressed air supply at 5–8 N/mm^2.

The equipment comprises a holder for the carbon electrode to which is supplied the first current with *electrode* +ve, and the compressed air which is controlled by a lever-operated valve on the electrode holder.

The jaws of the electrode holder are rotatable enabling the carbon electrode to be held in any position and twin jets of compressed air emerge from the head on each side of the carbon and parallel to it, the two jet streams converging at the point where the arc is burning. Air is also circulated internally in the holder to keep it cool.

Operation

The carbon electrode is placed in the holder with 75–150 mm projecting, with the twin jet holes pointing towards the arc end of the carbon.

The carbon is held at approximately 45° to the job to be cut while for shallow gouging the angle may be reduced to 20°. This angle together with the speed of travel affects the depth of cut. The carbons have been specially designed for the process and are a mixture of carbon and graphite covered over all with a thin sheath of copper. The copper coating prevents tapering

Fig. 7.13. Flame-gouged edge of plate, prepared for welding.

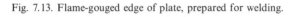
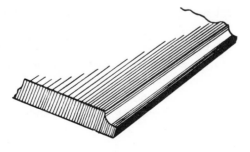

and ensures a cut of regular width in addition to enabling higher current to be used and consequently greater speeds of cutting to be obtained.

The carbons range in diameter from 4 mm (75–150 A) to 16 mm (550–700 A) and the higher the current density the more efficient the operation. Too high a current for a given size of electrode destroys the copper coating and burns the carbon at an excessive rate. Normally the copper coating burns away about 20 mm from the arc.

The process is applicable to work in all positions and is used for removing defects in castings, in addition to cutting and gouging. A reasonably clean surface of cut is produced with no adverse effect so that welding can be carried out on the cut surface without further grinding.

Cutting with the carbon arc. The carbon arc, owing to its high temperature, can be used for cutting steel. A high current is required, and the cut must be started in such a spot that the molten metal can flow away easily. The cut should also be wide enough so that the electrode (of carbon or graphite) can be used well down in it, especially when the metal is thick, so as to melt the lower layers. Cast iron is much more difficult to cut, since the changing of the iron into iron oxide is not easily performed owing to the presence of the graphite. Carbon arc cutting does not produce a neat cut, and because of this is only used in special circumstances.

Plasma cutting

This process gives clean cuts at fast cutting speeds in aluminium, mild steel and stainless steel. All electrically conducting materials can be cut using this process, including the rare earth metals such as tungsten and titanium and it has superceded powder cutting.

The arc is struck between the central tungsten electrode and a copper nozzle body of a water-cooled torch. A gas mixture passes under pressure through the restricted nozzle orifice and around the arc, emerging as a high-temperature (up to 25 000 °C) ionized plasma stream, and the arc is transferred from the nozzle and passes between the electrode and work (Fig. 7.14a).

Great improvements have been made in plasma cutting torches by greatly constricting the nozzle and thus narrowing the plasma stream, giving a narrower and cleaner cut with less consumption of power. In certain cases 'double arcing' may occur, in which the main arc jumps from tungsten electrode to nozzle and then to work, damaging both nozzle and electrode.

Power unit

A 24 kVA fan cooled unit has input voltages of 220, 380–415, and 500 and cutting currents ranging from 50 A minimum to 240 A maximum at 100% duty cycle and OCV of 200 V d.c. with heavy duty rectifiers for main and pilot arc control. An automatic pilot arc switch cuts off when the cutting arc is transferred. This size of unit enables metal of up to 30 mm thickness to be cut and edges prepared quickly and cleanly in stainless steel, alloy steels, aluminium and copper.

A 75 kVA unit of similar input voltage has cutting currents of 50–250 A with OCV of 300 V and cutting voltage of 170 V enabling thicknesses of 2.7 to 50 mm upwards, in the previously mentioned materials, to be cut and edges prepared at high speed (Fig. 7.14b). The cutting head of either unit can be fitted to a carriage to obtain high speeds of cut with greater precision than with manual cutting (Fig. 7.15).

The gases used are argon, hydrogen and nitrogen and the controls can be seen on the upper right-hand corner of the illustration in Fig. 7.14b. It should be noted that since high voltages of up to 300 V on open circuit and about 170 V during the cutting process, are encountered, great care should be taken to follow all safety precautions and avoid contact with live parts of the torch when the unit is switched on. All work to be done on the torch head should be with the power supply switched off.

Argon–hydrogen and argon–nitrogen are used for cutting, the controls for which are in the upper left-hand corner of the illustration. Any combination of the gases can be selected at will to suit the material and thickness and the actual ratio of the gases will depend upon the operator. The tungsten electrode may be of 1.6 or 2.0 mm diameter and gives a cut of

Fig. 7.14. (*a*) Plasma cutting.

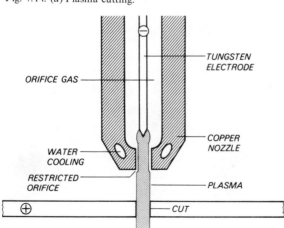

width about 2.5–3.5 mm with a torch-to-work distance of about 10 mm when cutting with about 250 A.

Since hydrogen is an explosive gas and nitrogen combines with oxygen of the atmosphere to form the oxides of nitrogen (NO, N_2O, NO_2 and N_2O_4) in the heat of the arc, cutting should be done in a well-ventilated shop (not enclosed in any way) and protective clothing and the correct eye lens protection always worn. Some units operate on the air–plasma system

Fig. 7.14. (*b*)

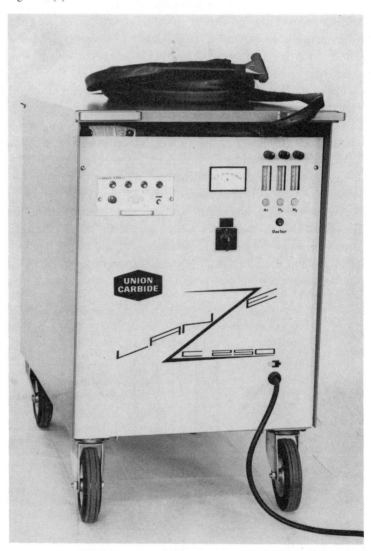

using dry compressed air at a pressure of 5 bar from either cylinders or a compressor. The air acts both as the plasma gas, blows away the products of the cut and cools the torch head. The 3-phase input power unit has 20 A and 40 A output tappings and all electrically conductive materials from 0.6 to 10 mm thickness can be cleanly cut. A flat zirconium electrode is used and these units can also be used with the usual argon, hydrogen, and nitrogen gas combinations.

Operation. The unit is switched 'on', current is selected on the potentiometer knob and checked on the ammeter, gases are adjusted for mixture and the torch switch pressed. The pilot arc gas flows and the pilot arc is ignited. The torch is now brought down to about 10 mm from the work (take care not to touch down) and, making sure that all protective clothing including head mask is in order, the torch switch is released and the main arc is transferred to the work. If this does not occur a safety cut-off circuit ensures that the pilot arc current circuit is de-energized after a few seconds.

Fig. 7.15. Mechanized plasma cutting.

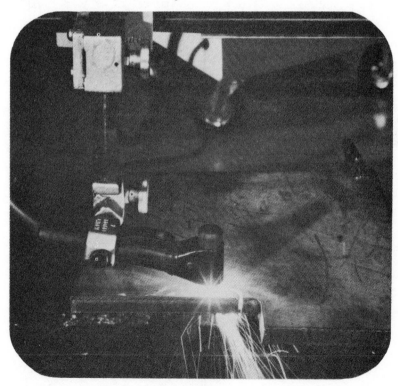

The arc commences with slow start current and when the thickness of work is pierced, the torch is moved along the plate and if cutting current and speed of travel are correct a steady stream of molten metal flows from the underside of the cut giving a smoothly cut surface. Average currents are in the region of 250 A.

The arc is extinguished automatically at the end of the cut or the torch can be pulled away from the work.

Using argon–hydrogen or argon–nitrogen gas mixtures the tungsten electrode can be 1.6 mm for thinner sections and 2 mm for thicker sections giving a cut width of 2.5–3.5 mm with a torch-to-work distance of about 10 mm.

Torch. Torches can be either hand or machine operated and are supplied with spare electrodes, cutting tips and heat shields (Fig. 7.16). In certain cases the presence of water can contribute to an electrolytic action in the

Fig. 7.16. Plasma cutting torch.

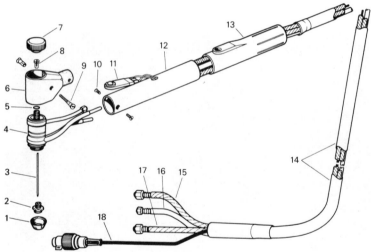

Key:
1. Heat shield.
2. Cutting tip.
3. Tungsten electrode 3.2 mm diameter.
4. Torch body.
5. O ring.
6. Torch head cover.
7. Torch cap.
8. Collet.
9. Cover retaining nuts and screw.
10. Screw.
11. Torch switch.
12. Handle.
13. Switch boot.
14. Sheath.
15. Main arc cable.
16. Gas hose.
17. Pilot arc cable.
18. Control cable with plug.

torch which quickly causes sufficient corrosion in the waterways of the torch to give water leaks, ruining the torch. This electrolytic action occurs as a result of the presence of dissolved minerals in the water, and potential differences between the electrode and nozzle. To prevent this a supply of de-ionized water is used (about 2 litres per minute) and this prevents the electrolytic action.

De-ionized water

Water taken directly from a tap contains dissolved carbon dioxide (CO_2) with possibly hydrogen sulphide (H_2S) and sulphur dioxide (SO_2) together with mineral salts such as calcium bicarbonate and magnesium sulphate which have dissolved into the water in its passage through various strata in the earth. These minerals separate into ions in the water, for example, magnesium sulphate $MnSO_4$ separates into Mg^{2+} ions (positively charged) and sulphate ions SO_4^{2-} which are negatively charged. Similarly with calcium bicarbonate ($Ca(HCO_3)_2$) which gives metallic ions, Ca^{2+} and bicarbonate ions HCO_3^-.

By boiling or distilling the water the minerals are not carried over in the steam (the distillate) so that the distilled water is largely free of mineral salts but may contain some dissolved CO_2. Distillation is however rather slow and costly in fuel and a more efficient method of obtaining de-ionized water is now available and uses ion exchange resins. The resins are organic compounds which are insoluble in water. Some resins behave like acids and others behave like alkalis and the two kinds can be mixed without any chemical change taking place.

When water containing dissolved salts is passed through a column containing the resins the metal ions change places with the hydrogen ions of the acidic resin so that the water contains hydrogen ions instead of metallic ions. Similarly the bicarbonate ions are replaced by hydroxyl (OH) ions from the alkaline resin. During this exchange insoluble metallic salts of the acid resin are formed and the insoluble alkali resin is slowly converted into insoluble salts of the acids corresponding to the acid radials previously in solution.

The water emerging from the column is thus completely de-ionized and now has a greater resistivity (resistance) than ordinary tap water so that it conducts a current less easily. A meter can be incorporated in the supply of de-ionized water to measure its resistivity (or conductivity) and this will indicate the degree of ionization of the water in the cooling circuit. When the ionization rises above a certain value the resins must be regenerated. This is done by passing hydrochloric acid over the acidic resin, so that the free hydrogen ions in the solution replace the metallic ions (Ca and Mg).

This is followed by passing a strong solution of sodium hydroxide (NaOH) through the column, when the hydroxyl ions displace the sulphate and bicarbonate ions from the alkaline resin.

This process produces de-ionized water, purer than distilled water, easily and quickly.

Water injection plasma cutting

In this process, water is injected through four small-diameter jets tangentially into an annular swirl chamber, concentric with the nozzle, to produce a vortex which rotates in the same direction as the cutting gas (Fig. 7.17). The water velocity is such as to produce a uniform and stable film around the high-temperature plasma stream, constricting it and reducing the possibility of double arcing. Most of the water emerges from the nozzle in a conical jet which helps to cool the work surface.

The cut produced is square within about 2° on the right-hand side (viewed in the direction of cutting), whilst the other side is slightly bevelled, caused in general by the clockwise rotation of the cutting gas, which is commercial purity nitrogen. The use of nitrogen reduces cutting costs as it is cheaper than gas mixtures.

The process gives accuracy of cut at high cutting speeds with very smooth

Fig. 7.17. The principle of water-injection plasma cutting.

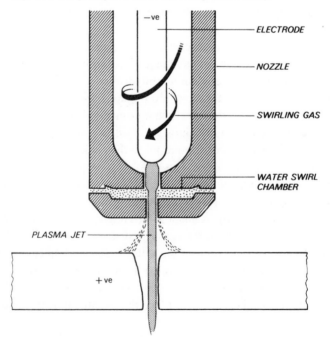

cut surfaces. There is little or no adherent dross and the life of the cutting nozzles is greatly increased; mild and carbon steels, alloy and stainless steels, titanium and aluminium are among the metals which can be efficiently cut, in thicknesses from 3 to 75 mm for stainless steel.

The noise and fumes associated with plasma cutting can be reduced by the use of a water muffler fitted to the torch to reduce the noise and by the use of a water table which replaces the normal cutting table and which removes up to 99.5% of the particles and fumes by scrubbing, using the kinetic energy of the hot gases and molten metal stream from the kerf.

This cutting equipment is currently fitted to existing cutting installations up to the largest sizes used in shipyards and including those with numerical control.

Powder injection cutting

The process of powder injection cutting causes a great deal of fumes to be given off and is one of the less liked operations. As a result it has, in many cases, been superceded by the plasma arc method of cutting of stainless steels, 9% nickel steel and non-ferrous metals. Powder cutting enables stainless steel to be cut, bevelled and profiled with much the same ease and facility as with the oxygen cutting of low-carbon steels.

In cutting low-carbon steels a pre-heating flame raises the temperature of a small area to ignition point. This is the temperature at which oxidation of the iron occurs, and iron oxide is formed when a jet of oxygen is blown on to the area. The heat of chemical combination, together with that of the pre-heating flame enables the oxidation, and hence the cut to continue, the oxide being removed by being blown away by the oxygen jet, resulting in a narrow cut.

For this sequence to occur, the melting point of the oxide formed must be lower than that of the metal being cut. This is the case in low-carbon steel. In the case of stainless steels and non-ferrous metals the oxides formed have a melting point higher than that of the parent metal.

When attempting to cut stainless steel with the ordinary oxy-acetylene equipment the chromium combines with oxygen at high temperature and forms a thin coating of oxide which has a melting point higher than that of the parent steel, and since it is difficult to remove, further oxidation does not occur and the cut cannot continue.

In the powder-cutting process a finely divided iron powder is sprayed by compressed air or nitrogen into the cutting oxygen stream on the line of the cut. The combustion of this iron powder so greatly increases the ambient temperature that the refractory oxides are melted, fluxed, and to a certain extent eroded by the action of the particles of the powder, so that a clean surface is exposed on to which the cutting oxygen impinges and thus the cut continues. The quality of the cut is very little inferior to that of a cut in a low-carbon steel.

Equipment

For hand or machine cutting the powder is delivered to the reaction zone of the cut by means of an attachment fitted to the cutting blowpipe (Fig. 7.18). The attachment consists of powder valve, powder nozzle and tubing. The nozzle is fitted over the standard cutting nozzle and the powder valve is clamped near the gas valves. The iron powder is carried down the outside of the cutting nozzle and after passing through inclined ports, is injected through the heating flame into the cutting stream of oxygen which it meets at approximately 25 mm below the end of the cutting nozzle.

The nozzle is normally one size larger than for cutting the same thickness of low-carbon steel and is held as for normal cutting except that a clearance of 25 to 35 mm is given between nozzle and plate to be cut to allow the powder to burn in the oxygen stream. The great heat produced makes pre-heating unnecessary on stainless steel and what may be termed a 'flying start' can be made.

The powder dispenser unit (Fig. 7.19) is of the injector type and is a pressure vessel which incorporates:

(1) A hopper for filling.
(2) An air filter.
(3) An air pressure regulator.
(4) A dryer.
(5) An injector unit.

The removable cover enables the hopper to be filled and is fitted with a pressure relief valve which lifts at 0.15 N/mm^2. A screen for removing over-large particles from the powder and a tray for the drying agent are fitted.

Compressed air flows into the dispenser, picks up powder and carries it

Fig. 7.18. Cutter equipped for powder cutting.

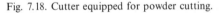
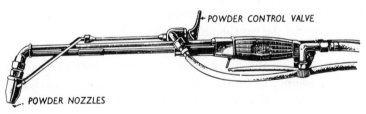

POWDER CONTROL VALVE

POWDER NOZZLES

along a rubber hose to the cutting nozzle. Nitrogen may be used instead of air but never use oxygen, as it is dangerous to do so. The powder flow from the dispenser is regulated by adjusting the nozzle or by varying the air pressure.

The dispenser should be fed from an air supply of 1.5 m³ per hour at a pressure of 0.3 N/mm². The usual pressure for operating is from 0.02 to 0.03 N/mm².

Since the whole process depends upon a uniform and smooth flow of powder every care must be taken to ensure this. Any moisture in either the powder or in the compressed air can cause erratic operation and affect the quality of the cut.

Silica gel (a drying agent) is incorporated in the dispenser to dry the powder but the amount is not sufficient to dry out the compressed air and a separate drying and filtering unit should be installed in the air line to ensure dry, clean air. The oxygen and acetylene is supplied through not less than 9 mm bore hose with 6 mm bore for the powder supply.

The single-tube attachment discharges a single stream of powder into the cutting oxygen and is used for straight line machine cutting of stainless steel. The multi-jet type (Fig. 7.18) has a nozzle adaptor which fits over a standard cutting nozzle and has a ring of ports encircling it. The powder is fed through these ports and passes through the heating flame into the

Fig. 7.19. Powder dispenser unit.

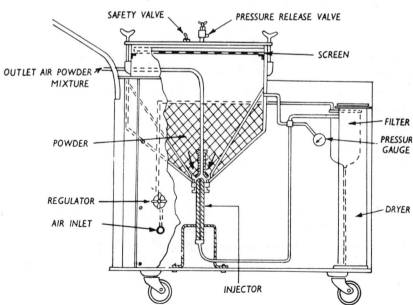

cutting oxygen. This type is recommended for hand cutting and for profile, straight and bevel machine cutting. Special cutting nozzles are available for this process. These give a high velocity parallel cutting oxygen stream from their bore being convergent–divergent. The pre-heater holes are smaller in number but more numerous and are set closer to the cutting oxygen orifice than in the standard nozzle. This gives a soft narrow pre-heating flame, giving a narrower cut and better finish with a faster cutting speed. Since the iron powder is very abrasive, wear occurs in ports and passages through which the powder passes. By using stainless tube where possible, avoiding sharp internal bends and reinforcing certain parts, wear is reduced to a minimum.

Technique

In cutting any metal by this process, correct dispenser setting, dry air and powder, clean powder passages and leak-tight joints all help towards ensuring a good quality cut. The rate of powder flow is first adjusted to the correct amount for the particular work in hand by trial cuts.

Stainless steel. Size of nozzle is one size larger than for the same thickness low-carbon steel. For thicknesses up to 75 or 100 mm the nozzle is held about 25 mm from the plate, for thicknesses up to 150 mm the distance is increased to about 35 mm, while for heavier sections it can be about 50 mm distant. No pre-heating is necessary – an immediate or flying start can be made. Scrap plate may be loosely put together and a single cut can be made, the great heat enabling this to be done in spite of gap between the plates.

Cast iron and high-alloy steels. Technique for cast iron is similar to that for stainless steel but cutting speed is up to 50% slower. With high-alloy steels for example, a 25/20 nickel–chrome steel takes 30–40% longer than for the 18/8 nickel–chrome steel. Pre-heating of high-alloy tool steels is advisable to prevent risk of cracking due to localized heat.

Round carbon steel bars are easily cut by laying them side by side. The cut is done from bar to bar, powder being switched on to start the cut on each bar giving no interruption in cut from bar to bar.

Nickel and nickel alloys. Pure nickel is extremely difficult to cut because it does not readily oxidize and it can only be cut at slow speed in thicknesses up to 25 mm section. Alloys like Inconel, Nimonic and Nichrome can be cut up to 125 mm thick, speeds being up to 60% slower than for the same thickness of stainless steel.

Copper and copper alloys. In this operation copper is melted (and not cut) the particles of iron powder removing the molten metal and eroding the cut.

Much pre-heating is required due to the high conductivity of copper and copper alloys – for example – a powder cutting blowpipe capable of cutting 180 mm thick stainless steel can only deal with copper up to 25 mm thick and brass and bronze up to 100 mm thick. A lower cutting oxygen pressure than normal is used and the nozzle moved in a forwards, upwards, backwards, downwards motion in the line of the cut, thus helping to remove the molten metal and avoiding cold spots.

Aluminium and aluminium alloys. The quality of the cut is poor and of ragged profile. Alloys containing magnesium such as MG7 develop a hard surface to the cut due to formation of oxide and this may extend to a depth of 16 mm. As a result powder cutting is limited to scrap recovery. Cutting (or melting) speeds on thinner sections are much the same as for stainless steel, but with thicker sections molten metal chokes the cut, so that large nozzles with reduced cutting oxygen pressures should be used.

Powder cutting can be incorporated on cutting machines and gouging can be carried out as with the normal blowpipe with the same ease and speed, the nozzle being again held further away from the work, to allow space for the combustion of the powder.

Underwater cutting

Underwater cutting of steel and non ferrous metals is carried out on the oxy-arc principle using the standard type of electrode which has a small hole down its centre through which the pressurized oxygen flows to make the cut. By using a thermic electrode, this oxygen stream causes intense oxidation of the material of the electrode, giving a greatly increased quantity of heat for the cut. In both cases the products of the cut are removed or blown away by the oxygen stream.

A d.c. generator is used with the torch negative and the special type earth connexion, connected to clean, non-rusted metal in good electrical contact with the cut, is positive polarity: a.c. is not suitable as there is increased danger of electric shock.

The torch head for ordinary or thermic electrodes is similar. It is tough rubber covered and has connections for oxygen and electric current. The electrodes, usually of 4 mm ($\frac{5}{32}''$) and 4.8 mm ($\frac{3}{16}''$) diameter are securely clamped in the head by a collet for the size electrode being used. A twist of the torch grip clamps and releases the collet and is used for stub ejection.

These collets can be changed to allow the torch head to be used for welding (Fig. 5.23).

The oxygen cylinder, fed from a manifold, has a high volume regulator which will give free flow and pressures from 8 bar (118 psi) at 12 m depth to 20 bar (285 psi) at 107 m. In general the oxygen supply to the torch should give about 6 bar (90 psi) over the bottom pressure at the electrode tip.

Ingress of water to the torch head is prevented by suitable valves and washers and replaceable flash arrester cartridge and screen are fitted.

The seal for the electrode in the torch head is made with washers and collet and tightened in place with the torch head locknut. When the electrode is fitted to the head it must bottom on to the washer and be held tight to prevent leakage of oxygen. Note that the thermic type of electrode will continue burning once it has been ignited, even when the electrical supply is switched off, due to the oxodizing action of the gas stream.

Use of equipment

An eyeshield with the correct lens fitted is attached to the outside of the diver's helmet. No oil nor grease should be used on the equipment and there should be no combustible nor explosive materials near to the point where cutting (or welding) is to be performed. Hose connexions should be checked for leaks and all electrical connexions tight, especially check the earth connexion and see that it is in good electrical connexion with the position of the cut. As electrolysis can cause rapid deterioration, all equipment should be continually inspected for signs of this. If any part of the work is above water level, connect the earth clamp to this after checking that there is a good return path from the cut to the earth clamp.

A double pole single throw switch of about 400 A capacity is connected in the main generator circuit as a safety switch. This should always be kept in the 'off' position except when the cut is actually taking place. To begin cutting, strike the arc and open the oxygen valve by pressing the lever on the torch.

When working, the diver-welder must always face the earth connexion so that the cutting is done *between* the diver and the earth connexion. When the electrode has burnt to a stub of about 75 mm in length the diver should call to the surface to shut off the current by opening the safety switch. This having been done the collet is loosened by a twist of the wrist, the oxygen lever is pressed and the stub is blown out of the holder. A new electrode is now fitted, making sure that it sits firmly against the sealing washer, and the handle twisted to lock it in place. Stubs should not be burnt close to the holder for fear of damage. When the safety switch is again placed in the 'on' position cutting can again commence.

When cutting, bear down on the torch so as to keep the cutting electrode against the work so that the electrode tip is in the molten pool and proceed with the cut as fast as possible consistent with good cutting. Spray back from the cut indicates that it is not through the work completely. Keep all cables and hoses from underfoot and where anything may fall on them. Pipes and tubes may be cut accurately to size under water by a hydraulically operated milling machine which is driven around a circumferential guide.

Safety precautions

The general safety precautions which should be observed when welding are listed at the end of the section on oxy-acetylene welding but special precautions should be taken when cutting processes using oxygen are used in the welding shop or confined spaces. Oxygen is present in the atmosphere at about 29.9% by volume and below 6–7% life cannot be sustained.

In the appendix article on fire precautions we state that for a fire to occur we need (1) oxygen, (2) fuel and (3) heat.

Oxygen rapidly oxidizes combustible materials and in the case of grease or oil the oxidation reaction produces so much heat that the material may be ignited, thus causing fire or explosion. When cutting with the oxy-acetylene or oxy-propane flame the oxygen (cutting) lever should not be kept open longer than necessary because not only is the compressed gas expensive but it also causes oxygen enrichment of the atmosphere in confined spaces. Should the operator be wearing gloves and clothing that are greasy or oily, the oxidation of this grease and oil may cause burning with evolvement of heat so that the operator may sustain burns and a fire or explosion may endanger life.

Never use compressed oxygen for any other purpose (e.g. in place of compressed air) other than cutting (or welding). Make sure that the ventilation of the area is good – cutting in confined spaces always presents a hazard. Oil, grease and other combustible material should never be in the vicinity where cutting is taking place.

Since oxygen has no smell or taste its presence is difficult to detect and no smoking should be allowed in areas where oxygen is being used.

Appendix 1

Welding symbols

A weld is indicated on a drawing by (1) a symbol (Fig. A1.1*a*) and (2) an arrow connected at an angle to a reference line usually drawn parallel to the bottom of the drawing (Fig. A1.1*b*). The side of the joint on which the arrow is placed is known as the 'arrow side' to differentiate it from the 'other side' (Fig. A1.1*c*). If the weld symbol is placed *below* the reference line, the weld face is on the *arrow-side* of the joint, while if the symbol is above the reference line the weld face is on the other side of the joint (Fig. A1.1*c*). Symbols on both sides of the reference line indicate welds to be made on both sides of the joint while if the symbol is across the reference line the weld is within the plane of the joint. A circle where arrow line meets reference line indicates that it should be a peripheral (all round) weld, while a blacked in flag at this point denotes an 'on site' weld (Fig. A1.1*d*). Intermittent runs of welding are indicated by figures denoting the welded portions, and figures in brackets the non-welded portions, after the symbol (Fig. A1.1*e*). A figure before the symbol for a fillet weld indicates the leg length. If the design throat thickness is to be included, the leg length is prefixed with the letter '*b*' and the throat thickness with the letter '*a*'. Unequal leg lengths have a × sign separating the dimensions (Fig. A1.1*f*). A fork at the end of the reference line with a number within it indicates the welding process to be employed (e.g. 131 is MIG, see Table A1.1) while a circle at this point containing the letters NTD indicates that non-destructive testing is required (Fig. A1.1*g*). Weld profiles, flat (or flush), convex and concave profiles are shown as supplementary symbols in Fig. A1.1*h*.

Figure A1.2*a* gives the elementary symbols and A1.2*b* typical uses of them. The student should study BS 499, Part 2, 1980, *Symbols for welding*, for a complete account of this subject.

Table A1.1 shows the numerical indication of processes complying with International Standard ISO 4063.

340

Fig. A1.1

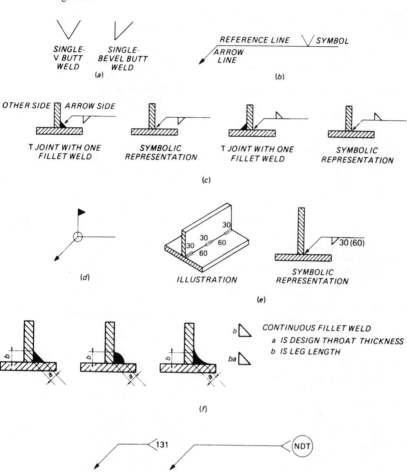

Table A1.1. *Numerical indication of process (BS 499, Part 2, 1980)**

No.	Process	No.	Process
1	**Arc welding**	43	Forge welding
11	Metal-arc welding without gas protection	44	Welding by high mechanical energy
		441	Explosive welding
111	Metal-arc welding with covered electrode	45	Diffusion welding
		47	Gas pressure welding
112	Gravity arc welding with covered electrode	48	Cold welding
113	Bare wire metal arc welding	**7**	**Other welding processes**
114	Flux cored metal-arc welding	71	Thermit welding
115	Coated wire metal-arc welding	72	Electroslag welding
118	Firecracker welding	73	Electrogas welding
12	Submerged arc welding	74	Induction welding
121	Submerged arc welding with wire electrode	75	Light radiation welding
		751	Laser welding
112	Submerged arc welding with strip electrode	752	Arc image welding
		753	Infrared welding
13	Gas shielded metal-arc welding	76	Electron beam welding
131	MIG welding	77	Percussion welding
135	MAG welding: metal-arc welding with non-inert gas shield	78	Stud welding
		781	Arc stud welding
136	Flux cored metal-arc welding with non-inert gas shield	782	Resistance stud welding
14	Gas-shielded welding with non-consumable electrode	**9**	**Brazing, soldering and braze welding**
		91	Brazing
141	TIG welding	911	Infrared brazing
149	Atomic-hydrogen welding	912	Flame brazing
15	Plasma arc welding	913	Furnace brazing
18	Other arc welding processes	914	Dip brazing
181	Carbon arc welding	915	Salt bath brazing
185	Rotating arc welding	916	Induction brazing
		917	Ultrasonic brazing
2	**Resistance welding**	918	Resistance brazing
21	Spot welding	919	Diffusion brazing
22	Seam welding	923	Friction brazing
221	Lap seam welding	924	Vacuum brazing
225	Seam welding with strip	93	Other brazing processes
23	Projection welding	94	Soldering
24	Flash welding	941	Infrared soldering
25	Resistance butt welding	942	Flame soldering
29	Other resistance welding processes	943	Furnace soldering
291	HF resistance welding	944	Dip soldering
		945	Salt bath soldering
3	**Gas welding**	946	Induction soldering
31	Oxy-fuel gas welding	947	Ultrasonic soldering
311	Oxy-acetylene welding	948	Resistance soldering
312	Oxy-propane welding	949	Diffusion soldering
313	Oxy-hydrogen welding	951	Flow soldering
32	Air fuel gas welding	952	Soldering with soldering iron
321	Air-acetylene welding	953	Friction soldering
322	Air-propane welding	954	Vacuum soldering
		96	Other soldering processes
4	**Solid phase welding; Pressure welding**	97	Braze welding
41	Ultrasonic welding	971	Gas braze welding
42	Friction welding	972	Arc blaze welding

* This table complies with International Standard ISO 4063.

Fig. A1.2. (*a*) Elementary welding symbols (BS 499, Part 2, 1980).

DESCRIPTION	SECTIONAL REPRESENTATION	SYMBOL
1. *BUTT WELD BETWEEN FLANGED PLATES (FLANGES MELTED DOWN COMPLETELY)*		⨆⨆
2. *SQUARE BUTT WELD*		‖
3. *SINGLE-V BUTT WELD*		V
4. *SINGLE-BEVEL BUTT WELD*		⌐/
5. *SINGLE-V BUTT WELD WITH BROAD ROOT FACE*		Y
6. *SINGLE-BEVEL BUTT WELD WITH BROAD ROOT FACE*		Y
7. *SINGLE-U BUTT WELD*		Y
8. *SINGLE-J BUTT WELD*		ⱶ
9. *BACKING OR SEALING RUN*		⌣
10. *FILLET WELD*	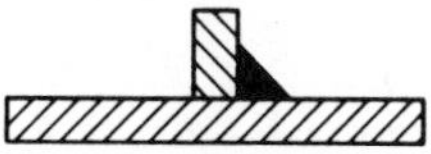	◿
11. *PLUG WELD (CIRCULAR OR ELONGATED HOLE, COMPLETELY FILLED)*	*ILLUSTRATION* 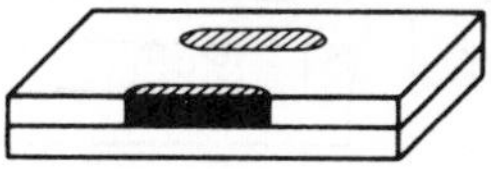	⊓
12. *SPOT WELD (RESISTANCE OR ARC WELDING) OR PROJECTION WELD*	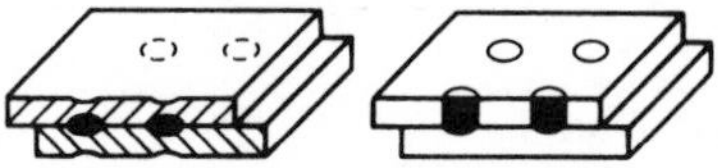 (a) *RESISTANCE* (b) *ARC*	◯
13. *SEAM WELD*	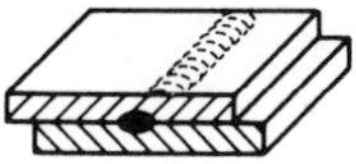	 ⊖

Fig. A1.2. (*b*) Examples of uses of symbols (BS 499, Part 2, 1980).

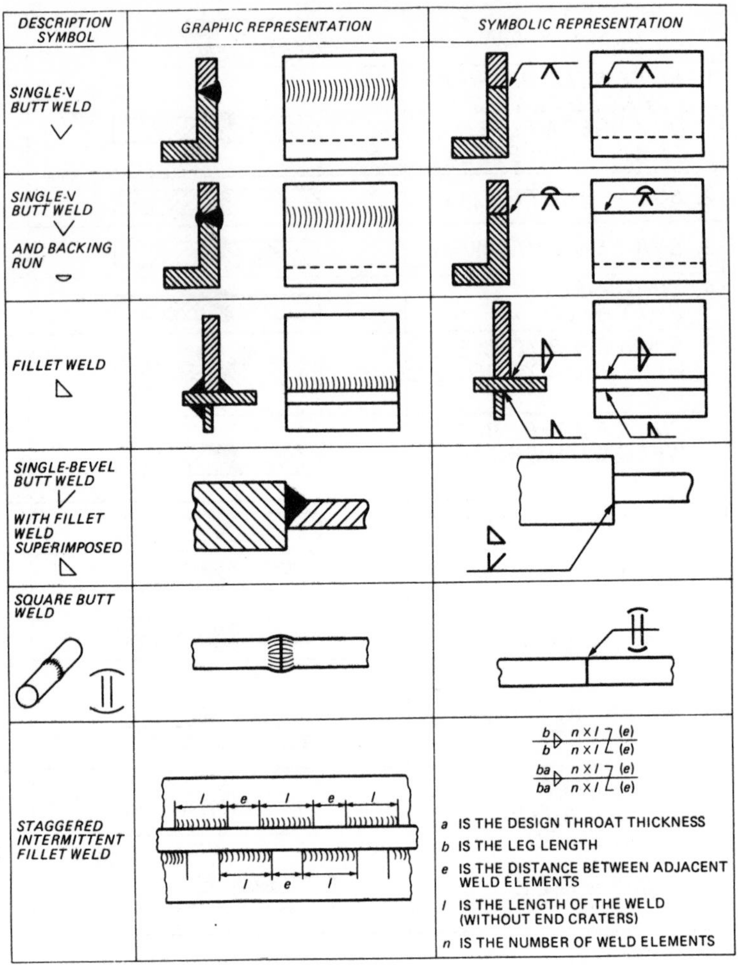

DESCRIPTION SYMBOL	GRAPHIC REPRESENTATION	SYMBOLIC REPRESENTATION
SINGLE-V BUTT WELD ∨		
SINGLE-V BUTT WELD ∨ AND BACKING RUN ⌣		
FILLET WELD △		
SINGLE-BEVEL BUTT WELD ∨ WITH FILLET WELD SUPERIMPOSED △		
SQUARE BUTT WELD ‖		
STAGGERED INTERMITTENT FILLET WELD		a IS THE DESIGN THROAT THICKNESS b IS THE LEG LENGTH e IS THE DISTANCE BETWEEN ADJACENT WELD ELEMENTS l IS THE LENGTH OF THE WELD (WITHOUT END CRATERS) n IS THE NUMBER OF WELD ELEMENTS

AMERICAN WELDING SYMBOLS

From *Standard Welding Symbols and Rules for their Use*, published by The American Welding Society

Note: A groove weld is a weld made in the groove between two members to be joined.

Fig. A1.3. Location of information on welding symbols.

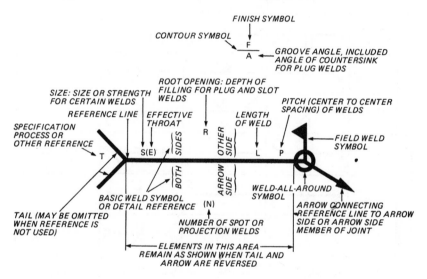

Fig. A1.4. Welding symbols.

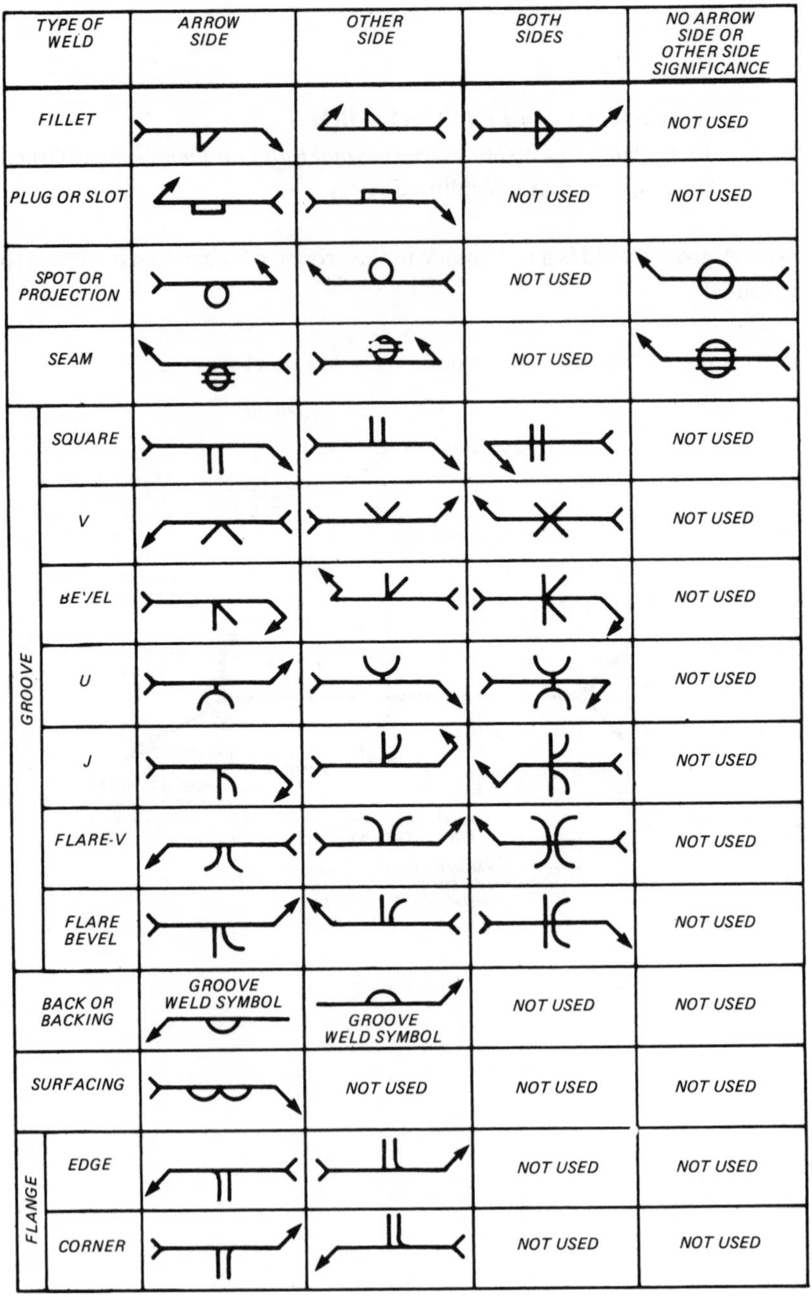

TYPE OF WELD		ARROW SIDE	OTHER SIDE	BOTH SIDES	NO ARROW SIDE OR OTHER SIDE SIGNIFICANCE
FILLET					NOT USED
PLUG OR SLOT				NOT USED	NOT USED
SPOT OR PROJECTION				NOT USED	
SEAM				NOT USED	
GROOVE	SQUARE				NOT USED
	V				NOT USED
	BEVEL				NOT USED
	U				NOT USED
	J				NOT USED
	FLARE-V				NOT USED
	FLARE BEVEL				NOT USED
BACK OR BACKING		GROOVE WELD SYMBOL	GROOVE WELD SYMBOL	NOT USED	NOT USED
SURFACING			NOT USED	NOT USED	NOT USED
FLANGE	EDGE			NOT USED	NOT USED
	CORNER			NOT USED	NOT USED

Fig. A1.5. Supplementary welding symbols.

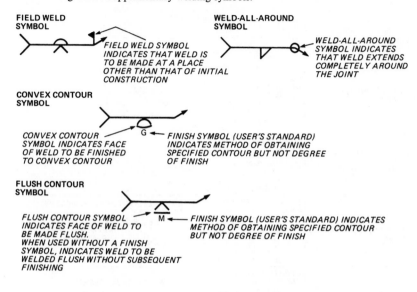

FIELD WELD SYMBOL

FIELD WELD SYMBOL INDICATES THAT WELD IS TO BE MADE AT A PLACE OTHER THAN THAT OF INITIAL CONSTRUCTION

WELD-ALL-AROUND SYMBOL

WELD-ALL-AROUND SYMBOL INDICATES THAT WELD EXTENDS COMPLETELY AROUND THE JOINT

CONVEX CONTOUR SYMBOL

CONVEX CONTOUR SYMBOL INDICATES FACE OF WELD TO BE FINISHED TO CONVEX CONTOUR

FINISH SYMBOL (USER'S STANDARD) INDICATES METHOD OF OBTAINING SPECIFIED CONTOUR BUT NOT DEGREE OF FINISH

FLUSH CONTOUR SYMBOL

FLUSH CONTOUR SYMBOL INDICATES FACE OF WELD TO BE MADE FLUSH. WHEN USED WITHOUT A FINISH SYMBOL, INDICATES WELD TO BE WELDED FLUSH WITHOUT SUBSEQUENT FINISHING

FINISH SYMBOL (USER'S STANDARD) INDICATES METHOD OF OBTAINING SPECIFIED CONTOUR BUT NOT DEGREE OF FINISH

CONTOUR		
FLUSH	*CONVEX*	

Fig. A1.6. Other typical welding symbols.

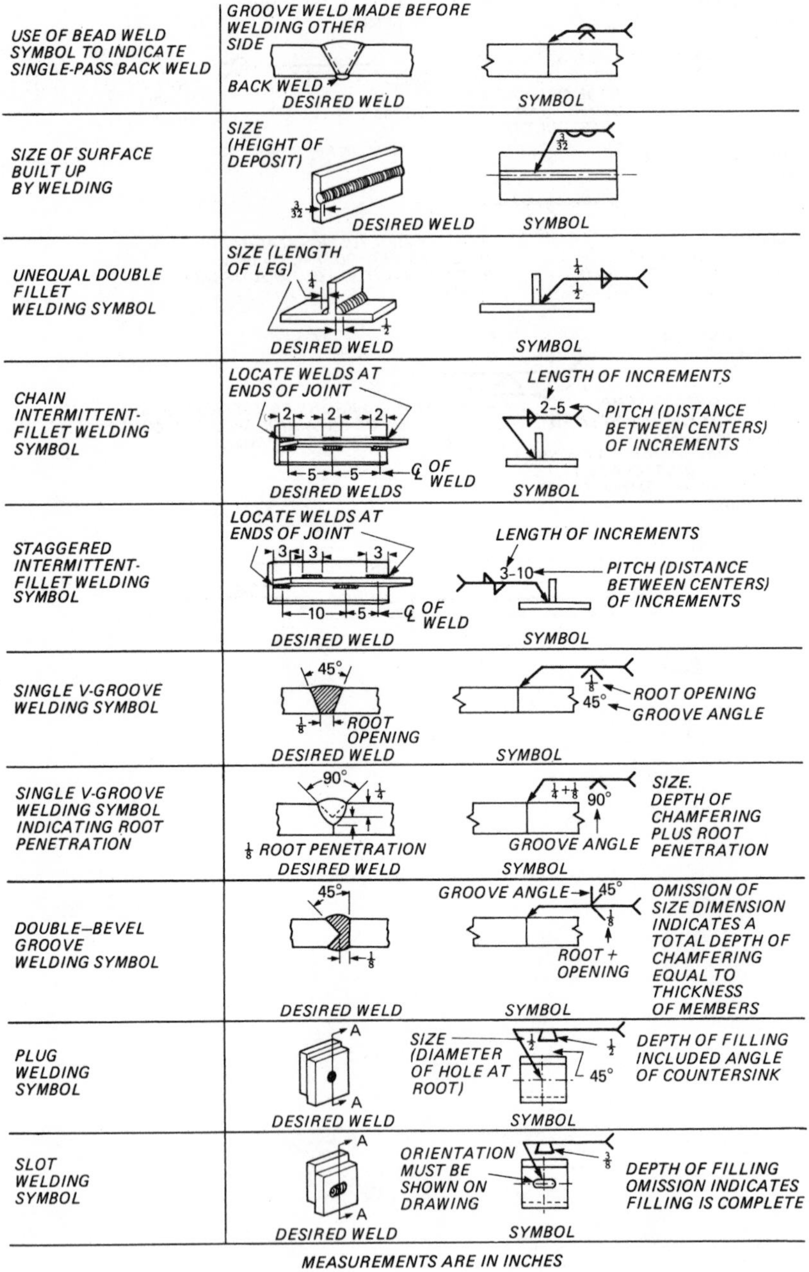

MEASUREMENTS ARE IN INCHES

Appendix 2

Simplified notes on the operation of an SCR (thyristor)

An 'n' type silicon wafer or element has impurities such as antimony, arsenic or phosphorus added to it to increase the number of negatively charged electrons available, termed 'doping'.

A 'p' type has impurities such as indium, gallium or boron added to increase the number of positive carriers or 'holes' available.

Both types have increased conductivity compared with the pure silicon.

When 'n' and 'p' types are placed in contact this forms a diode (two elements) and this will pass a current in one direction but not in the other (subject to certain temperature considerations), converting the a.c. input into half-wave pulsating d.c. The electrons flow from the 'n' type where there are free electrons to the 'p' type where there are positive 'holes'. By connecting four diodes in bridge connection full wave rectification is obtained.

The silicon controlled rectifier (SCR) has four elements with three junctions and connection is made to a further terminal called the 'gate', which is connected to the 'p' element on the cathode side as shown. The

Fig. A2.1

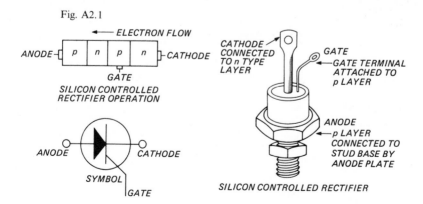

349

device can be turned on or triggered by applying a low-intensity signal to the gate and it continues to operate with no further control. The timing of this signal to the gate controls the output of the device and thus controls the welding current. The 'firing angle' of the thyristor is the point on the wave at which the gate is pulsed. (Each half-wave is 180° or π radians.)

By altering the angle (or position on the graph) at which the gate is pulsed or fired, the current in the circuit (the welding current) can be varied by a single continuous control.

If the gate is not pulsed the SCR acts as a solid state contactor and has no moving parts.

Fig. A2.2

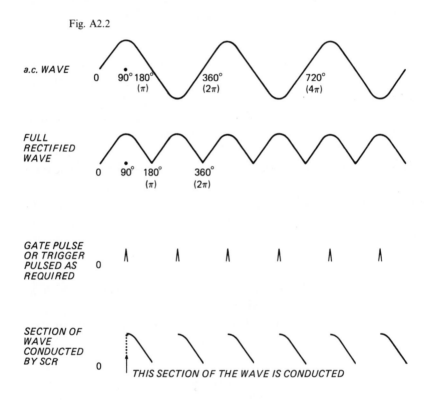

THIS SECTION OF THE WAVE IS CONDUCTED

Fig. A2.3. Simplified schematic diagram for one knob control of welding current by silicon controlled rectifier (SCR).

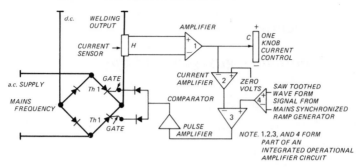

The value of welding current is set by the single knob current control *C* on the front panel. This value is compared by the control circuits with the value of output current received from *H* and the firing angle of the thyristor is altered to bring the output current to that value set by the current control.

Appendix 3

Proprietary gases and mixtures

Red colour on the shoulder of the cylinder indicates a flammable gas. Yellow colour on the shoulder of the cylinder indicates a toxic gas. Cylinders of mixed gases have the contents labelled on the shoulder of the cylinder and they have a body cylinder colour together with bands of colour around the cylinder. In many cases the composition of the contents together with the formula is stencilled on the cylinder body side. Colour codes are a guide only. Average cylinder contents 8 m³ (300 cu ft).

Proprietary gases
Air Products Ltd

Name	Cylinder contents	Cylinder colours
Argon	Argon	Peacock blue
Helium	Helium	Brown
Coogar 5	Argon 93%, CO_2 5%, oxygen 2%	Peacock blue with pastel purple neck
Coogar 20	Argon 78%, CO_2 20%, oxygen 2%	Peacock blue with pastel purple neck
Astec 15	Helium 75%, argon 23.5%, CO_2 1.5%	Pale blue. Contents stencilled in black
Astec 25	Helium 75%, argon 25%	Pale blue. Contents stencilled in black
Astec 30	Argon 70%, helium 30%	Pale blue. Contents stencilled in black
Hytec	Argon, hydrogen	Pastel purple with red neck
Apachi	Propylene	Orange with red shoulder

British Oxygen Gases Ltd

Name	Cylinder contents	Cylinder colours
Argon	Argon	Peacock blue
Helium	Helium	Brown
Argonox 1	Argon 99%, oxygen 1%	Peacock blue with broad black band round centre of cylinder
Argonox 1	Argon 98%, oxygen 2%	Peacock blue with broad black band round centre of cylinder
Argoshield 5	Argon 94%, CO_2 5%, with a trace of oxygen	Peacock blue with brunswick green band round centre of cylinder
Argomix	Argon 80%, CO_2 20%	Peacock blue with brunswick green band round centre of cylinder
Argoshield 20	Argon 79%, CO_2 20%, with trace of oxygen	Peacock blue with brunswick green band round centre of cylinder
Hyplas	Argon 95%, hydrogen 5%	Peacock blue with signal red band round centre of cylinder
Helishield 1	Helium 83%, argon 15%, CO_2 1.5%, with a trace of oxygen	Brown with blue band round centre of cylinder
Helishield 2	Helium 75%, argon 25%	Brown with blue band round centre of cylinder
Hylite	Argon 99%, hydrogen 1%	Peacock blue with signal red band round centre of cylinder

Distillers Company (Carbon Dioxide) Ltd

Name	Cylinder contents	Cylinder colours
Carbon dioxide	Carbon dioxide	CO_2 commercial liquid withdrawal – black. With 25 mm white stripe down cylinder length CO_2 commercial vapour withdrawal – black.

Appendix 4

Tests for wear-resistant surfaces

The following tests for wear are used

(1) *Adhesive wear test* in which a test specimen in the form of a block is pressed or loaded against a rotating alloy wear test specimen in the form of a ring or of a comparison standard. The block and ring may be of similar or dissimilar alloy and the ring may be lubricated (optional). The volume loss for various loads is measured for a given number of revolutions.

(2) *Abrasive wear test* in which dry sand is fed between the surfaces of a wear test specimen which is pressed or loaded against a rotating rubber wheel. The volume loss of the specimen for a given number of revolutions and load is measured.

(3) *Galling test* in which a rotating cylindrical pin made from undiluted weld metal (e.g. TIG) is pressed (loaded) against a standard ground block and the stress in kg/mm² at which visual sign of material transfer (galling) occurs after one revolution of the pin is noted.

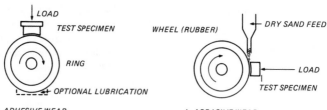

a ADHESIVE WEAR   b ABRASIVE WEAR

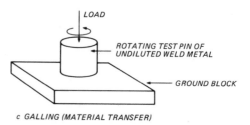

c GALLING (MATERIAL TRANSFER)

Conversion factors

inch	mm	Preferred mm	inch	mm	Preferred mm
$\frac{1}{64}$	0.4	—	$\frac{5}{16}$	8.0	8
$\frac{1}{32}$	0.8	—	$\frac{3}{8}$	9.5	10
$\frac{1}{16}$	1.6	—	$\frac{7}{16}$	11.1	11
$\frac{3}{32}$	2.4	—	$\frac{1}{2}$	12.7	12
$\frac{1}{8}$	3.2	3	$\frac{9}{16}$	14.4	—
$\frac{5}{32}$	4.0	4	$\frac{5}{8}$	16.0	15
$\frac{3}{16}$	4.8	5	$\frac{11}{16}$	17.6	18
$\frac{1}{4}$	6.4	6	$\frac{3}{4}$	19.0	20
			$\frac{7}{8}$	22.4	22
			1	25.4	25

SWG	mm	SWG	mm
28	0.38	16	1.6
26	0.46	14	2.0
24	0.56	12	2.5
22	0.71	10	3.25
20	0.91	8	4.0
18	1.22	6	5.0
		4	5.9

1 kgF = 9.8 N (10 N approx.).
1 bar = 14.5 lbf/in^2 = 0.1 N/mm^2.
1 tonf/in^2 = 15.4 N/mm^2 = 1.54 hbar.
1000 mb = 1 bar = 14.5 lbf/in^2 = 1 kgf/cm^2.
1 hbar = 100 bar.
1 cu. ft = 0.028 m^3 = 283 litres.

Appendix 6

Low hydrogen electrode, downhill pipe welding

A d.c. power source is used as with cellulosic rods (p. 71) and is usually engine driven in the field, the electrode being connected to the $-$ ve pole. Weld metal deposition is faster than with cellulosic or uphill low hydrogen methods and a short arc is used with no 'pumping' of the electrode. Since shrinkage is important, if internal clamps are used there should be more spacers. The preparation of a joint using different diameter pipes is show in Fig. A6.1 and the tack welds should be tapered at each end.

For the root run the electrode should lightly touch the joint and the weld should continue through 6 o'clock about 2 mm towards 7 o'clock to facilitate link up with the run to be deposited on the other side. There is no weave and keeping a short arc and high travel rate the slag is kept above the

Fig. A6.1. Joint preparation.

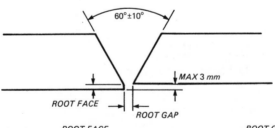

ROOT FACE	
ELECTRODE DIAMETER	FACE
2.5 mm	
3.25 mm	

ROOT GAP	
ELECTRODE DIAMETER	GAP
2.5 mm	2.5 mm + 1.0 mm
3.25 mm	2.5 mm + 1.0 mm

Note: further reduction of the root gap is possible in thinner wall thicknesses, for experienced pipeline welders using the 3.25 *mm* electrode.

Fig. A6.2   Fig. A6.3

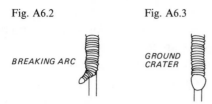

arc at a distance of about the coating diameter. The arc is broken by turning into the bevels and the metal is tailed off as shown in Fig. A6.2.

The finishing crater should be ground to a pear-shaped taper (Fig. A6.3) and the next electrode is started about 10 mm above the top end of the tapered crater. Slag is removed from each run with power brush and chipping hammer and grinding is used to remove any excess metal generally from 4 to 6 o'clock. The penetration bead may be slightly concave but has a smooth tie-in with the pipe metal and is acceptable. Too much heat gives increasingly convex penetration and is due to too slow travel or too high a current.

Avoid burn through between 12 and 1 o'clock when using a leading arc by keeping the arc as short as possible and directing it alternately to each side of the joint. Fig A6.4 shows angles of electrode, electrode diameters and currents.

For the filling and capping runs the arc is kept short with a weave not exceeding 3 to 4 times the core wire diameter and the slag–electrode distance

Fig. A6.4. Angles of electrode, electrode diameters and currents, root run.

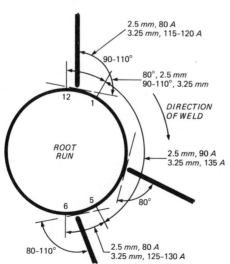

should be about 10 mm (Fig. A6.6). Fig A6.5 shows angles of electrode, electrode diameter and currents, Fig. A6.6 the slag–electrode tip distance and Fig A6.7 the restrike and electrode movement.

Low hydrogen electrodes are supplied in sealed air-tight tins and should be stored as on p.13, the hydrogen content being 5 ml per 100 g of weld metal (IS03690). If the electrodes are left unused for 8 hours or more they should be rebaked at 280 °C + 20°C for 1 hour and stored in a holding oven at 120–150 °C until used. They should not be exposed to any damp conditions

Fig. A6.5. Angles of electrode, electrode diameters and currents, filling and capping runs.

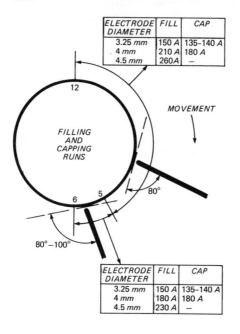

ELECTRODE DIAMETER
FILL
CAP
3.25 mm
150 A
135–140 A
4 mm
210 A
180 A
4.5 mm
260A
–
12
MOVEMENT
FILLING AND CAPPING RUNS
80°
6
5
80°–100°
ELECTRODE DIAMETER
FILL
CAP
3.25 mm
150 A
135–140 A
4 mm
180 A
180 A
4.5 mm
230 A
–

Fig. A6.6

MOVEMENT
SLAG
10 mm
ELECTRODE TIP

Fig. A6.7

MOVEMENT
STRIKE

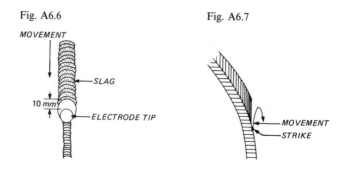

during use and the joint should be completely free of all moisture. The weld metal has high resistance to cracking and is of composition C 0.06–0.09%, Mn 1.00–1.40%, Si 0.03–0.07% and classification AWS A5.5 E8018-G BS 639 E5154B 120 90(H).

Appendix 7

The manufacture of extruded MMA electrodes

As previously stated, almost all electrodes are produced by the extrusion process. The present day electrode consists of a core wire, centrally placed in a coating, usually silicate bonded (Vol. 1, p. 60). Double coated electrodes are also produced.

A semi-automatic production line may consist of the following:

(1) *Wire drawing.* Non-slip hydraulic machines draw the wire through dies to the correct size, after which the wire is stored on reels.

(2) *Wire cutting.* The wire is then straightened and cut to electrode length, giving burr-free wire rods of any required length which are stacked in bins.

(3) *The coating.* The raw materials of the coatings are kept in raw material silos and from these the materials are sieved and dosed. It is the type and composition of these materials that determines the characteristic of the particular electrode. The dosing installations consist of a number of silos each fitted with a discharge system, dosing control and weighing facilities.

(4) *Silicate storing* consists of tanks that are temperature controlled and from which the silicate is pumped to the mixer.

(5) *Dry blending and wet mixing.* The batch is first thoroughly mixed in the dry state and then a measured quantity of silicate is added from a dosing unit. The wetted batch is then roughly mixed, emptied and removed to the slug press, where it is pressed into slugs ready for extrusion.

(6) *Extrusion.* The slugs are placed into the charging trough and when this is full the load of slugs is thrust into the extrusion cylinder. The breech of the hydraulic extrusion cylinder is then locked by a simple lever movement. The cut and deburred wires are stored adjacent to the wire feed from which, by opening a slide in the bin

side, the wire rods drop into the feed magazine. The extruder piston is then operated and when the flux flow pressure is reached the wire feed commences. Accelerator and drive rolls take the rods dropping from the magazine and push them axially through the press to the extrusion head where they receive their coating.

(7) *Concentricity check and brushing.* It is important that the core wire should be centrally placed in the covering (pp. 5–6). This is checked electronically and adjustments made if required. A transfer belt then takes the electrodes to a brushing station where the end for the electrode holder and the striking tip are formed, after which the number or type of electrode is printed on the coating by a printing machine.

(8) *Drying and baking.* The electrodes are air dried to prevent cracking of the coating and are then loaded into baking ovens, after which they are ready for packing.

Fig. A7.1. Simplified diagram of electrode production process.

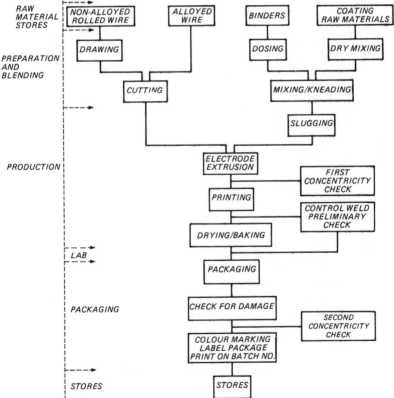

Where high capacity is required, two extruders are placed in parallel. When one is being charged the other is working, so that the same brushing and drying ovens are used and delay in production is minimised. In many cases the high speed wire drawing and cutting machines are fully automatic and the drying ovens are of the bucket type with conveyor chains up to 1000 m long and temperatures up to 450 °C.

Appendix 8

Notes on fire extinguishing (BS 5306, Part 3 (1980))

For a fire to occur we need *oxygen, fuel* and *heat*.

Oxygen is present in the atmosphere and in the compressed form contained in cylinders used for oxy-acetylene or oxy-propane cutting.

Fuel may be any combustible substance such as petrol, diesel oil, paper, cleaning materials, rubber or litter etc. – anything that will burn.

Heat may be supplied by discarded cigarette ends, match sticks and, in the welding shop, by sparks from the arc or bad connections such as those at the earth or return terminal to the set. The arc is an intense source of heat and, together with the sparking, will rapidly cause ignition of any combustible solids or liquids in the immediate vicinity.

If any one of these above causes is removed, the fire will be contained. For example the oxygen supply from the atmosphere may be contained by smothering the fire with a blanket or dry sand or by turning off the oxygen supply if it is from a cylinder. The heat can be controlled by ceasing to use the arc and switching off all power supplies to prevent further arcing at poor contacts.

Classes of fire with suitable extinguishing media

(*A*) Solid materials (carbonaceous fires forming glowing embers). Use water type extinguisher (cooling method) except on electrical apparatus.

(*B*) Liquids or liquifiable solids (petrol, paraffin, wax, etc.). Use foam, dry powder, CO_2, vapourizing liquid, fire blanket, dry sand. (smothering method).

(*C*) Flammable gases -liquid petroleum gas including propane and butane, acetylene). Turn off the supply (starvation method).

(*D*) Metals (magnesium, metallic powders). Special powders containing sodium or potassium chloride are required. Such fires should be dealt with by the fire brigade.

Type of extinguishers

(1) *Carbon dioxide CO_2*. This contains liquid CO_2 under pressure, which is released as a gas when the trigger or plunger on the extinguisher is operated.

(2) *Extinguishing powder (gas cartridge)*. The powder is expelled by the pressure from a gas cartridge.

(3) *Extinguishing powder (stored pressure)*. The powder is expelled by pressure stored within the body of the extinguisher.

(4) *Foam (chemical)*. Chemicals stored in the body of the extinguisher are allowed to mix and react, producing foam, which is expelled from the extinguisher.

(5) *Foam (gas cartridge)*. Foam is expelled from the extinguisher by the pressure from a gas cartridge.

(6) *Foam (mechanical stored pressure)*. Pressure stored in the body of the extinguisher expels the foam.

(7) *Halon (stored pressure)*. The body of the extinguisher contains a halon which is expelled by pressure stored within the body.

(8) *Water (gas cartridge)*. Water is expelled from the extinguisher by the pressure from a gas cartridge.

(9) *Water (stored pressure)*. Water is expelled by pressure stored within the extinguisher body.

(10) *Water (soda-acid)*. An acid–alkaline reaction within the body produces the pressure to expel the water.

Extinguishers containing dry powder, CO_2 and vaporizing liquids
These extinguishers are suitable for fires involving flammable liquids and live electrical apparatus (as are met with under welding conditions). For fires in liquids (containers or spillage) the jet or horn should be directed at the near side of the fire, which should be driven away from the operator to the far side with a side-to-side motion of the jet. If falling liquid is alight begin at the bottom and move upwards. On electrical apparatus direct the jet directly at the fire after first switching off the mains supply. If the latter is not possible, contain the fire and summon the fire brigade.

If there is no shut-off valve on the extinguisher continue discharging it over the fire area until empty. If the discharge is controlled, shut off the extinguisher discharge when the fire is extinguished and hold it ready for any further outbreak which may occur when the atmosphere has cleared.

Do not use vaporizing liquid extinguishers in confined spaces where there is poor ventilation and a danger of fumes being inhaled.

Foam extinguishers give semi-permanent protection against contained fires and will partially extinguish a fire, which will not gain its full intensity until the foam covering is destroyed. They may be used on liquids, in tanks to protect them from ignition or from giving off flammable vapours. Dry powder and foam should not be used together unless a specially compatible powder/foam are used.

BCF extinguishers

When this type is used the bromochlorodifluoromethane, which is a stable compound, appears as a heavy mist that blankets the fire and excludes the atmospheric oxygen. It is more efficient than chlorobromomethane (CB), methyl bromide (MB) or carbon tetrachloride (CTC). It does not damage normal materials and leaves no deposit. It decomposes slightly in the heat of a flame but the amount of harmful gas produced (which has a pungent smell) has no effect on adjacent personnel even when excess BCF is used. It has high insulating properties and can be used on electrical equipment and is particularly suitable for the welding or fabrication shop since it can be used on Class *A* fires in addition to those of Class *B* and *C*.

Fire prevention where welding processes are involved

General precautions

(1) All areas must be free of litter and combustible waste with all combustible material stored in metal bins, with lids, well away from the welding processes.

(2) Make certain that all electrical cables are not chafing and connections are secure and will not cause arcing. This applies particularly to the earth or return connection to the welding set.

(3) All chemicals including cleaning fluids should be kept away from the welding area, in a store room.

(4) Oxygen cylinders should be kept in a room apart from fuel gases such as acetylene and propane (LPG), which should be kept in another room. All cylinders and manifolds should be turned off except when in use. Keep cylinder keys with the appropriate cylinders in case of urgent need.

(5) No smoking signs should be prominently displayed and should be strictly adhered to.

(6) Defects in plant and equipment should be reported at once and indicated. Faulty electrical equipment is always a fire hazard.

(7) Always read the instructions for operation of the extinguisher situated in your shop.

Notes on procedures to be followed when using welding apparatus.

Leakage of oxygen, as for example from a cylinder used in the cutting process, may lead to the ignition of any glowing or red-hot material due to the rapid oxidizing action of the escaping oxygen. The gas must be turned off at the cylinder valve (leave the cylinder key attached to the cylinder so that it is never misplaced). Remember that should flames play on a cylinder containing oxygen, there is a danger of explosion due to the high pressure involved – although oxygen is non-combustible it is a powerful supporter of combustion and may cause re-ignition.

Fuel gas cylinders

LPG, acetylene, etc. and accompanying apparatus and appliances

(1) Gas leak on fire. If the fire is away from regulators and cylinders first turn off the gas supply at the cylinder or appliance valve. If this is not possible and the fire is spreading to nearby combustible material, extinguish the flames and then seal the leak, taking care that re-ignition is not possible.

If the gas is on fire near a regulator or cylinder valve first extinguish the flames if this is possible within two or three minutes of the beginning of the combustion and use a cloth or gloves (e.g. welders) to turn off the valve and guard against re-ignition.

If the valve cannot be shut, evacuate the area and call the fire brigade.

If the gas is on fire and the flames are impinging on the gas cylinders, first extinguish the flames if this is possible within two to three minutes of the conflagration. If this is not possible, divert the flames from the affected cylinder away from the other cylinders and combustible material, evacuate the area and call the fire brigade.

Any cylinders exposed to the heat of fire should be removed within 2–3 minutes. If this is not possible, evacuate the area and call the fire brigade.

A burning gas leak can be extinguished by a dry powder, BCF or water type fire extinguisher in this order of preference but small fires may be smothered with a wet blanket or cloth.

A water type fire extinguisher should be used on any adjacent burning material and whilst dry powder or BCF types can be used in this case they will have only a limited effect.

Do not continue to fight a fire if there is a danger of gas cylinders exploding or if the fire continues to grow in spite of your efforts to contain it. Remember that extinguishers are only intended to deal with small fires so that the quicker the extinguisher is brought into use the better as fires spread quickly. Position yourself between the fire and the exit from the shop so that your escape will not be cut off.

If there is any possibility of the conflagration getting out of control in the slightest degree, call the Fire Brigade whose personnel are trained in fire fighting. They must be met on arrival and the officer in charge advised of the presence and position of all gas cylinders.

(2) Gas leak. No fire.

(*a*) The danger is from an explosion if the fuel gas is ignited. Keep all sources of ignition away from the gas. Do not smoke, switch on electric lights, electrical apparatus or telephones, etc. and do not drive vehicles in the vicinity.

(*b*) Shut off the cylinder valve or pipeline. If this is impossible move the cylinder(s) to the open air so as to dissipate the gas.

(*c*) Do not re-occupy the shops until they are properly ventilated and pay attention to pockets of gas that may exist in ducting, sumps and pits and see that they are completely purged.

Appendix 9

Tables of brazing alloys and fluxes

General purpose silver brazing alloys: cadmium bearing

Description	Nominal composition	Melting range (°C)		BS 1845	Remarks
		Solidus	Liquidus		
Easy-flo	50% Ag:Cu:Cd:Zn	620	630	AG1	Cadmium-bearing alloys generally offer the best combination of melting range, flow characteristics and mechanical properties. Thus, where the presence of cadmium is acceptable, they would normally be recommended as the most economical alloys.
Easy-flo No. 2	42% Ag:Cu:Cd:Zn	608	617	AG2	
DIN Argo-flo	40% Ag:Cu:Cd:Zn	595	630	—	
Argo-flo	38% Ag:Cu:Cd:Zn	608	655	—	Easy-flo, Easy-flo No. 2 and DIN Argo-flo are the lowest melting point alloys available. All three alloys have excellent flow characteristics. The highest joint strengths are obtained with Easy-flo.
Mattibraze 34	34% Ag:Cu:Cd:Zn	612	668	AG11	
Argo-swift	30% Ag:Cu:Cd:Zn	607	685	AG12	
Metalflo	25% Ag:Cu:Cd:Zn:Si	606	720	—	Alloys with 30–40% silver are normally considered as general-purpose alloys. They still offer good flow characteristics and melting ranges and are suitable for mechanized brazing operations.
Argo-bond	23% Ag:Cu:Cd:Zn	616	735	—	
Metalsil	20% Ag:Cu:Cd:Zn:Si	603	766	—	
Metalbond	17% Ag:Cu:Cd:Zn:Si	610	782	—	The low silver alloys have good fillet-forming characteristics and are widely used in fabricating brass fittings and furniture. Owing to their wide melting ranges they are not generally used as preforms due to the danger of liquation.
Metaloy	13% Ag:Cu:Cd:Zn:Si	605	795	—	

General purpose silver brazing alloys: cadmium free

Description	Nominal composition	Melting range (°C)		BS 1845	Remarks
		Solidus	Liquidus		
Silver-flo 55	55% Ag:Cu:Zn:Sn	630	660	AG14	The range of cadmium-free alloys has been developed for applications where the presence of cadmium is not allowed due to the brazing environment or the service conditions of the assembly, i.e. food-handling equipment.
Silver-flo 55	45%Ag:Cu:Zn:Sn	640	680	—	
Silver-flo 452	45%Ag:Cu:Zn:Sn	640	680	—	
Silver-flo 45	45% Ag:Cu:Zn	680	700	—	
Silver-flo 44	44%Ag:Cu:Zn	675	735	—	
Silver-flo 40	40%Ag:Cu:Zn:Sn	650	710	AG20	Silver-flo 55 is the lowest melting-point alloy in the group and has been widely adopted as a substitute for Easy-flo No. 2.
Silver-flo 38	38% Ag:Cu:Zn:Sn	660	720	—	
Silver-flo 34	34%Ag:Cu:Zn:Sn	630	740	—	
Silver-flo 33	33%Ag:Cu:Zn	700	740	—	
Silver-flo 302	30%Ag:Cu:Zn:Sn	665	755	AG21	Silver-flos 30 to 45 are general-purpose alloys suitable for use on most common engineering materials.
Silver-flo 30	30%Ag:Cu:Zn	695	770	—	
Silver-flo 25	25%Ag:Cu:Zn	700	800	—	
Silver-flo 24	24% Ag:Cu:Zn	740	780	—	The alloys with less than 24% silver are widely used on copper and steel. They give a good colour match on brass, but require close temperature control due to their high melting temperature.
Silver-flo 20	20% Ag:Cu:Zn:Si	776	815	—	
Silver-flo 18	18%Ag:Cu:Zn:Si	784	816	—	
Silver-flo 16	16%Ag:Cu:Zn	790	830	—	
Silver-flo 12	12%Ag:Cu:Zn	820	840	—	Unlike the cadmium-bearing range, this group contains low silver content alloys such as Silver-flo 33, 24 and 16, with relatively high melting points, which make them ideal for step brazing.
Silver-flo 4	4% Ag:Cu:Zn	870	890	—	
Silver-flo 1	a% Ag:Cu:Zn:Si	880	890	—	

General purpose silver brazing alloys: phosphorus bearing

Description	Nominal composition	Melting range (°C)		BS 1845	Remarks
		Solidus	Liquids		
Sil-fos	15% Ag:Cu:P	644	700**	CP1	These alloys are recommended for fluxless brazing of copper. They may also be used on brass and bronze with the application of Easy-flo Flux or Tenacity 4A Flux. They should not be used on ferrous- or nickel-base materials or nickel-bearing copper alloys.
Sil-fos 5	5% Ag:Cu:P	650	710**	CP4	
Silbralloy	2% Ag:Cu:P	644	740**	CP2	
Copper-flo	7% P:Cu	714	800	CP3	
Copper-flo No. 2	6% P:Sb:Cu	690	800	CP5	
Copper-flo No. 3	6% P:Cu	714	850	CP6	The silver bearing alloys are more ductile than the Copper-flos and are recommended where joints will be subjected to significant stress levels.

** A small percentage of higher melting point material remains at this temperature.

Silver brazing alloys for tungsten carbide

Description	Nominal composition	Melting range (°C)		BS 1845	Remarks
		Solidus	Liquidus		
Easy-flo No. 3	50% Ag:Cu:Zn:Cd:Ni	634	656	AG9	General-purpose cadmium-containing alloy with excellent mechanical properties.
Argo-braze 38	38% Ag:Cu:Zn:Cd:Ni	615	655	—	An economical alternative to Easy-Flo No. 3 which has proved adequate for many applications.
Argo-braze 49H	49% Ag:Cu:Zn:Ni:Mn	680	705	—	A cadmium-free alternative to Easy-flo No. 3. The addition of manganese enhances wetting on difficult carbides.
Argo-braze 49LM	49% Ag:Cu:Zn:Ni:Mn	670	710	—	A free-flowing alloy giving good wetting. Suitable for small carbides.
Argo-braze 40	40% Ag:Cu:Zn:Ni	670	780	—	Economical alloys with good wetting characteristics. Their application is restricted by their relatively high melting points.
Argo-braze 25	25% Ag:Cu:Zn:Ni:Mn	710	810	—	
Easy-flo Tri-foil 'C'	Easy-flo bonded to both sides of a copper shim	620	630	—	For brazing large pieces of cemented carbide. The copper ensures that a thick stress-absorbing joint is achieved.
Easy-flo Tri-foil 'CN'	Easy-flo bonded to both sides of a copper-nickel shim	620	630	—	Recommended for joints subjected to high compressive stresses during service.
AB49 Tri-foil	Argo-braze 49LM bonded to both sides of a copper shim	670	710	—	A cadmium-free alternative to Easy-flo Tri-foil 'C'.

Brazing fluxes

Fluxes are normally selected on the basis of two main criteria: 1. The melting range of the brazing alloy – both solidus and liquidus should fall well within the quoted working range of the flux. 2. The parent metals to be joined – some alloys, such as aluminium bronze, require special fluxes. Copper, brass and mild steel are effectively cleaned by all fluxes.

Description	Working range °C	DIN 8511	Availability	Remarks	Residue removal
Easy-flo Flux	550–800	F-SH1	Powder/Paste	A general-purpose flux with good fluxing activity and long life at temperature. The powder has good hot-rodding characteristics.	Residues are generally soluble in hot water. Where difficulty is encountered, immersion in 10% caustic soda is suggested.
Easy-flo Flux Dipping Grade	550–775	F-SH1	Paste	A highly active and fluid flux which exhibits a minimum of bubbling, making it ideal for induction brazing. This paste is formulated to give a thin, stable consistency suitable for dipping.	
Easy-flo Flux Stainless Steel Grade	550–775	F-SH1	Powder	More active than Easy-flo Flux. Recommended for use on stainless steel with the lower melting-point brazing alloys.	
Easy-flo Flux Aluminium Bronze Grade	550–800	F-SH1	Paste	Similar for Easy-flo Flux Paste but modified for use on alloys containing up to 10% aluminium.	When finished components are heavily oxidized, cleaning and flux removal may be accomplished by the use of 10% sulphuric acid.

Brazing fluxes (cont.)

Description	Working range °C	DIN 8511	Availability	Remarks	Residue removal
Tenacity Flux No. 6	550–800	F-SH1	Powder/Paste	An opaque flux due to the addition of elemental boron. Recommended for tungsten carbide, refractory metals and stainless steel.*	
Tenacity Flux No. 14	550–750	F-SH1	Powder	An extremely fluid principally developed to prevent red staining on brass.	
Tenacity Flux No. 4A	600–850	F-SH/1/2	Powder	A general-purpose flux with good resistance to overheating, used with higher melting temperature.	
Tenacity Flux No. 5	600–900	F-SH1/2	Powder	Recommended for stainless steel and heavy assemblies, where flux exhaustion is likely to occur due to prolonged heating	Residues are virtually insoluble in water.
Tenacity Flux No. 125	750–1200	F-SH2/3	Powder/Paste	A general-purpose high temperature flux for use with the bronze alloys.	Immersion in 10% caustic soda or mechanical removal is recommended.
Tenacity Flux No. 12	800–1300	F-SH3	Powder	A fluoride-free, high temperature flux for use with 'B' Bronze and 'G' Bronze.	
Silver-flo	550–775	F-SH1	Paste	A general purpose flux paste which combines exceptional activity with low fluxing temperature and melt viscosity.	
Mattiflux 100	550–800	F-SH1	Paste	Combines a relatively low melt viscosity with a good resistance to overheating and is particularly well suited to brazing mild steel assemblies.	

* This flux is unsuitable for use on stainless steel where crevice corrosion is likely to be a hazard in service.

Specialised silver brazing alloys

Description	Nominal composition	Melting range (°C)		BS 1845	Remarks
		Solidus	Liquidus		
Silver-flo 60	60% Ag:Cu:Zn	695	730	AG13	Principally used for brazing assemblies which will be exposed to sea water. Easy-flo No. 3 and Silver-flo 55 are also suitable for marine service.
Silver-flo 43	43% Ag:Cu:Zn	690	775	AG5	Specifically developed to prevent crevice corrosion, which may result in rapid failure of joints in stainless steel when exposed to water.
Argo-braze 56	56% Ag:Cu:In:Ni	600	711	—	
Argo-braze 35	35% Ag:Cu:Sn:Ni	685	887	—	An economical alloy suitable for fluxless brazing. Principally used on stainless steel and nickel silver.
Silver-copper eutectic	72% Ag:Cu	778	778	AG7 and AG7V	Ideal for fluxless brazing of copper, nickel and metallized ceramics. Available in two grades: Grade 1 (AG7V), a high purity material for brazing assemblies which will operate in vacuum. Grade 3 (AG7) for general engineering.
RTSN	60% Ag:Cu:Sn	602	718	—	Suitable for fluxless furnace brazing. Relatively fast heating rates necessary to avoid liquation.
15% Manganese-Silver	85% Ag:Mn	951	960	AG19	A copper-free alloy for brazing assemblies which will be in contact with ammonia.

Flux coated rods

Alloy	Nominal composition	Melting range (°C)	BS 1845	Remarks
Cadmium bearing				
Easy-flo No. 2	42% Ag:Cu:Zn:Cd	608–617	AG2	A very fluid alloy, ideal for small components.
DIN Argo-flo	40% Ag:Cu:Zn:Cd	595–630	—	A general-purpose alloy with good flow characteristics.
Mattibraze 34	34% Ag:Cu:Zn:Cd	612–668	AG11	Fillet-forming alloys with limited flow characteristics.
Argo-swift	30% Ag:Cu:Zn:Cd	607–685	AG12	
Cadmium free				
Silver-flo 55	55% Ag:Cu:Zn:Sn	630–660	AG14	A free-flowing alloy generally suitable for all applications where a neat appearance is required.
Silver-flo 40	40% Ag:Cu:An:Sn	650–710	AG20	A range of cadmium-free alloys with fillet-forming properties.
Silver-flo 33	33% Ag:Cu:Zn	700–740	—	
Silver-flo 30	30% Ag:Cu:Zn	695–770	—	An economical alloy suitable for brazing mild steel and copper.
Silver-flo 20	20% Ag:Cu:Zn:Si	776–815	—	

Copper base brazing alloys

Description	Nominal composition	Melting range (°C)		BS 1845	Remarks
		Solidus	Liquidus		
'B' Bronze	97% Cu:Ni:B	1081	1101	CU7	Ideal substitute for pure copper in furnace brazing where joint gaps range from interference to 0.5 mm.
'C' Bronze	86% Cu:Mn:Ni	965	995	—	Suitable for furnace brazing. Requires a dewpoint of better than − 40°C, or a small addition of flux.
'D' Bronze	86% Cu: Mn: CO	980	1030	—	Recommended for brazing cemented carbide into rock drills and similar percussively loaded joints.
'F' Bronze	58% Cu:Zn:Mn:Co	890	930	—	
'G' Bronze	97% Cu:Ni:Si	1090	1100	—	The group offers a range of brazing characteristics to match different material and heat-treatment requirements.
Argentel	49% Cu:Zn:Ni:Si	913	930	CZ8	Low cost alloys ideal for use on tubular steel furniture and display racks. The high strength of Argentel makes it particularly suitable for butt joints while Argentel B offers improved flow characteristics and a lower working temperature.
Argentel B	57% Cu: Zn: Ni: Mn: Si	890	910	—	

Silver-bearing soft solders

| Description | Nominal composition | Melting range (°C) | | Remarks |
		Solidus	Liquidus	
Plumbsol	2.5% Ag:Sn	221	225	Lead-free alloys ideal for use in plumbing and the fabrication of assemblies where toxicity is a problem.
P35	3.5% Ag:Sn	221	221	Good colour match on stainless steel.
P5	5% Ag:Sn	221	235	P35 and P5 are stronger than tin-lead alloys at elevated temperatures.
Ceramic No. 1	9% Ag:Sn:Pb	183	250	For soldering of silver-metallized components in the electronics industry.
Comsol	1.5% Ag:Sn:Ph	296	296	Good creep resistance at elevated temperatures. Suitable for soldering armature windings and radiators.
A25	2.5% Ag:Pb	304	304	For use in 'hard chlorinated waters where tin-bearing alloys suffer corrosion.
A5	5% Ag:Pb	304	370	
LM10A	10% Ag:Sn:Cu	214	275	A free-flowing alloy with above average tensile strength and creep resistance.
LM15	5% Ag:Cd:Zn	280	320	The strongest alloy in the soft solder range. Used where temperature limitations do not permit silver brazing.
LM5	5% Ag:Cd	338	390	Can be considered for applications where good strength at moderate temperatures is necessary.

Soft solder fluxes

Description	Flux residue removal	Working range (°C)	Remarks
Soft Solder Flux No. 1	Corrosive – can be removed with cold water	150–400	A general-purpose acid flux suitable for copper, brass, mild steel and stainless steel.
Soft Solder Flux No. 2	Non-corrosive – can be removed with cold water	150–400	Recommended where flux residues cannot be removed. Suitable for copper, bass and most plated surfaces.

Appendix 10

Synergic MIG welding

The output of the power source is modified to give a square wave that is also pulsed (pp. 145–6). The power unit is directly linked to the wire feed mechanism and a p.c.b. enables the one-knob control to match the burn off rate with the arc voltage and the wire feed speed. In effect, with correct pulse parameters, one droplet is projected with each pulse, the background current being kept low with high peak current values so that positional welding can be performed. This results in, it is claimed, less spatter, burnback and stubbing and less pre-heat is required. Welders are easily trained in this process which can be usefully employed in heavy section fabrication, and in narrow gap and stainless steel sections, for example. Low current values are obtainable for thinner sections.

The following terms are also used by certain manufacturers.

SAM (servo-adjusted MIG). This uses constant average arc current which provides constant heat output, depending upon wire size and type and shielding gas. It is maintained by feedback from the welding arc.

SPM (synchro-pulsed MIG). This gives uniform detachment of the droplet by automatic frequency regulation when wire size and type and shielding gas are programmed for any required wire feed speed. It is maintained by feedback regulation.

City and Guilds of London Institute
Examination questions

Note: All dimensions are given in millimetres, unless otherwise stated.

Manual metal arc

1 State *four* precautions which should be taken to protect the eyes and skin of the welder and other workers exposed to radiation from arc welding operations.

2 Sketch and label *three* defects that may occur in a double-vee butt welded joint which would affect fatigue strength.

3 Give *three* reasons why hydrogen-controlled electrodes are often used in preference to general purpose electrodes for welding low-alloy steels.

4   (*a*) Explain why the degree of penetration is greater using a cellulose-covered electrode compared to a rutile-covered electrode using the same heat input.

  (*b*) An electrode used for manual metal-arc welding has over 100 per cent metal recovery. State

    (1) the addition which must have been made to the covering,

    (2) *two* advantages of using this type of electrode.

5 The figure shows two views of a fractured grey cast-iron bracket which is to be repaired by the manual metal-arc process.

    (*a*) Make a pictorial sketch of the bracket.

    (*b*) List *eight* items of information that should be included on a welding procedure sheet for the repair of the bracket.

    (*c*) State *four* additional difficulties that would occur if the bracket has been made from 3% silicon alluminium alloy.

6 Give *three* reasons why hot cracking may be a problem when manual metal-arc welding austenitic stainless steel.

7 List *four* factors that should be considered when determining the pre-heating temperature to be used for welding a steel fabrication.

Question 5

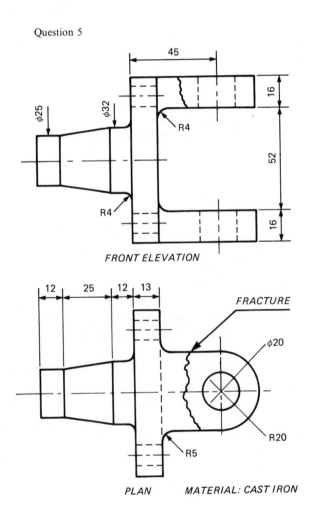

8 (*a*) A 30 mm thick alloy of the composition shown has to be manual metal-arc welded using rutile-coated electrodes.
Composition: 0.2% carbon; 2% chromium; 0.5% molyb denum; 0.8% manganese; 0.4% silicon; 0.35% sulphur; 0.035% phosphorus; remainder iron.

$$\text{Carbon Equivalent} = \%C + \frac{\%Mn}{20} + \frac{\%Ni}{15} + \frac{\%Cr + \%Mo + \%V}{10}$$

(1) Using the formula given calculate the carbon equivalent of the alloy.
(2) From the information given in the table below construct a graph and from it determine the pre-heating temperature of the alloy.

Carbon equivalent %	0.3	0.35	0.4	0.45	0.5	0.55	0.6
Pre-heat temp. (°C)	100	150	200	250	300	350	400

(3) If a hydrogen-controlled electrode is used to weld the material explain why the pre-heating temperature may be lowered by 100°C.

(*b*) Two of the hazards which may arise in welding are explosions and fumes. For any *one* of these state *four* possible causes and the precautions that should be taken to avoid them.

9 Brittle fracture may occur in welded structures in low-alloy steels.
(1) List *four* possible causes.
(2) State *four* precautions that should be taken in order to reduce its occurrence.
(3) Describe one test procedure which may be used to give information on the parent plate's susceptibility to this type of failure.

10 (*a*) Explain what is meant by percentage metal recovery in manual metal-arc welding.
(*b*) Electrode manufacturers claim that over 100% metal recovery is possible when using certain of their electrodes.
(1) State the addition which must be made to the coatings of these electrodes.
(2) State *two* advantages.

11   (*a*) State *two* factors that should be considered before selecting a filler electrode for a hard surfacing application.

(*b*) State *three* precautions which may be taken by the welder to control the lever of dilution when surfacing using an arc welding process.

12 Steel components are to be hardfaced by the manual metal-arc process.

(*a*) Give *two* reasons why cracking may occur.

(*b*) State *three* precautions that can be taken in order to reduce the possibility of cracking.

(*c*) Name *one* other problem that may arise when hardfacing.

13   (*a*) With the aid of sketches, show *two* examples of distortion caused by manual metal-arc welding.

(*b*) For *each* example shown, suggest a suitable method of controlling the distortion.

14 The figure shows a typical beam-to-stanchion connection used in structural steelwork. Copy the drawing as shown and insert the appropriate welding symbols to indicate the following:

(*a*) the gusset to be fillet welded on each side to the stanchion in the shop;

(*b*) the beam to be added on site, and fillet welded all round to the stanchion;

(*c*) the gusset to be fillet welded each side to the beam.

Question 14

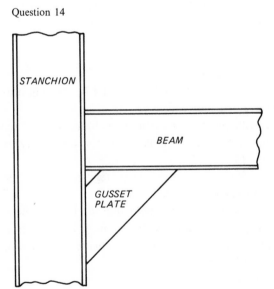

15 (a) Name two functions of the filter glasses used when metal-arc welding.

(b) Why should the cable used for the welding lead connexion of an arc welding circuit be flexible?

(c) Name two types of arc welding processes in which non-consumable electrodes are used.

(d) Give one example of a semi-automatic arc welding process.

16 Name *four* different types of metal-arc welding plant which fulfil the electric power supply requirements for welding. Describe, with aid of a sketch, one of the plants named.

17 What would be the practical effect of each of the following during metal arc welding:

(a) loose circuit connexions;

(b) variations in the mains supply voltage?

18 Describe in detail, with the aid of sectional sketches, three defects which may be produced during the metal-arc welding in the horizontal–vertical position of close-square-tee joints in mild steel plate. In *each* case state the cause and explain how the defects should be avoided.

19 Explain briefly the effect of any two of the following upon the production of an efficient joint in a partially chamfered butt weld:

(a) depth of root face,

(b) the angle of the vee,

(c) the gap setting.

20 Describe, with the aid of a sketch, one manual metal-arc welding technique used for making butt welds in mild steel pipelines.

21 Explain, by means of sketches, what is meant by *each* of the following:

(a) The *slope and tilt* of the filler rod and blowpipe *when making* a butt weld in low-carbon steel, 2 mm thick, in the vertical position by the oxy-acetylene process.

(b) The *slope and tilt* of the cutting electrode *when making* a straight cut in 8 mm low-carbon steel plate in the flat position by the oxygen-arc process.

22 (a) State *three* factors which may influence slag control during manual metal-arc welding.

(b) Give *one* example of when each of the following are used in manual metal-arc welding:

(1) tong test ammeter,

(2) voltmeter.

23   (*a*) What is meant by (1) open circuit voltage, and (2) arc voltage in manual metal arc welding?

(*b*) State the likely effect on root penetration and weld deposit when manual metal-arc welding low-carbon steel 6 mm thick with too long an arc.

24 State *two* safety precautions which should be observed with *each* of the following for manual metal-arc welding:

(*a*) treating components prior to welding by the use of tri-chloroethylene degreasing plant,

(*b*) welding in confined spaces,

(*c*) welding in close proximity to glossy finished surfaces,

(*d*) preparing vessels which have contained liquids with flammable vapours for repair by welding.

25   (*a*) Sketch the joint set-up and state the diameter of electrodes and current values to be used when making butt welds in the flat position in *each* of the following thicknesses of low-carbon steel:

(1) 3 mm,

(2) 6 mm,

(3) 10 mm.

(4) 14 mm,

(*b*) Discuss the effect of *each* of the following factors on depth of root penetration when making butt welded joints in the flat position:

(1) current,

(2) arc length,

(3) speed of travel,

(4) angle of electrode (slope and tilt).

26 State why pre-heating is to be recommended when welded joints are to be made in *each* of the following:

(*a*) low-alloy, high-tensile steel in cold weather,

(*b*) 50 mm thick low-carbon steel,

(*c*) 6 mm thick copper plate.

27 The figure shows two views of a cast iron support bracket which has fractured in service.

(*a*) If the casting is to be replaced by a low-carbon steel welded fabrication:

(1) make a sketch showing the complete bracket assembled ready for welding. By the use of weld symbols (BS499) indicate the type and location of the weld joints;

(2) detail freehand the steel parts that you would require.

(b) If the casting is to be repaired by manual metal arc welding state:

(1) all preparations necessary to be made before welding;

(2) the weld procedure, including details of electrode and current values;

(3) any precautions to be carried out after welding.

28 Explain briefly why nickel is liable to crack when metal-arc welded. State briefly how this defect may be avoided.

29 With the aid of sketches, give two examples of distortion of welded work that may result from the application of metal-arc welding.

30 (a) Explain what is meant by the term 'arc eye'.

(b) What precautions should be taken to avoid 'arc eye'?

(c) What action should be taken in the case of severe 'arc eye'?

31 When each of the following materials is welded by the manual metal-arc process state, in each case, two difficulties which may arise:

(1) copper, (2) grey cast iron.

(a) State two difficulties which may be encountered when manual-metal arc welding dissimilar metals.

(b) Explain how these difficulties can be overcome.

32 Give *three* precautions which should be taken to produce acceptable joints when manual metal-arc welding austenitic stainless steel. Butt welds are to be made in the construction of a pipe line.

(a) Give *two* reasons why excessive penetration should be avoided.

Question 27

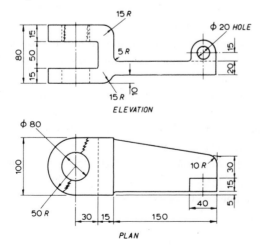

(*b*) Describe briefly *two* methods of controlling root penetration.

33 Describe, with the aid of sketches, any *four* of the following. In *each* case state *one* typical application, *one* advantage and *one* limitation:

(*a*) stove pipe technique,

(*b*) tongue bend test,

(*c*) manipulators,

(*d*) arc on gas techniques,

(*e*) 'studding' when welding cast iron.

34 Describe, with the aid of sketches, each of the following:

(*a*) back-step welding,

(*b*) backing strips,

(*c*) backing bars.

35 The figure shows details of a pipe assembly. Answer the following:

(*a*) Recommend a suitable method in each case for producing to correct shape and size (1) the flanges, (2) the main pipe and the branch pipe.

(*b*) Describe, with the aid of sketches, a suitable edge preparation that may be used to ensure satisfactory penetration at the following joints: (1) a flange to the main pipe, (2) branch pipe to the main pipe.

(*c*) Show, with the aid of sketches, the procedure for welding both of these joints, stating the sequence of runs, the current value and diameter of electrode used in each case.

(*d*) Describe briefly how the finished welds may be tested for surface defects.

36 The figure shows details of a pipe assembly. Answer the following:

(*a*) Show, by means of a sketch, how the main pipe and the flanges may have distorted as a result of welding.

(*b*) Sketch a simple fixture that would be suitable for locating the branch to the main pipe and assist in controlling distortion.

(*c*) Describe briefly a suitable method for pressure testing the completed assembly.

(*d*) Calculate the total weight of the three flanges; 1 m³ of low-carbon steel weighs 7750 kg.

37 Briefly explain the function of *each* of the following in manual metal-arc welding:

(*a*) a low-voltage safety device,

(*b*) a rectifier.

38   (*a*) What are the *three* basic types of wear to which hard-faced
       components are subjected?

       (*b*) Select *one* of these types and suggest a suitable type of
       electrode to be used for surfacing to meet requirements.

       (*c*) State *two* precautions which should be taken to minimize
       cracking.

39 An aluminium alloy casting is to be repaired by the manual metal-
    arc welding process. Give *each* of the following:

       (*a*) a workshop method of indicating the pre-heating
       temperature,

       (*b*) the type of current to be used.

       (*c*) *three* factors which make this material more difficult to weld
       than low-carbon steel.

40 Select a suitable electrode for welding *each* of the following low-
    carbon steel joints:

       (*a*) a severely stressed single-vee butt joint welded in the flat
       position,

       (*b*) a lightly stressed fillet welded joint in the vertical position,

       (*c*) fillet welds in the flat position where a high metal recovery
       rate is required.

41 State one important safety precaution that must be observed for
    each of the following:

       (*a*) before commencing welding repair work on a two-
       compartment tractor fuel tank, having one compartment
       used for petrol and the other for fuel oil;

Question 35 and 36

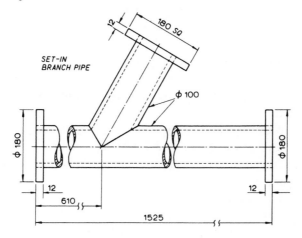

　　　(*b*) when flame gouging in the vertical position a defective
　　　　vertical joint out of a heavy fabrication in preparation for
　　　　rewelding.

42　State *two* advantages which may be obtained by the use of a U
　　preparation instead of a V preparation for welded butt joints in
　　thicker materials.

43　Sketch the joint preparation and set-up, and state the size of the
　　electrodes, number of runs and current values to be used for
　　making
　　　(1) butt welds, without backing bar in the flat position,
　　　(2) tee fillet welds in the horizontal–vertical position, in each of
　　　　the following thicknesses of mild steel: 6.4 mm and 12.7 mm.

44　State the plate thickness limitations of the upward–vertical and the
　　downward–vertical metal arc welding techniques.

45　　(*a*) Describe, with the aid of sketches, what effect each of the
　　　　following would have on the depth of root penetration and
　　　　the quality of the deposited metal:
　　　　(1) the use of too high current value,
　　　　(2) incorrect angle of electrode slope,
　　　　(3) too fast a speed of travel,
　　　　(4) incorrect arc length.
　　　(*b*) Explain how measuring equipment could be used to check
　　　　current and voltage values available in a metal arc welding
　　　　circuit during welding.

46　State one safety precaution that must be carried out:
　　　(*a*) before commencing welding repair work on a tank which
　　　　has contained acids,
　　　(*b*) when metal arc welding from a multi-operator set is carried
　　　　out during the construction of multiple-storey steel framed
　　　　structures.

47　State what is meant by:
　　　(*a*) deep-penetration coated electrode,
　　　(*b*) non-consumable electrode.

48　What is the purpose of a choke reactance used in manual metal arc
　　welding?

49　　(*a*) What is meant by arc blow in manual metal arc welding?
　　　(*b*) Arc blow may arise when manual metal arc welding with
　　　　either direct current or alternating current supply. Give *two*
　　　　possible causes.

50　　(*a*) State what is meant by:
　　　　(1) rutile-coated electrode,
　　　　(2) hydrogen-controlled coated electrode.

(b) Give *one* defect which may arise when manual metal arc welding low-carbon steel, using an eccentrically coated electrode.

51 Describe with the aid of sketches the *technique* required when cutting low-carbon steel plate in the flat position by the use of *each* of the following arc cutting processes:

(1) air-arc,

(2) oxygen-arc.

52 Sketch and label *two* different types of edge preparation for butt joints other than close-square butt, suitable for manual metal arc welding.

53 Give *one* possible cause of *each* of the following defects when manual metal arc welding close-square-tee fillet joints in low-carbon steel in the vertical position by the upwards technique:

(a) incomplete root penetration,

(b) undercut,

(c) unequal leg-length.

54 (a) Explain what is meant by 'percentage metal recovery' in metal arc welding.

(b) If over 100% is claimed, what additions could have been made to the electrode coating?

55 Manufacturers of metal arc welding electrodes take care to specify a minimum and maximum current for each size and type of mild steel electrode. Discuss in detail the important consequences of using

(a) insufficient current,

(b) excessive current.

56 (a) State *three* reasons for using welding fixtures in welding fabrication.

(b) State *three* basic factors that must be considered for the effective operation of a welding fixture.

(c) Describe briefly, with the aid of a sketch, the principle of operation of any welding fixture with which you are acquainted.

57 A number of butt welded joints have to be made between the ends of 100 mm diameter low-carbon steel pipe. Sketch a simple jig for holding them in position for tack welding.

58 Give *four* important functions of a slag.

(a) Briefly describe one test, other than visual inspection, used to reveal surface defect in welded joints.

(b) Give one limitation of such a test.

(c) Name four constituents used in electrode coatings.

(*d*) State the most important function of any one of these constituents.

59 Give *four* differences in the behaviour of the metal that a welder will find between the metal arc welding of (*a*) aluminium and (*b*) low-carbon steel.

60   (*a*) Describe in detail the metal arc welding process commonly used for manually welding mild steel.

(*b*) Give *two* examples showing how the process in (*a*), using standard electrodes, may be partially automated so as to reduce costs.

61 State *three* factors that may influence metal transfer phenomena when using a metal arc welding process.

62 Give *three* reasons why stray-arcing is undesirable.

63 Give *three* reasons why damp flux-coated electrodes should not be used for welding mild steel.

64 With the aid of a sketch, show the principle of the buttering technique.

65   (*a*) Construct a table showing the ranges of welding currents typically used with 2.5 mm, 3.2 mm, 5.0 mm and 6.0 mm general-purpose, mild steel electrodes.

(*b*) Plot a graph from the figures in the table in part (*a*).

(*c*) From the graph in part (*b*) determine the probable current range to be used with a 4.0 mm electrode.

66 State *three* safety precautions to be observed when using engine-driven metal arc welding equipment.

67 Give any *two* advantages of using multi-operator a.c. metal arc welding equipment.

68 The figure shows two views of a bracket which has to be fabricated by manual metal arc welding.

(*a*) Make a pictorial sketch of the bracket and indicate, by the use of weld symbols, according to BS499, the location and the type of weld joints required.

(*b*) List *ten* items of information that could be included on a Welding Procedure Sheet for the bracket shown.

69 The figure shows two views of a cast iron bracket which has fractured in service.

(*a*) If the casting is to be repaired by manual metal arc welding,

(1) state *all* preparations necessary to be made before welding,

(2) describe a suitable procedure, including details of the type of electrodes and the current values,

(3) state *two* precautions carried out after welding.

Question 68

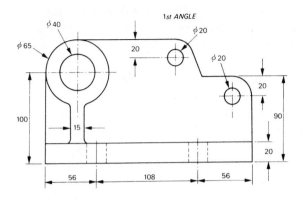

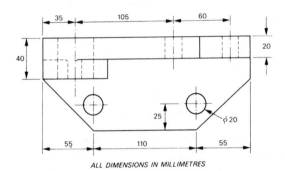

ALL DIMENSIONS IN MILLIMETRES

Question 69

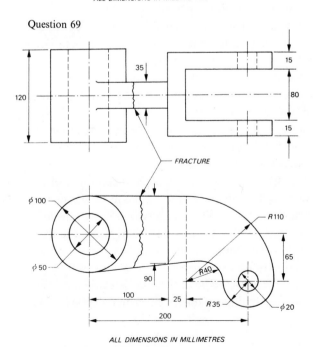

ALL DIMENSIONS IN MILLIMETRES

(*b*) Make a sketch to show the complete bracket assembled ready for welding, if the casting is to be replaced by a low-carbon steel welded fabrication. By the use of weld symbols (BS499), indicate the type and location of the weld joints to be used.

70 Give *one* reason why, in particular situations, it might be difficult to control the arc when arc welding mild steel.

71 Give what you consider to be *three* important factors likely to affect the cost of the welded fabrication of a very large one-off component.

72   (*a*) By means of a labelled sketch show what is meant by buttering.

(*b*) Name *one* material on which buttering could helpfully be used to make an effective arc fusion welded joint.

73 Explain how the difficulties which may arise from the presence of hydrogen may be overcome during the welding of low-alloy steels.

74 What is an alloy? Give *two* reasons why an alloying addition might be necessary in the composition of a metal intended for use as weld metal for joints to be made by gas-shielded arc welding.

Gas shielded metal arc MIG and TIG

1   (*a*) List *four* welding problems to be overcome when gas-shielded metal-arc welding austenitic heat-resisting steels to be used for high-temperature service.

(*b*) State *three* features which should be present in a manipulator intended for extensive gas-shielded metal-arc welding of joints in a large assembly.

(*c*) Outline *three* functions of jigs designed to be used for the metal-arc gas-shielded welding of joints in components during flow production.

2   (*a*) State *five* factors which influence the cost of metal-arc gas-shielded welding in the batch production of welded fabrications in non-ferrous metals.

(*b*) A butt welded joint in a flat, low-carbon steel bar L mm wide and t mm thick is to be made by the metal-arc gas-shielded process. When in service a static tensile load of P newtons will be carried by the butt welded joint. Show by means of a simple formula how the tensile stress in the butt weld may be calculated.

(*c*) Describe, with the aid of sketches, a technique which may be used for making a fillet welded part-lapped close outside corner joint in the flat position in 6 mm thick aluminium plate by the metal-arc gas-shielded process.

3 (*a*) With the aid of a labelled block diagram to show the arrangement, describe the equipment necessary to fulfil the electrical gas-shielding and ancillary requirements when one operator is required to carry out effective tungsten-arc gas-shielded welding with

(1) alternating current,

(2) direct current as required for different types of material.

(*b*) State the purpose of *each* item of equipment in the system and give details of current control together with a labelled sketch to show the power source characteristic.

4 (*a*) State *four* different-types of material suitable for tungsten-arc gas-shielded welding.

(*b*) In the case of *two* of these materials state

(1) *two* difficulties that may arise during the welding of the material,

(2) the methods used to overcome the difficulties encountered.

5 (*a*) List *four* gases *or* gas mixtures used in tungsten-arc welding and give *one* typical application of *each*.

(*b*) Describe with the aid of sketches, a technique used for making a close-square butt joint in the vertical position by the upward method in 2.5 mm thick austenitic stainless steel by the tungsten-arc process.

(*c*) State *four* possible causes of incomplete penetration when making butt welds by the tungsten-arc process in the flat position.

6 (*a*) Outline *two* essential functions of the gas shield required for effective gas-shielded metal-arc welding.

(*b*) List *four* gases *or* gas mixtures, other than carbon dioxide, used in metal-arc gas-shielded welding and give *one* typical application of *each*.

(*c*) State *two* possible causes of *each* of the following defects when encountered in welds made by the metal-arc gas-shielded process in low alloy steel:

(1) porosity,

(2) cracking.

(*d*) Explain why care should be taken to avoid leakage at joints in the shielding gas supply lines when gas-shielded metal-arc welding aluminium alloys.

7 (*a*) Show, by means of a labelled diagram, the voltage wave-
        form obtained when using *either* a high-frequency unit *or*
        a surge injector for gas-shielded tungsten-arc welding with
        alternating current supply.
   (*b*) State the function of each of the regularly alternating high
        and low current levels used in controlled spray (pulse) mode
        of metal transfer in gas-shielded metal-arc welding.

8 (*a*) Outline *one* method of arc length control which may be used
        in metal-arc gas-shielded welding.
   (*b*) Show, by means of labelled sketches, the stages in spray
        mode of metal transfer when metal-arc gas-shielded
        welding.

9 (*a*) Explain, with the aid of a sketch, how the level of dilution of
        a butt-welded joint may be determined.
   (*b*) Show by means of a labelled sketch *one* type of edge
        preparation used to control pick-up effects when making a
        butt-welded joint in clad steel.

10 The following gas shielding mixtures are used in gas-shielded arc
   welding:
   (*a*) argon,
   (*b*) argon and hydrogen,
   (*c*) argon and oxygen,
   (*d*) argon and nitrogen,
   (*e*) argon and helium.
   State, in *each case*, a material and the welding process for which
   these mixtures are best suited.

11 (*a*) Name the type of current and electrode required for gas-
        shielded tungsten-arc welding *each* of the following
        materials:
        (1) low-carbon steel,
        (2) aluminium alloy,
        (3) copper alloy.
        (4) austenitic stainless steel.
   (*b*) State *two* factors, other than thickness of material, which
        should be considered when selecting the filler wire diameter
        to be used for tungsten-arc welding.

12 (*a*) Outline the procedure for rectifying a fault which has been
        indicated in a metal-arc gas-shielded welding circuit by
        failure to strike an arc and a falling voltage reading at the
        power source.
   (*b*) Show, by means of a sectional sketch, the contact tube

setting and electrode extension required for dip transfer welding of low-carbon steel.

13 (a) Sketch a tungsten-arc welding torch fitted with a gas lens and increased electrode extension, indicating the pattern of gas flow which would be obtained from the torch.

(b) State *two* reasons why rectification must be controlled when alternating current is used for the gas-shielded tungsten-arc welding of aluminium.

14 (a) State *two* safety precautions, other than ventilation, which are necessary when using gas-shielded arc welding processes.

(b) State how

(1) ozone,

(2) carbon monoxide

may be formed during gas-shielded arc welding.

15 (a) State the purpose of a choke reactance when used for gas-shielded metal-arc welding.

(b) State the function of

(1) a surge injector,

(2) a high frequency unit,

when tungsten arc welding with alternating current.

16 Show by means of a labelled diagram of output curves the essential differences between the characteristics of the power source suitable for

(a) manual tungsten-arc gas-shielded welding,

(b) semi-automatic metal-arc gas-shielded welding with a self-adjusting arc.

17 (a) State what current values are being shown by the ammeter when welding with controlled spray (pulse) mode of metal transfer.

(b) Show, by means of labelled sketches, the stages in metal transfer when metal-arc gas-shielded welding by means of dip mode of metal transfer.

18 State, in *each* case, *two* precautions which must be taken in the storage of

(a) electrode wires for metal-arc gas-shielded welding,

(b) filler wires for tungsten-arc gas-shielded welding,

to ensure effective welding when required.

19 (a) During gas-shielded tungsten-arc welding, porosity has appeared in the weld face and metal deposition has become difficult. State *two* possible causes that may produce this condition.

(*b*) Give *one* suitable material in *each* case which may be used for a backing bar insert when tungsten-arc gas-shielded welding
    (1) ferrous metals,
    (2) non-ferrous metals.

20   (*a*) Give *two* undesirable effects which may occur when using excessive wire feed speed during gas-shielded metal-arc welding.

    (*b*) State the shielding gas or gas mixture which is best suited to obtain the required modes of metal transfer for the efficient metal-arc gas-shielded welding of *each* of the following metals:
       (1) low-carbon steel and low-alloy steel by spray transfer in the flat position,
       (2) austenitic stainless steel and aluminium by controlled spray (pulse) transfer in the vertical position.

21   (*a*) State how gas flow rate is controlled and measured to enable the production of efficient welded joints in tungsten-arc gas-shielded welding.

    (*b*) Name *four* factors upon which the selection of the correct current value will depend in tungsten-arc welding.

22   (*a*) Give *two* advantages which may be obtained by the use of the metal-arc gas-shielded spot welding process instead of the electric resistance spot welding process.

    (*b*) State *one* typical application in each case for the following welding process:
       (1) electron beam,
       (2) friction.

23   (*a*) Outline the procedure for rectifying a fault which has been indicated in a metal-arc gas-shielded welding circuit by failure to strike an arc and a falling voltage reading at the power source.

    (*b*) Show, by means of a sectional sketch, the contact tube setting and electrode extension required for dip transfer welding of low-carbon steel.

24   (*a*) Sketch a tungsten-arc welding torch fitted with a gas lens and increased electrode extension, indicating the pattern of gas flow which would be obtained from the torch.

    (*b*) State *two* reasons why rectification must be controlled when alternating current is used for the gas-shielded tungsten-arc welding of aluminium.

25 Explain why the presence of moisture should be avoided when gas-shielded arc welding low-alloy steels.

26   (*a*) State in *each* case *one* probable cause of the following weld defects in joints made by tungsten-arc welding:
      (1) lack of root penetration,
      (2) porosity.
  (*b*) State *two* reasons why abrupt changes of section should be avoided in the contours of a welded joint.

27   (*a*) Outline *two* essential functions of the gas shield required for tungsten-arc welding.
  (*b*) State *two* factors which influence the selection of the flow rate of argon gas necessary to obtain efficient welded joints.
  (*c*) State in *each* case *two* difficulties that may arise during the tungsten-arc welding of
      (1) stainless steel,
      (2) aluminium,
      (3) magnesium alloy.

28   (*a*) Describe, with the aid of a sketch, the distortion effects which are likely to be produced when an unrestrained 450 mm long single-vee butt joint between 6 mm thick austenitic stainless steel plates 225 mm wide is made by the tungsten-arc process in two runs from one side.
  (*b*) Give *two* methods used to control distortion.
  (*c*) Outline, with the aid of sketches, *two* methods of providing gas backing for the butt welding of pipe joints.

29   (*a*) Outline the shut-down procedure to be followed at the end of work with tungsten-arc welding plant in order to ensure safety and continued efficiency of equipment.
  (*b*) Show, by means of a labelled sketch, the gas paths and the function of *each* of the main components of a water-cooled tungsten-arc welding torch.
  (*c*) Describe, with the aid of sketches, a tungsten-arc technique used for making a close-square butt-welded joint in the vertical position by the upward method in 1.5 mm thick austenitic stainless steel.

30   (*a*) Describe, with the aid of a labelled block arrangement diagram, the purpose of each part of the equipment essential for argon-shielded metal-arc welding aluminium alloys.
  (*b*) State *one* precaution which must be taken with the temporary backing bars used in welding aluminium alloys in order to ensure the production of efficient joints.

    (*c*) State the shielding gas or gas mixture which is most suited to obtain controlled spray (pulse) mode of metal transfer for welding
        (1) low alloy steel,
        (2) aluminium.

31   (*a*) List *four* welding problems to be overcome when welding steel containing 9% nickel, which is to be used for low-temperature service.

    (*b*) State *three* features which should be present in a jig designed for use in the welding of joints made in components during batch production.

    (*c*) Describe, with the aid of sketches, a gas-shielded metal-arc technique which may be used for making a lap-fillet welded joint in the vertical position by the downward method in 3 mm thick low-carbon steel plate.

32   (*a*) List *three* advantages and *three* disadvantages in *each* case which may be obtained by the use of
        (1) carbon dioxide,
        (2) argon,
    for metal-arc gas-shielded welding.

    (*b*) Outline *two* effects of the chromium content on the weldability of low-alloy high-tensile steel.

    (*c*) State *two* essential factors which must be considered before making welded joints between dissimilar metals by the metal-arc gas-shielded welding process.

33   (*a*) Give *two* reasons why it is necessary to be particularly careful with the insulation of components in the welding circuit when using high-frequency current in tungsten-arc gas-shielded welding operations.

    (*b*) Give *two* safety precautions which should be taken when preparing to use the metal-arc gas-shielded welding process for making welds on site in an exposed position which is 12 metres above ground level.

34   (*a*) State the purpose of a drooping voltage transformer-rectifier when used for tungsten-arc gas-shielded welding.

    (*b*) Outline *two* methods used to control the current surges at the short-circuitings which take place when metal-arc gas-shielded welding with dip mode of metal transfer.

35   (*a*) Explain what is meant by *each* of the following in fusion welding:
        (1) dilution,
        (2) pick-up.

(b) Show by means of a labelled sketch *one* type of joint preparation used to control pick-up effects when making a butt-welded joint in clad steel.

36 (a) State *two* reasons why the d.c. component must be suppressed when alternating current is used for gas-shielded tungsten-arc welding.

(b) Outline, with the aid of a sketch, *one* method of providing back purging for the butt welding of a joint in flat plate by the tungsten-arc process.

37 (a) State *one* undesirable effect in *each* case which is likely to be produced when metal-arc gas-shielded welding with wire feed drive roll pressure adjustment which is
   (1) insufficient,
   (2) excessive.

(b) Give *four* routine checks which should be made on the electrode wire before fitting a new spool into a gas-shielded metal-arc welding plant.

38 Distortion may result from gas-shielded arc welding operations. State *two* possible causes and give *two* methods of control.

39 (a) Give *one* reason in *each* case for the use of
   (1) a remote control switch,
   (2) a welding contactor,
   in tungsten-arc welding.

(b) State the likely effect of allowing cables in the welding circuit to coil when using high-frequency currents in gas-shielded tungsten-arc welding.

40 (a) Explain the technique to be followed to avoid contamination of the electrode when tungsten-arc welding butt joints in thicker section material in the flat position.

(b) Give *two* advantages which may be obtained by the use of the two-operator upward-vertical technique for tungsten-arc welding.

41 (a) Give *two* reasons why flux-cored electrode wire is used with carbon dioxide shielding for the metal-arc gas-shielded welding of steel.

(b) When using a gas-shielded arc welding process, on an exposed site, which *two* problems are most likely to be encountered as a result of weather conditions?

42 (a) What advantageous feature is provided by a constant potential power source when the arc is initiated to start the weld in metal-arc gas-shielded welding?

(b) Outline the procedure for rectification when a fault condition has been indicated in a metal-arc gas-shielded welding circuit, by falling voltage reading at the power source, when the end of the electrode wire is touched on to the work with the torch control trigger actuated.

43 Name *four* factors which may affect the cost of metal-arc gas-shielded welding a fabrication in a non-ferrous alloy.

44 With the aid of a sketch, explain briefly the principles of the tungsten-arc spot welding process.

45   (*a*) Describe, with the aid of a sketch, the distortion effects which are likely to be produced when an unrestrained 300 mm long single-vee butt joint between 5 mm thick austenitic stainless steel plates, 150 mm wide, is made in two runs from one side by the tungsten arc process.

   (*b*) Outline *two* methods used to control distortion in tungsten-arc welding operations.

46 List *four* of the main welding problems encountered when tungsten-arc gas-shielded welding magnesium alloys in thicknesses ranging from 2 mm to 10 mm, indicating how *each* of these problems is overcome, and mentioning the effects of surface preparation.

47 Twenty 2 metre pipe lengths are to be fabricated from existing stock of 1 metre lengths of 150 mm internal diameter low-carbon steel pipe of 10 mm wall thickness by metal-arc gas-shielded welding. Describe, with the aid of sketches, *each* of the following for the welding of one joint:

   (*a*) a suitable joint preparation and set-up,
   (*b*) the assembly and tack welding procedure,
   (*c*) the mode of metal transfer, shielding gas and electrode wire size to be used,
   (*d*) an effective welding procedure.

48   (*a*) Compare the use of carbon dioxide with the use of argon for metal-arc gas-shielded welding by listing *two* relative advantages and *two* relative disadvantages which may be obtained by the use of *each* gas.

   (*b*) State *two* advantages which may be obtained by the use of a manipulator as an aid to fabrication by metal-arc gas-shielded welding.

49 State the sequence which could be used to complete the joints numbered 1, 2, 3, and 4 in the figure of a welded fabrication made from five parts as shown. Give reasons for your answer.

50 For joint 1 in the figure give details of (*a*) the type of preparation and (*b*) the welding procedure to be used.

51 For joint 2 in the figure give details of (*a*) the type of preparation and (*b*) the welding procedure to be used.

52 Describe briefly four problems which may be encountered in the welding of the fabrication shown in the figure.

53 State why a self-adjusting arc is 'not likely to operate effectively in CO_2 shielded metal-arc welding'.

54 By means of a labelled block diagram show clearly the equipment required to give effective control of gas flow to the torch or welding head during gas-shielded arc welding. Assume that a suitable gas is being used.

55 By means of a labelled sketch indicate any *three* types of weld defect likely to occur in gas-shielded arc welded joints.

56 What is the purpose of a suppressor unit in a gas-shielded arc welding circuit?

57 State *three* different types of material suitable for tungsten arc gas-shielded welding. In *each* case state

  (*a*) two difficulties that may arise during the welding of the material,

  (*b*) the methods used to overcome the difficulties encountered.

58 With the aid of a simple labelled block diagram show the name, location and purpose of each part of the equipment essential for the effective tungsten-arc gas welding of aluminium.

Questions 49–52. Section through a heater shield. Material: 18% Cr, 8% Ni austenitic steel, Ti stabilized. Process: gas-shielded tungsten arc welding.

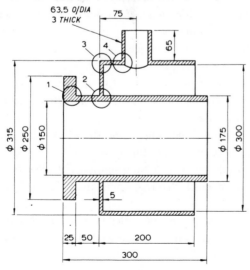

59 (*a*) Sketch a tungsten-arc welding torch fitted with gas lens, indicating the pattern of gas flow which would be obtained from the torch.

(*b*) State *two* advantages which may be obtained by the use of a gas lens in tungsten-arc welding.

60 (*a*) Describe briefly the recommended procedure to follow for clearing a 'burn back' in gas-shielded metal-arc welding.

(*b*) State *four* causes, other than joint preparation, of lack of root penetration in butt-welded joints made by gas-shielded metal-arc welding.

61 (*a*) Outline the typical main functions of a jig or fixture suitable for use in tungsten-arc gas-shielded welding.

(*b*) Give *two* situations in which the use of a jig or fixture would be considered essential, in the manufacture of a component by tungsten arc gas-shielded welding.

62 Explain what is meant by gas-backing in tungsten-arc welding and show *two* ways in which it may be applied.

63 List *two* advantages and *two* limitations of gas-shielded metal-arc welding as a process for general-purpose welding repair work.

64 With the aid of a simple labelled block diagram show the name, location and purpose of each part of the equipment essential for the effective CO_2 shielded metal-arc welding of low-carbon steels.

65 State *four* features which should be present in a manipulator intended for extensive gas-shielded metal-arc welding of joints in a large component.

66 (*a*) State the type of filter which should be used to provide eye and skin protection against radiation effects when gas-shielded arc welding with:

(1) the tungsten arc process,

(2) the metal arc process.

(*b*) Give *one* reason, in *each* case, for the type of filter used.

67 (*a*) What is the purpose of a choke reactance when used for gas-shielded metal-arc welding?

(*b*) State *two* effects which may be produced by using excessive electrode wire extension during welding.

68 (*a*) Give *two* advantages of tungsten-alloyed electrodes over plain tungsten electrodes when used for tungsten-arc welding.

(*b*) Show by means of simple labelled sketches what is meant by the electrode tip (vertex) angle when tungsten-arc welding with:

(1) alternating current,

(2) direct current.

69 Show by means of a single labelled sketch the essential differences between welding power source characteristics suitable for:

    (*a*) manual tungsten-arc welding,

    (*b*) welding with a self-adjusting arc.

70   (*a*) State *two* factors which influence the amount of dilution of the weld deposit in gas-shielded arc welding.

    (*b*) With the aid of an outline sketch, show how the level of dilution of the welded joint may be determined.

71   (*a*) State *two* modes of metal transfer, other than spray, used for gas-shielded metal-arc welding.

    (*b*) Give *one* application of *each* mode of transfer named.

72   (*a*) Give *two* important physical properties of ceramic gas nozzles.

    (*b*) State *two* of the factors which govern the gas-shielding necessary to obtain efficient welded joints by tungsten-arc welding.

73   (*a*) State the type of current and tungsten alloyed electrode recommended to be used for tungsten-arc welding *each* of the following materials:

      (1) low-carbon steel up to 2 mm thick,

      (2) austenitic stainless steel up to 5 mm thick,

      (3) aluminium over 3 mm thick,

      (4) magnesium alloy up to 5 mm thick.

    (*b*) List *three* gases or gas mixtures in tungsten-arc welding and give one typical use of each.

74   (*a*) State *four* factors that have to be taken into account when costing for tungsten-arc gas-shielded welding.

    (*b*) Explain what is meant by:

      (1) arcing time,

      (2) floor to floor time.

75   (*a*) List the main welding problems when tungsten-arc gas-shielded welding copper plates in thicknesses ranging from 3 to 10 mm.

    (*b*) Outline with the aid of sketches how each of these problems may be counteracted.

76   (*a*) Show by means of a labelled sketch *one* method of providing gas backing for the tungsten-arc welding of butt joints in plate in the flat position.

    (*b*) Why should syphon-type cylinders be used to supply carbon dioxide for shielding gas in welding?

77 With the aid of simple sketches show how you would prepare the joints for welding the top central part of the beam shown in the

figure, where the central boss and the top flange join with each other and where the vertical web joins with the top plate.

78 State the welding sequence, which should be used for welding the beam shown in the figure. Give reasons for each step.

79 Describe four of the problems which are likely to arise in welding the ends of the beam shown in the figure.

80 On a sketch of the figure give all the information needed to show the type and location of each welded joint, using the system given in BS 499 (Part 2).

81 Give *three* reasons why a suitable shade of filter glass should be used for viewing the arc during gas-shielded arc welding.

82 The following gas-shielding mixtures are used in gas-shielded arc welding:

   (*a*) argon + carbon dioxide + oxygen,
   (*b*) argon + oxygen,
   (*c*) arbon + hydrogen,
   (*d*) helium + argon,
   (*e*) helium + argon + carbon dioxide.

State in *each* case:

   (1) the material for which the mixture is best suited,
   (2) the approximate percentages of gases in the mixture.

83 State *five* desirable features in jigs intended for gas-shielded metal-arc welding of joints in components to be mass produced.

Questions 77–80. Support beam. Material: high yield stress structural steel (BS 4360 Grade 50) welded by CO_2 shielded metal arc process.

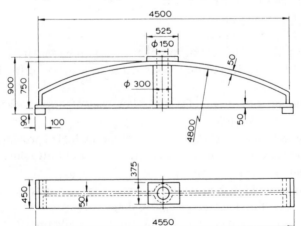

84 (*a*) With the aid of a labelled block diagram to show the arrangement, describe the purpose of *each* part of the equipment necessary for the effective argon-shielded metal-arc welding of aluminium alloys.

(*b*) State *one* precaution which must be taken in *each* case, with:
  (1) the shielding gas supply lines,
  (2) temporary backing bars,
in order to assist in the production of efficient joints.

85 (*a*) Give *one* safety precaution to be taken before tungsten-arc gas-shielded welding when the gas cylinder is located near the welding area.

(*b*) Give *two* hazards present when metal arc gas-shielded welding equipment is inadequately earthed.

86 (*a*) State the purpose of a combined transformer-rectifier when used for tungsten-arc welding.

(*b*) Give *two* important reasons for the use of a welding contactor in tungsten-arc welding.

87 (*a*) State the type of current and tungsten alloyed electrode recommended to be used for tungsten-arc welding *each* of the following materials:
  (1) low-carbon steel up to 2 mm thick,
  (2) austenitic stainless steel up to 5 mm thick,
  (3) aluminium over 3 mm thick,
  (4) magnesium alloy up to 5 mm thick.

(*b*) List *three* gases or gas mixtures in tungsten-arc welding and give one typical use of each.

88 (*a*) State the purpose of a surge injector unit as used in arc welding.

(*b*) State the purpose of a suppressor unit as used in arc welding.

89 A cylindrical vessel 620 mm diameter by 2.5 metres long with low-carbon steel flanged ends 25 mm thick is to have two austenitic stainless steel pipes coming out at right angles to the vessel axis and to each other midway along the vessel axis. The pipes are 100 mm inside diameter with a wall thickness of 9.5 mm and the vessel is made of 12.5 mm thick low-carbon steel, clad inside with austenitic steel to a further 1.6 mm thickness. The pipes and flanges are to be joined to the vessel by gas-shielded tungsten-arc welding.

(*a*) Give particulars of (1) the type of joint, (2) the joint preparation and (3) the welding procedure that you recommend for the welding of flanges to the vessel.

(*b*) Give particulars of (1) the type of joint, (2) the joint preparation and (3) the welding procedure that you recommend for welding the pipes into the vessel.

90 A special I-section girder 6 metres long has the following sectional dimensions: flanges 450 mm wide by 50 mm thick, web 19 mm thick, and overall depth 1 metre. Stiffening ribs 150 mm wide by 12.5 mm thick, running from flange to flange, are located opposite each other against the web at right angles to the rider axis at 1.2 m intervals. The girder is to be fabricated from high-tensile constructional steel of welding quality by gas-shielded metal-arc welding.

(*a*) Give particulars of (1) the type of joint, (2) the joint preparation, and (3) the welding procedure that you would use for joining the web to the flanges.

(*b*) Give particulars of (1) the type of joint, (2) the joint preparation, and (3) the welding procedure that you would use for attaching the stiffening ribs.

91 There are particular economic and metallurgical problems associated with the effective use of gas-shielded metal-arc welding in production applications. Outline these problems and explain how they may be overcome.

92 Austenitic heat-resisting steels containing higher nickel and chromium contents are used for fabricating structures for high-temperature service.

(*a*) State *five* main welding problems that are likely to arise in the metal-arc gas-shielded welding of this type of material.

(*b*) Explain how *each* of these problems may be effectively overcome to ensure efficient welded joints for service at high temperatures.

93 (*a*) When a d.c. arc is operating, what proportion of the heat may be generated at each side of the arc?

(*b*) Would the heating situation be the same when tungsten-arc gas-shielded welding aluminium with the electrode positive?

94 A gas pressure regulator and a flowmeter are each essential for the successful operation of a gas-shielded arc welding process.

(*a*) With the aid of an outline diagram show the usual location of each in a conventional gas-shielded tungsten-arc welding arrangement.

(*b*) Why is it necessary to have a gas pressure regulator?

95 In gas-shielded tungsten-arc welding each of the following plays an important part: (1) length of electrode projecting beyond the nozzle, (2) length of arc, (3) angle of electrode relative to the workpiece.

Give *two* important reasons why small diameter filler wire is used for gas-shielded metal-arc welding.

96 (*a*) State one difficulty with the dip-transfer mode of metal deposition as it is used in CO_2 shielded metal-arc welding.

(*b*) State briefly how this difficulty is overcome.

97 On a simple labelled block diagram show the name and purpose of each essential part of the plant and equipment needed for making tungsten-arc gas-shielded welds in a variety of metals and a range of thicknesses.

98 Give two checks you should make on the electrode wire before fitting a new spool into a gas-shielded metal-arc welding machine.

99 Why is a double-bevel or a double J butt-welded tee joint preferable to a double fillet welded close-square-tee joint for welding two highly stressed members?

100 (*a*) Explain why, in spite of the overall efficiency of argon as a shielding gas, so much effort is spent in developing the use of gases such as carbon dioxide.

(*b*) Name two difficulties likely to be encountered in using carbon dioxide as a shielding gas in arc welding.

(*c*) For each of the difficulties given in your answer to (*b*) name one method used for overcoming the problem.

101 State the main difficulty you would expect to have to overcome in joining each of the following types of material by gas-shielded tungsten-arc welding:

(*a*) malleable cast iron,

(*b*) solution-treatable aluminium alloy.

102 (*a*) Name any *two* limitations of the gas-shielded tungsten-arc welding process.

(*b*) Name any *two* limitations of the gas-shielded metal-arc welding process.

(*c*) Why is a gas shield needed in the two types of arc processes in which it is used?

103 (*a*) In tungsten-arc welding how should the welding arc be initiated without contact between electrode and parent metal?

(*b*) With the aid of a simple line diagram show all the essential equipment for the conditions that you outline in (*a*) and clearly label each part.

104 (*a*) With the aid of sketches show how a single-vee butt joint preparation for tungsten-arc welding should differ from that of gas-shielded metal-arc welding in the same thicknesses of similar materials.

(*b*) Give the main reason why they should differ.

105 The support column shown in the figure is to be made up from 25 mm thick plates of high-strength low-alloy structural steel mounted on a square mild steel base plate. The sizes are given on the diagram.

 (*a*) Outline the main material problems to be overcome in welding this construction.

 (*b*) Indicate where you would locate the four transverse joints required to make up the vertical member from the two plates.

106 For the support column shown in the figure and the materials quoted in question, give

 (*a*) the details of the appropriate longitudinal joint preparation used to complete the hollow section,

 (*b*) the number and sequence of deposition of the runs required to complete the section,

 (*c*) any precautions necessary to ensure sound welds.

107 For the support column shown in the figure and the materials quoted in question, give

 (*a*) the details of the appropriate joint preparations used to attach the vertical member to the base,

 (*b*) the number and sequence of deposition of runs used to complete this joint,

 (*c*) any precautions necessary to ensure sound welds.

Questions 105–108. Overall height 5.6 m. Material available for main stem: plate 3.5 m × 1.2 m × 25 mm, plate 2.0 m × 1.2 m × 25 mm.
Note. Gas shielded metal arc welding to be used for all except tack welds.

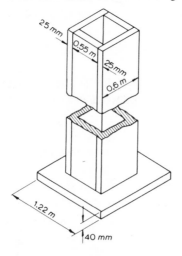

108 Discuss the main factors affecting the desirability of making up a special welding fixture for use when making the longitudinal welds in the vertical member of the support column shown in the figure.

109   (*a*) Give *two* reasons why it is necessary to be particularly careful of plant insulation when using HF current.

    (*b*) Why may it be dangerous to weld with an exposed arc near to a tank filled with non-flammable degreasing liquid?

110   (*a*) Give the principle reason why carbon dioxide gas is used for the gas-shielded metal-arc welding of mild steel.

    (*b*) Give *one* specific difficulty to be overcome when using carbon dioxide as a shielding gas for arc welding mild steel.

111 With the aid of an outline sectional sketch show the relative positions and proportions of (*a*) the electrode, (*b*) the collet, (*c*) the gas entry and (*d*) the nozzle, in a typical gas-shielded tungsten-arc welding torch head.

112 On a simple outline sketch locate and name the function of any *three* parts essential to the operation of the welding head of a fully automatic gas-shielded metal-arc welding plant.

113 Give any *three* possible differences between the respective preparation and deposition techniques for making a flat single V butt weld in 12 mm thick material by (*a*) gas-shielded tungsten-arc welding and (*b*) gas-shielded metal-arc welding.

Other welding processes

1   (*a*) Explain *four* principles involved in the production of a weld.

    (*b*) State *four* variables which influence the production of friction welds.

2 The electroslag process is used for a wide range of welded work.

    (*a*) Outline the principles of this process.

    (*b*) Sketch a sectional view of the weld area during welding showing all the essential parts.

    (*c*) Explain how the joint may be set up to retain its shape and minimize transverse shrinkage.

    (*d*) State *three* reasons why run-on and run-off plates are required.

    (*e*) Describe a suitable method of testing the weld metal to assess its toughness.

    (*f*) Describe a heat treatment process that can be used to improve the toughness of the completed welded joint.

3   (*a*) Explain how heat is produced in the resistance spot welding process.

(b) The figure shows a time, pressure, current diagram which illustrates the sequence of operations for the formation of a resistance spot weld.

(1) Calculate the percentage of the total time during which the current flows to form the weld.

(2) Calculate the total heat energy generated in joules using the formula $J = I^2 Rt$, where

I = current in amperes,

$R = 0.001$ ohm resistance,

t = time in seconds,

J = heat energy.

4 (a) Explain *four* principles involved in the production of a friction weld.

(b) State *four* variables which influence the production of friction welds.

5 For each of three of the following give a description of (1) the principles of operation, (2) a typical application:

(a) submerged arc welding.

(b) electric resistance spot welding,

(c) electroslag welding,

(d) arc stud welding.

6 With the aid of sketches show how resistance projection welding differs from resistance spot welding.

Question 3

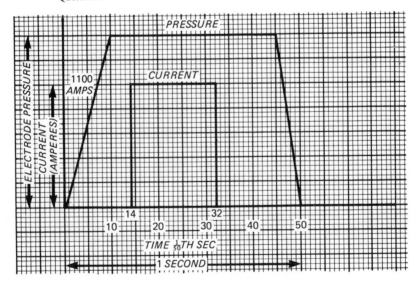

7 Describe briefly the principles of operation, and give *one* suitable industrial application, for *three* of the following welding processes:
  (*a*) electroslag,
  (*b*) electric resistance flash butt,
  (*c*) electron beam,
  (*d*) friction.

8 The figure shows a fixture arrangement for the automatic arc welding of copper floats using the gas-shielded tungsten-arc process.
  (*a*) State the influence the thermal conductivity of the metal would have on weld penetration.
  (*b*) State *three* factors other than thermal conductivity which would influence weld profile.

9 (*a*) List the following welding processes in their order of usefulness for welding steel components 50 mm thick, in the vertical position:
    (1) arc welding with coated electrodes,
    (2) electron beam welding,
    (3) electroslag welding,
    (4) laser beam welding.
  (*b*) Give a brief outline of friction welding.

10 (*a*) Explain why a flux is not required for resistance spot welding.
  (*b*) Give *three* possible causes of defective resistance spot welds.

11 The submerged arc welding process is used for a wide range of welded work.
  (*a*) Outline the principles of this process.
  (*b*) Explain why run-on and run-off plates are used.
  (*c*) State the purpose of vacuum recovery.

Question 8

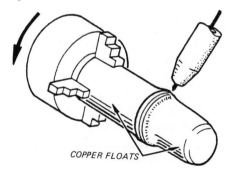

COPPER FLOATS

(*d*) State *four* precautions to be taken in order to avoid porosity of the welded joint.

(*e*) State *four* advantages this process has over the manual metal-arc process.

Cutting processes

1   (*a*) State *two* methods used for gas cutting bevels on plate edges in preparation for welding.

(*b*) Describe, with the aid of a sectional sketch indicating the gas paths, the operation of an oxy-fuel gas cutting blowpipe.

2   (*a*) Make labelled sketches to show the essential difference between the nozzle assemblies used with
   (1) acetylene,
   (2) propane,
   for the oxy-fuel gas cutting of low-carbon steel.

(*b*) After manual oxy-fuel gas cutting 50 mm thick low-carbon steel, it is required to cut 6 mm thick low-carbon steel plate. State the adjustment which will need to be made for this operation to be efficiently carried out.

3 Explain the action which produces the severance of the metal when oxy-fuel gas cutting low-carbon steel and state the difference between the basic principles of gas cutting and flame gouging.

4   (*a*) Give three operations which are essential during the assembly of a high-pressure oxy-acetylene cutting plant to ensure safe and effective service.

(*b*) Show by means of labelled sketches the angles of slope and tilt of the electrode when making a straight cut in 8 mm thick low-carbon steel plate in the flat position by the air-arc process.

5   (*a*) What is meant by 'oxygen lance cutting'? Give *two* examples of its use.

(*b*) Describe with the aid of sketches the technique required when cutting low-carbon steel-plate in the flat position by the use of each of the following arc cutting processes:
   (1) air-arc,
   (2) oxygen arc.

6 Unsuitable variations during oxy-fuel gas cutting of steel may lead to faults along the cut face. Show, by means of labelled sketches of the cut face, the faults which would be caused by either:
   (1) the speed of travel being too fast, or

(2) the nozzle being too high above the work surface.

7 (*a*) Explain how the correct size of cutting nozzle to be used in the oxy-fuel gas cutting process should be determined.

(*b*) Discuss the influence on the oxy-fuel gas cutting operation of any two alloying elements which may be present in the steel.

(*c*) Describe how cast iron can be cut by the oxy-fuel gas cutting process.

8 (*a*) Explain the action which causes the severance of the metal when using the oxy-fuel gas cutting process for the cutting of austenitic stainless steel.

(*b*) What is meant by stack cutting? List *two* advantages of the use of stack cutting.

9 (*a*) Give *two* advantages in *each* case which may be obtained when oxy-fuel gas cutting:

(1) manually,

(2) by machine.

(*b*) Describe, with the aid of a sketch, a suitable arc cutting process for the cutting of ferrous and non-ferrous metals.

10 (*a*) What is arc plasma?

By means of a simple labelled outline diagram, show one method of using arc plasma for cutting purposes.

(*b*) Explain the differences between

(1) a single port nozzle,

(2) a multi-port nozzle as used with plasma arc cutting.

(*c*) Show, by means of diagrams, the finished cut that is produced by each of these nozzles after cutting 4.5 mm thick austenitic stainless steel.

11 The following thermal cutting equipment is available for use:

oxy-acetylene manual cutting equipment,

arc plasma profile cutting machine,

oxy-acetylene cutting equipment complete with lance,

oxy-acetylene cutting equipment complete with dispenser unit.

From the cutting equipment available select, with reasons, which cutting process should be used for cutting each of the following materials to the shape or form specified. Do not use the same process for cutting both materials.

(1) 200 mm diameter discs to be cut from 20 mm thick austenitic stainless steel plate.

(2) The removal of excess material from a newly cast grey iron casting.

(3) Outline the principles of the thermal cutting process used.

(4) State *two* aspects of safety that are particularly appropriate to each process used.

12 State why the arc plasma process would be used in preference to oxygen and fuel gas cutting of austenitic stainless steel.

13 (*a*) Why is it necessary to pre-heat medium- and high-carbon steels before cutting by the oxy-fuel gas cutting process?

(*b*) State the factors that prevent the use of natural gas as an oxy-fuel cutting gas.

14 (*a*) Discuss the health hazards that can be encountered when cutting galvanized steel sheet.

(*b*) State the safety precautions that should be adopted to ensure that workshop personnel can be protected against these hazards.

15 (*a*) Describe two suitable thermal cutting processes which could be used to cut 1 metre diameter blanks from 12 mm thick austenitic steel plate.

(*b*) Compare the *two* methods selected in (*a*) with regard to:
    (1) cost,
    (2) speed of cutting,
    (3) distortion,
    (4) treatment after cutting.

16 (*a*) List the following cutting process in sequence with regard to best quality of cut face for the thermal cutting of austenitic stainless steel 38 mm thick in the flat position:
    (1) oxygen-arc,
    (2) air-arc,
    (3) oxy-fuel-gas powder,
    (4) arc-plasma.

(*b*) Explain, with the aid of a sketch, how the heat is produced for cutting by the plasma cutting process.

17 (*a*) Explain briefly the principles of *laser cutting*.

(*b*) List the advantages of laser cutting over conventional methods of cutting such as oxy-fuel gas and oxy-arc cutting.

Oxy-acetylene

1 (*a*) State *four* problems associated with the welding of copper.

(*b*) State what type of flame setting should be used and how it is attained.

(*c*) Describe in detail the welding procedure.

2 (*a*) Unsuitable variations during oxy-fuel gas cutting of steel may lead to faults along the cut face. Show, by means of labelled sketches of the cut face, the faults which would be caused by *either*

(1) the speed of travel being too fast, *or*

(2) the nozzle being too high above the work surface.

(*b*) State *two* methods which may be used for back gouging the root, ready for a sealing run, on the reverse side of welded butt joints in low-carbon steel.

3 Describe briefly the fundamental differences between the following: (*a*) the technique for the bronze welding of cast iron, and (*b*) the technique for the fusion of welding of cast iron.

4 (*a*) Describe, with the aid of sketches, the all-position rightward technique to be used when making a butt joint in low-carbon steel pipe 100 mm diameter by 5 mm wall thickness, the pipe axis to be in the fixed vertical position throughout.

(*b*) Compare the respective advantages and limitations of the leftward and the rightward techniques when used for the butt welding of low-carbon steel plate, 5 mm thick, in the flat position.

5 (*a*) Describe briefly, with the aid of a sectional sketch and indicating the gas paths, the mode of operation of *either*

(1) a non-injector type welding blowpipe, *or*

(2) a single-stage gas pressure regulator.

(*b*) What precautions are necessary during the assembly of a high-pressure oxy-acetylene cutting plant in order to ensure safe and effective operation?

(*c*) Explain why the specified discharge rate of a dissolved acetylene cylinder must not be exceeded when in use.

6 (*a*) Describe in detail, with the aid of sectional sketches, *four* defects which may occur during the oxy-acetylene welding of low-carbon steel, stating in *each* case its cause and explaining how it may be avoided.

7 Describe the technique necessary to form a tee-fillet joint in 1.6 mm commercially pure aluminium sheet by the flame brazing process.

State the type of filler wire and flame setting needed. You may use a sketch to illustrate your answer.

8 (*a*) Name three impurities found in acetylene gas immediately after generation.

(*b*) Explain how acetylene gas may be tested for purity.

9   (*a*) Summarize the problems encountered in the oxy-acetylene repair welding of each of the following cast materials: (1) zinc base die cast alloy, (2) magnesium alloy.

(*b*) Describe the preparation, welding technique and post-weld treatment necessary for each.

10   (*a*) When would a carburizing flame be used for joining metals?

(*b*) What purpose is served by the use of such a flame adjustment?

11 Describe, with the aid of a simple outline sketch, one example of a repair welding operation where *studding* could be employed with advantage.

(*a*) Name two advantages of using a high silicon content in cast iron, or in cast iron filler rods for welding purposes.

(*b*) Give the approximate percentage of silicon contained in a 'super-silicon' cast iron filler rod.

12   (*a*) What is meant by the term 'hard-facing'?

(*b*) Name two types of filler rod that may be used for hard-facing operations.

13   (*a*) Name the fuel gas generally used for underwater flame cutting operations.

(*b*) State one good reason why this gas is used.

14   (*a*) Explain, in detail, four safety measures that should be taken before carrying out welding repairs on a 150 litre (30 gallon) low-carbon steel petrol tank.

(*b*) Describe the hazards encountered when welding galvanized metals. State the precautions to be taken in the interests of health and safety.

15 A 6 mm thick magnesium alloy casting is cracked for a length of 150 mm and is to be repaired by oxy-acetylene welding.

(*a*) Outline the preparation which may be required before welding.

(*b*) Give *two* methods of indicating the correct pre-heat temperature.

(*c*) Describe a suitable method of stress relieving after welding.

(*d*) Explain how the flux residue should be removed.

16 A cobalt-based, hard-facing alloy is required to be deposited on to a low-carbon steel component.

(*a*) State how the component could be prepared.

(*b*) Name the type of flame setting required for a single layer deposit and give *two* reasons for your choice.

(c) State *two* precautions to be taken to avoid defects.

17 What is the included angle of preparation necessary for single-vee butt welded joints to be made by the oxy-acetylene process using (1) the leftward technique, and (2) the rightward technique?

18   (a) What is meant by 'all-position rightward welding'?

      (b) Give two important advantages which may be obtained by the use of this technique for the welding of pipe joints.

19   (a) Explain briefly how (1) carburizing and (2) oxidizing oxy-acetylene flame settings are obtained.

      (b) Give one important application of each flame setting.

20 What is the purpose of (a) portable cylinder couplers in oxy-acetylene welding, (b) a brushless d.c. generator for metal-arc welding?

21   (a) Make a labelled section-sketch to show the arrangement of the nozzle assembly used for progressive flame gouging.

      (b) State two methods used for gas-cutting bevels on plate edges in preparation for welding.

22 Name one possible cause of each of the following, when oxy-acetylene welding butt joints in mild steel: (a) excessive penetration, (b) incomplete penetration, (c) adhesion.

23   (a) What is the difference between a backfire and a flashback?

      (b) Name five possible causes of backfiring when using a gas welding blow-pipe. In each case explain how backfiring could have been avoided.

      (c) State two difficulties which may be experienced in the efficient operation of a gas pressure regulator.

      (d) State the purpose of each part of the high-pressure oxy-acetylene welding system.

24 Sketch the joint preparation and set-up, and state the nozzle size (in cubic feet or litres per hour) and filler rod size to be used, for making: (1) butt welds in the vertical position in each of the following thickness of mild steel: 3.2 mm ($\frac{1}{8}$ in.) and 5 mm ($\frac{3}{16}$ in.); (2) a tee fillet-weld in the horizontal–vertical position in 5 mm ($\frac{3}{16}$ in.) thick mild steel plate.

Describe in detail, with the aid of sketches, the fusion welding of a butt joint in cast iron 10 mm ($\frac{3}{8}$ in.) thick.

25 Explain what is meant by gas velocity and outline how this is controlled in welding practice.

26 Describe, with the aid of a sectional sketch indicating the gas paths, the operation of an oxy-fuel gas cutting blowpipe.

27   (a) Sketch the joint preparation and set-up and state the nozzle
       and filler rod sizes to be used for making:
       (1) a butt weld in the vertical position in 3 mm thick low-
           carbon steel,
       (2) a close-square tee fillet weld in the horizontal–
           vertical position in 5 mm thick low-carbon steel,
       (3) a fusion welded butt joint in the flat position in
           10 mm thick cast iron.
       (b) Describe in detail, with the aid of sketches, the procedure
       and the technique required for making any *one* of the joints
       in part (a) above.

28 Outline one method which may be used during the oxy-acetylene
   welding of vertical butt joints, in 4 mm thick low-carbon steel, to
   ensure freedom from weld defects when
   (a) starting the weld at the beginning of the joint,
   (b) restarting the weld at a stop-point along the joint.

29 State *two* safety precautions which should be taken for *each* of the
   following:
   (a) storage or use of dissolved acetylene cylinders,
   (b) using gas pressure regulators,
   (c) oxy-acetylene cutting operations in a confined space,
   (d) using non-injector type gas welding blowpipes.

30   (a) Explain the action which produces the severance of the
       metal when oxy-fuel gas cutting low-carbon steel and state
       the difference between the basic principles of gas cutting and
       flame gouging.
       (b) Describe, with the aid of a sectional sketch indicating the
       gas paths, the operation of an oxy-fuel gas cutting blowpipe.

31 Explain what is meant by capillary attraction and describe how
   this affects the making of certain brazed joints.

32 The figure shows a cast iron wheel with two broken spokes. With
   the aid of a sketch, indicate how partial pre-heating may be used to
   minimize the risk of cracking.

33   (*a*) Describe with the aid of sketches the technique for making a
            lap fillet weld in the horizontal–vertical position by the oxy-
            acetylene process.
      (*b*) Compare the respective advantages and limitations of the
            leftward and the Lindewelding techniques when butt weld-
            ing mild steel pipes.

Question 33

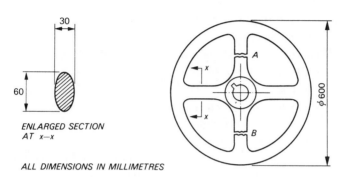

ENLARGED SECTION
AT x–x

ALL DIMENSIONS IN MILLIMETRES

Multiple choice

Note: The following are examples of the multiple choice type of question
but may not be representative of the entire scope of the examination either
in content or difficulty.

1 Because of the possibility of explosions, acetylene line fittings
  should *not* be made from
    (*a*) steel
    (*b*) copper
    (*c*) aluminium
    (*d*) cast iron.

2 One reason why low-carbon steel may be successfully welded by
  oxy-acetylene without the use of a flux, is that the oxide
    (*a*) is under the surface
    (*b*) has a higher melting point than the parent metal
    (*c*) has a lower melting point than the parent metal
    (*d*) melts at the same temperature as the parent metal.

3 An undesirable property of an aluminium flux residue is that it
   (*a*) is corrosive
   (*b*) obstructs the vision of the molten pool
   (*c*) decreases fluidity
   (*d*) requires great heat to melt it.

4 When a low-alloy steel has a hard and brittle structure it may be rendered soft and malleable by
   (*a*) recrystallization
   (*b*) cold working
   (*c*) lowering its temperature
   (*d*) hot quenching.

5 What happens to the mechanical properties of steel if the carbon content is increased to 0.5%?
   (*a*) The material becomes softer.
   (*b*) Malleability is increased.
   (*c*) The tensile strength is increased.
   (*d*) Ductility is increased.

6 The main reason for pre-heating medium- and high-carbon steels before cutting by the oxy-fuel gas technique is to
   (*a*) improve the quality of cut
   (*b*) increase the cutting speed
   (*c*) refine the grain structure
   (*d*) prevent hardening and cracking.

7 Which one of the following factors restricts the use of town gas as an oxy-fuel cutting gas?
   (*a*) Its low calorific value.
   (*b*) Its tendency to cause rapid melting.
   (*c*) Its unsitability for cutting plates less than 12 mm thick.
   (*d*) Its relatively high cost.

8 A suitable filler wire for brazing pure aluminium would consist of:
   (*a*) aluminium bronze
   (*b*) aluminium alloy containing 10/13% silicon
   (*c*) aluminium alloy containing 5% magnesium
   (*d*) pure aluminium.

9 Columnar growth takes place when a metal is
   (*a*) cold
   (*b*) losing heat
   (*c*) being heated
   (*d*) being rolled.

10 Difficulty may be encountered when welding aluminium because
   (*a*) the weld metal expands during solidication

(b) its coefficient of expansion is low compared to steel

(c) no colour change takes place to indicate its melting points

(d) its thermal conductivity is low compared to steel.

11 One purpose of a microscopic examination of a weld to establish the

(a) strength of the weld

(b) number of alloying elements

(c) grain size

(d) number of runs used.

12 Which one of the following components is employed to control amperage in an a.c. arc-welding circuit?

(a) rheostat

(b) choke

(c) voltmeter

(d) resistor.

13 When carrying out welds in low-carbon steel, using the carbon dioxide welding process, one purpose of the inductance control is to reduce

(a) porosity

(b) penetration

(c) undercut

(d) spatter.

14 One purpose of a reactor (choke) when manual metal arc welding is to

(a) change alternating current to direct current

(b) allow the correct amperage to be selected

(c) allow the desired arc voltage to be selected

(d) enable the correct polarity to be chosen.

15 Which shielding gas is generally recommended when butt-welding 6 mm nickel alloy sheet by the metal-arc gas-shielded process?

(a) Argon

(b) CO_2

(c) Hydrogen

(d) Nitrogen.

16 When TIG welding using a.c. output, which one of the following is essential in the circuit to stabilize the arc?

(a) A surge injector

(b) An open circuit voltage of 100 volts

(c) A flow meter

(d) An amperage regulator.

17 In manual metal-arc welding the flux coating to give deep penetration characteristics would contain
   (*a*) iron oxide
   (*b*) manganese
   (*c*) cellulose
   (*d*) calcium carbonate.

18 Which element is used as a deoxidant in copper filler rods?
   (*a*) Aluminium
   (*b*) Tin
   (*c*) Sulphur
   (*d*) Phosphorus.

19 An oxygen cylinder regulator being used in a flame-cutting supply may freeze up if the
   (*a*) gas withdrawal rate is exceeded
   (*b*) cylinder content is too low
   (*c*) cylinder is on its side
   (*d*) needle valve on the regulator is not fully open.

20 To test a component part for a vibrational loading, a suitable mechanical test would be
   (*a*) impact
   (*b*) tensile
   (*c*) compressive
   (*d*) fatigue.

21 The principal advantage of arc-on-gas welding is that it
   (*a*) allows controlled penetration of initial bead
   (*b*) requires less operator skill
   (*c*) entirely eliminates distortion
   (*d*) improves surface finish.

22 One reason why a grey cast iron casting should be slowly cooled after welding is to keep it
   (*a*) soft
   (*b*) spheroidal
   (*c*) hard
   (*d*) brittle.

23 An iron casting has a crack in it. Before oxy-acetylene fusion welding it may be necessary to drill the ends of the crack. One reason for this is to
   (*a*) balance out any shrinkage stresses
   (*b*) stop the crack from spreading
   (*c*) prevent the ends of the crack from being carburized
   (*d*) prevent grain growth.

24 Which one of the following metals may require the studding

techniques to be used when being repaired by manual metal arc welding?

(*a*) Low-carbon steel

(*b*) Aluminium

(*c*) Nickel

(*d*) Cast iron.

25 During the deposition of a manual metal-arc electrode, a certain percentage of the core wire is lost. This is due to

(*a*) voltage drop across the arc

(*b*) short arc length

(*c*) spatter

(*d*) excessive build-up.

26 Which one of the following can be welded by d.c. using the tungsten arc gas-shielded process?

(*a*) Copper

(*b*) Commercial pure aluminium

(*c*) Silicon–aluminium

(*d*) Magnesium alloys.

27 Backing bars for manual metal-arc welding of low-carbon steel should be made from

(*a*) copper

(*b*) low-carbon steel

(*c*) tool steel

(*d*) cast iron.

28 Peening may be carried out when manual metal-arc welding cast iron in order to

(*a*) reduce the effects of contraction

(*b*) make the bond more firmly adhering

(*c*) refine the grain structure

(*d*) speed up the welding.

29 The fillet welds on the support brackets in the figure should have 5 mm leg length with 50 mm intermittent welds as shown. The symbol at *A* to communicate this information should be

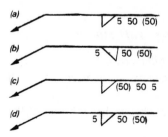

30 What is the volume of deposited metal in a fillet weld indicated by the symbol in the figure neglecting reinforcement?

   (*a*) 14000 mm³

   (*b*) 14900 mm³

   (*c*) 15000 mm³

   (*d*) 15 100 mm³.

Question 29. The figure shows a component fabricated from stabilized austenitic stainless steel sheet 3 mm thick.

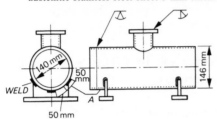

Question 30

Answer key

1 – *b*	11 – *c*	21 – *a*
2 – *c*	12 – *b*	22 – *a*
3 – *a*	13 – *b*	23 – *b*
4 – *a*	14 – *b*	24 – *d*
5 – *c*	15 – *a*	25 – *c*
6 – *d*	16 – *a*	26 – *a*
7 – *a*	17 – *c*	27 – *b*
8 – *b*	18 – *d*	28 – *a*
9 – *b*	19 – *a*	29 – *d*
10 – *c*	20 – *d*	30 – *c*

Welding engineering craft studies

All dimensions are in millimetres

1 Spot welds are to be made by the tungsten arc gas-shielded welding process.

   (*a*) Make a sectional sketch through one of these spot welds.

(*b*) State *one* advantage that this process has over the resistance spot welding process for making this type of weld.

(*c*) Name the additional equipment that would be necessary for making the spot welds using standard tungsten arc gas-shielded welding equipment.

2 (*a*) How is the size of a fillet weld with normal penetration determined in accordance to BS 499 in
   (1) a convex fillet weld,
   (2) a concave fillet weld?

(*b*) Make a sectional sketch of a mitre fillet weld.

3 State *five* factors that would influence the pre-heating temperature to be used for a welded steel fabrication.

4 (*a*) Give *two* reasons why hydrogen-controlled electrodes are preferred for manual metal-arc welding restrained joints in low-alloy steel.

(*b*) Explain the influence of the cellulose in the coating of an electrode on
   (1) voltage,
   (2) penetration.

5 A circular low-carbon steel plate of 750 mm in diameter and 10 mm thick has to be fitted and welded into a deck plate.

(*a*) State why hot cracking is likely to occur in this particular type of assembly.

(*b*) Outline *either* a suitable weld sequence *or* a change in the form of the plate insert that could be used to avoid the occurrence of hot cracking.

6 A worn press die has to be built up using the oxy-acetylene flame powder spraying process.

(*a*) From the table select the powder to be used for this repair.

(*b*) List *three* advantages that powder spraying has over arc welding for this type of repair.

(*c*) Name the type of flame setting to be used.

Powder no.	Resistance to:		Machinability
	Abrasion	Impact	
1	Fair	Excellent	Very good
2	Very good	Excellent	Very good
3	Excellent	Poor	Grind only

7 From the information given below, list in the correct order the welding sequence that should be carried out when friction welding two 20 mm diameter low-carbon steel bars.

(*a*) Place parts lightly in contact.

(*b*) Load machine.

(*c*) Apply axial force.

(*d*) Apply upset force.

(*e*) Rotate chuck and close gap.

(*f*) Release upset force.

(*g*) Arrest chuck movement.

(*h*) Remove specimen.

8 Steel cylindrical tanks 5 m long and rolled to 2 m internal diameter from 50 mm thick low-carbon steel plate are to be welded, using the electroslag welding process for the longitudinal seams.

(*a*) With the aid of a sketch outline how the weld area is protected from atmospheric contamination.

(*b*)   (1) State *three* advantages that the electroslag welding process has over arc welding processes for welding process for the longitudinal seams.

(2) Give *two* limitations of the electroslag welding process.

(*c*)   (1) What is the purpose of run-on and run-off plates when electroslag welding?

(2) After electroslag welding has commenced small additions of flux must continue to be added to the weld pool. Why is this?

(*d*) If it was considered that the cylinder would be working under conditions that may cause stress corrosion cracking:

(1) State what is meant by the term stress corrosion cracking,

(2) State *two* precautions that could be taken to reduce the occurrence of failure from this form of attack.

9 The figure shows a low-carbon steel nut to be resistance projection welded to 2 mm thick low-carbon steel sheet.

(*a*) Explain the principles involved in making this welded connection using the resistance welding process.

(*b*) Make a sectional sketch through *A–A* of the welded assembly shown in the figure.

(*c*) State *two* advantages of resistance welding over manual metal-arc welding for making this welded connection.

(*d*) List *three* defects that may be found when resistance *spot* welding, and in each case state the probable cause of the defect.

10 As part of a welding sequence, stiffeners require to be welded to the aluminium alloy plate shown in the figure.

   (*a*) Select a suitable process for welding the stiffeners.

   (*b*) Sketch and label the essential components of the welding circuit for the process selected.

Question 9

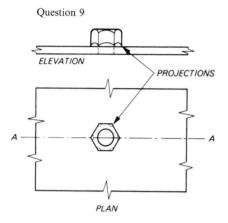

ELEVATION

PROJECTIONS

A ———— A

PLAN

Question 10

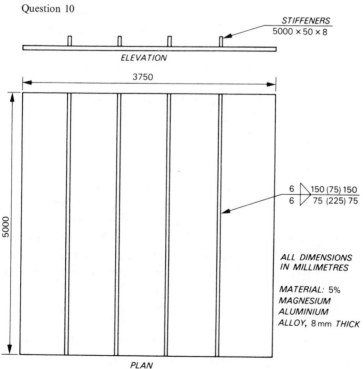

STIFFENERS
5000 × 50 × 8

ELEVATION

3750

5000

6 ⟩ 150 (75) 150
6 ⟩ 75 (225) 75

*ALL DIMENSIONS
IN MILLIMETRES*

MATERIAL: 5%
*MAGNESIUM
ALUMINIUM
ALLOY,* 8 mm *THICK*

PLAN

(c) Using a graph to illustrate your answer, give a brief outline of the output characteristics of the power source needed for the process used.

(d) What information is indicated by the weld symbol shown in the drawing?

11 Alloy steel of the following % composition has to be manual metal-arc welded.

Carbon (C)	Chromium (Cr)	Molybdenum (Mo)	Manganese (Mn)	Silicon (Si)	Nickel (Ni)	Vanadium (V)
0.16	1.5	0.5	0.8	0.4	0.0	0.0

Remainder: iron with acceptable limits of impurities.

(a) State *three* advantages of using alloying elements in steel.

(b) (1) Using the carbon equivalent formula given below determine the carbon equivalent of the alloy steel outlined above.

Carbon equivalent (%) =

$$C\% + \frac{Mn\%}{20} + \frac{Ni\%}{15} + \frac{Cr\% + Mo\% + V\%}{10}$$

(2) Indicate how the information produced may be used to determine the welding procedure used.

(c) Explain why pre-heating this alloy may reduce the occurrence of underbead cracking.

(d) Welds made on low-alloy steels are generally heat treated by normalizing.

(1) What is meant by normalizing?

(2) State *three* advantages produced by normalizing welded fabrications.

12 State *five* factors that will influence dilution during fusion welding.

13 (a) Give *three* reasons for pre-heating air-hardenable steels before or during oxy-fuel gas cutting.

(b) Give the main constituents in an air-hardenable steel.

14 Briefly explain how a friction weld is produced.

15 (a) State *three* advantages of using fully automatic welding processes in preference to manual welding processes.

(b) Under what circumstances may manual welding be preferred to fully automatic welding?

16 (a) Explain how heat is produced during cutting when using the oxygen lance.

(b) State *two* safety precautions that should be taken when using the oxygen lance for cutting.

17 List *five* problems which may be encountered when fusion welding aluminium.

18 Explain briefly why single-pass welds made by the arc plasma process may have the form shown in the macrograph in the figure.

19 One thousand circular containers as shown in the figure have been fabricated from low-carbon steel by using the metal-arc gas-shielded welding process with carbon dioxide as shielding gas.

   (*a*) Name the main type of stress, in *each* case, that weld *A* and *B* would be subjected to during hydraulic testing.

   (*b*) During testing some welds were found to contain porosity. State *three* probable causes of this defect.

20   (*a*) The figure shows a time and current graph making an electric resistance spot weld. The following two parts of this question refer to this diagram.

      (1) Calculate the weld time as a percentage of the welding cycle.

      (2) Explain why it is generally necessary for the forging time to be longer than the squeeze time.

   (*b*) A resistance spot welding machine was set up for welding low-carbon steel. If austenitic stainless steel of the same

Question 18

Question 19

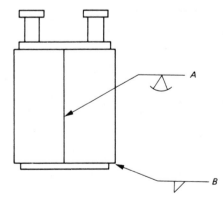

thickness is to be welded, what alterations would require to be made to

(1) the welding current,

(2) the welding time?

Explain why these alterations are necessary.

(*c*) Spot welding of sheet metal may also be carried out by fusion welding.

(1) Select a fusion welding process and describe how a spot weld is made using this process.

(2) What additional equipment would be necessary for making spot welds using standard equipment?

21 The figure shows a low-alloy steel rotary shear blade, used for trimming steel plate. The cutting edge has become worn due to severe abrasion during service and is to be repaired by using the manual metal-arc welding process.

(*a*) The general composition of three electrodes is shown below. Select the most suitable electrode for use in the building up of the cutting edge, and give a reason for your selection.

(1) Austenitic stainless steel.

(2) Medium-carbon low-alloy steel.

(3) Pure nickel.

(*b*) Outline the welding procedure that should be used to carry out this repair.

Question 20

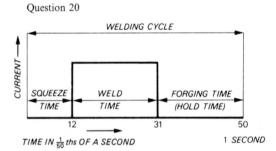

WELDING CYCLE

CURRENT →

SQUEEZE TIME | WELD TIME | FORGING TIME (HOLD TIME)

12   31   50

TIME IN $\frac{1}{50}$ ths OF A SECOND   1 SECOND

Question 21

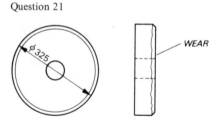

ø 325

WEAR

(c) Explain the influence that the dilution of the weld metal by the parent metal would have on the weld's mechanical properties.

(d) Under what circumstances would the oxy-acetylene welding process be preferred to arc-welding processes for hard surfacing?

22 The figure shows a low-carbon steel shaft to be fabricated by welding using the submerged arc welding process.

(a) With the aid of a sketch, explain how the weld area is protected from atmospheric contamination.

(b) State *three* advantages of using a backing strip for this joint.

(c) Sketch in detail the weld preparation that would be used for the butt weld shown at *A*.

(d) Why is the filler used in submerged arc welding copper coated?

(e) When submerged arc welding, explain why it is an advantage to use a multipower source which has one electrode using alternating current and the other direct current.

23 50,000 shear connectors are to be welded to the low-carbon steel box girder shown in the figure, using the drawn arc stud welding process.

(a) In what respect does the purpose-built power source used for this type of equipment differ from the power source used for metal-arc welding?

(b) Sketch in section, and label the main parts of, a ceramic ferrule used for stud welding.

(c) If the shear connectors are to be positioned at 100 mm centres, explain how this could be best achieved.

Question 22

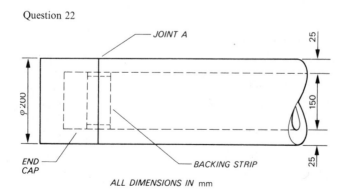

JOINT A

Φ 200

END CAP

BACKING STRIP

25

150

25

ALL DIMENSIONS IN mm

(*d*)   (1) Describe a method of testing the stud welds.

      (2) Welds having the section shown in *d*(2) below failed during testing due to the defects shown. State *four* possible causes of these defects.

24 Power souces used for electron beam welding may be rated as 30 kV.

(*a*) Explain the term 30 kV.

(*b*) Give *two* reasons why electron beam welding is generally carried out in a vacuum.

25   (*a*) State *four* of the variables involved in the production of friction welds.

(*b*) State the range or temperature necessary for the production of friction welds in low-carbon steel.

26 Sketch in section, and label the parts of, the head of an arc plasma torch for welding.

27   (*a*) State *three* advantages of introducing iron powder into the flame when oxy-acetylene powder cutting.

(*b*) State *two* hazards that the operator should guard against when oxy-acetylene powder cutting.

28 Low-carbon steel plate 25 mm thick has to be surfaced by the submerged arc welding process using stabilized austenitic stainless steel filler. List *five* variables that would influence the degree of dilution found in the weld.

29 Explain why it is recommended that materials containing sulphur should be removed from the weld area before welding nickel and nickel alloys.

Question 23

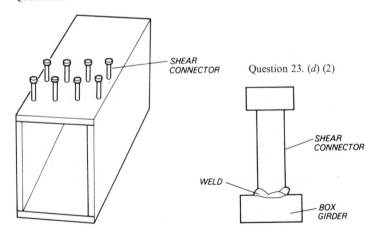

SHEAR CONNECTOR

Question 23. (*d*) (2)

SHEAR CONNECTOR

WELD

BOX GIRDER

30 State *five* factors that would need to be considered before deciding the pre-heating temperature to be used on low-alloy steel plate.

31   (*a*) With the aid of a graph, explain why a constant potential power source is often preferred for automatic arc welding processes.

    (*b*) List *four* variables, under the control of the welder, that can influence weld quality when using the submerged arc welding process.

    (*c*) Explain why the quality of submerged arc welds made in low-alloy steels is particularly high compared with the quality of welds made by other arc welding processes.

    (*d*) Inspection authorities may specify that welds made by the submerged arc welding process should be heat treated. Outline a suitable heat treatment.

32 The grey cast iron angle plate shown in the figure has to be repaired by manual metal-arc welding and then machined flush. Rutile-covered low-carbon steel and nickel alloy electrodes are available for making the weld.

    (*a*) Select the most suitable electrode to carry out this repair.

    (*b*) State *two* advantages and *one* disadvantage of the electrode selected.

33 The aluminium–magnesium alloy components shown in the figure are to be welded using the metal-arc gas-shielded welding process.

    (*a*) Explain why an argon/carbon dioxide gas mixture would be unsuitable as the gas shield to be used for welding the components.

Question 32

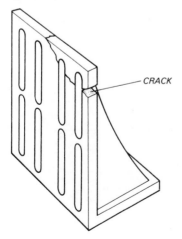

CRACK

(b) State *three* factors that would influence the weld profile.

(c) Explain why the oxy-acetylene fusion welding process would be unsuitable for welding the components.

(d) Outline the influence that welding would have on the metal's mechanical properties and grain structure.

(e) If 1000 components are to be fabricated, calculate the total length of welding carried out to complete the contract. Each joint is to be made by means of a single-run weld deposit.

34 (a) Calculate the volume of metal deposited in the welded joint shown in the figure if the length of the joint is 10 m, and the total weld reinforcement is taken as one quarter of the total volume of the joint gap.

(b) Outline the influence of *each* of the following factors on the cost of welded fabrication:

    (1) current density,

    (2) joint set-up.

(c) Outline how *four* factors in the welding procedure can influence welding costs.

(d) Sketch an example of a non-load bearing fillet weld, indicating by an arrow the direction of the applied load on the component when in service.

Question 33

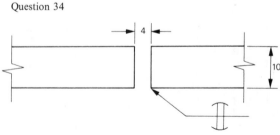

ϕ100

ALL DIMENSIONS IN mm

Question 34

4

10

ALL DIMENSIONS IN mm

35 A fabricated tank is to be manufactured from stabilized austenitic stainless steel. The weld preparation and holes are to be thermally cut using the arc plasma process.

   (*a*) Explain why the arc plasma process would be used in preference in oxygen and fuel gas for cutting this material.

   (*b*) Sketch in detail the design of the welded butt joint shown at *A* in the figure.

   (*c*) State *two* problems which may be encountered during the arc welding of this joint.

   (*d*) Explain why the depth of penetration would be greater in welds made in austenitic stainless steel, compared with similar welds made in low-carbon steel, assuming the energy input to be the same.

   (*e*) Explain, with the use of a graph, why a drooping characteristic power source should be used when manual metal-arc welding, rather than a constant potential power source.

36 Component parts for an oil rig, having the form shown in the figure, are to be fabricated and joined by welding.

   (*a*) Calculate the carbon equivalent of the steel given the following:

     Material composition: Carbon, 0.22%; Silicon, 0.5%;

Question 35

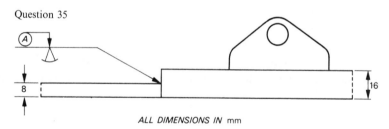

ALL DIMENSIONS IN mm

Question 36

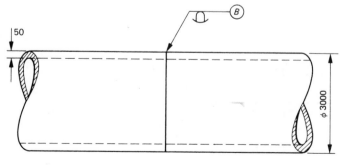

ALL DIMENSIONS IN mm

Manganese, 1.5%; Niobium, 0.1%; Vanadium, 0.1%; Sulphur, 0.05%; Phosphorus, 0.05%; remainder iron.

Carbon equivalent =

$$C\% + \frac{Mn\%}{6} + \frac{Cr\% + Mo\% + V\%}{5} + \frac{Ni\% + Cu\%}{15}$$

(*b*) Sketch in detail the butt joint preparation shown at *B*.

(*c*) Outline a welding procedure that could be used to make the joint.

(*d*) Since this component will be subjected to fatigue conditions during service, state *three* precautions which should be taken when welding has been completed.

37 (*a*) The figure shows a cruciform welded joint. Using the formula $l = 1.4\,t$ calculate the throat size (*t*) of a fillet weld if the leg length (*l*) is 21 mm.

(*b*) Show by means of a labelled sectional sketch a fillet weld leg length of 21 mm.

(*c*) A load is applied in the direction of the arrows indicated in the figure. State whether the welds are load bearing or non-load bearing.

(*d*) The material shown in the figure has a relatively high carbon equivalent and pre-heating temperatures have not been maintained during welding so that underbead cracking has resulted. Explain how the underbead cracking would take place.

(*e*) What is meant by the term 'fatigue fracture'?

(*f*) State *four* factors under the control of the welder which may adversely affect the fatigue life of a single-vee butt-welded joint.

Question 37

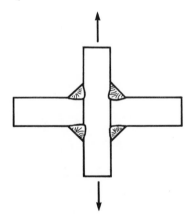

(g) State *three* precautions that could be taken to reduce the occurrence of a brittle fracture failure in a welded joint.

38 The figure shows an austenitic stainless steel hopper which has to be fabricated from 5 mm thick sheet.

    (a) State *four* methods which may be used to ensure a low heat input procedure for arc welding the hopper.

    (b)   (1) State the value of the included angle *X* for the joint shown at *B*.

        (2) State *two* reasons why the included angle *X* should be increased if welding has to be carried out.

    (c) Four different welding procedures could be used for making the butt-welded joints shown at *A* and *B*. State *one* advantage and *one* disadvantage for *each* of the following welding procedures.

        (1) Completely weld the joints using the tungsten arc gas-shielded process.

        (2) Completely weld the joints using the gas-shielded metal-arc process.

        (3) Weld the first run using the tungsten arc gas-shielded process and complete the weld using the manual metal-arc process.

Question 38

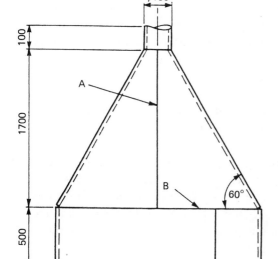
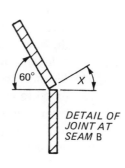

DETAIL OF JOINT AT SEAM B

(4) Completely weld the joints using the manual metal-arc process.

(*d*) (1) Explain why distortion is likely to be a problem when welding this material.

(2) Describe a method that could be used to control the distortion when welding joint *A*.

39 The figure shows a fabrication which is to be made from 25 mm thick low-carbon steel using the gas-shielded metal-arc welding process with flux cored electrode wire.

(*a*) Give *three* reasons why flux cored electrode wire may be preferred for welding this fabrication in preference to solid wire.

(*b*) State *two* advantages of using a gas shield in addition to the flux cored electrode wire when using this process.

(*c*) A root crack was observed when welding the single bevel butt weld joint shown in (*b*). State *three* possible causes for its formation.

(*d*) State *three* factors which may cause porosity during welding, assuming the joint surfaces were clean before welding was carried out.

(*e*) Calculate the area of the side plate shown as *A* on (*a*).

40 (*a*) State *one* reason why

(1) some welding specifications permit a small amount of slag inclusion in a welded joint,

(2) it would not be realistic for a specification to state that a welded joint must be free from defects,

(3) welding specifications do not permit welded joints containing cracks to be accepted.

(*b*) Show, with the aid of a sketch, how the design size of a butt weld is measured.

41 (*a*) The figure shows butt and fillet joints which are to be joined by manual metal-arc welding. Explain why the welding specification may state that for the same heat energy input the butt-welded joint would not require to be pre-heated, although the fillet welded joint should be.

(*b*) The hardness of the gas cut edge shown in '*X*' in (*b*) was found to be unacceptably high. State *three* changes in the oxy-fuel gas cutting procedure that could have been taken to reduce the hardness value.

42 The figure shows a low-carbon steel oil storage tank which is to be fabricated from 18 mm thick low-carbon steel plate using the submerged arc welding process.

Question 39

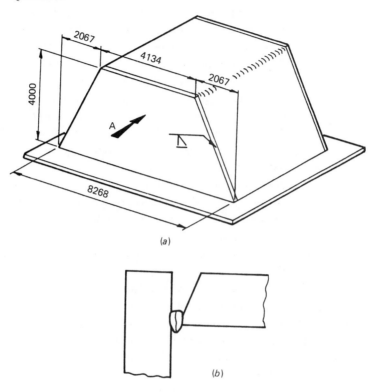

(a)

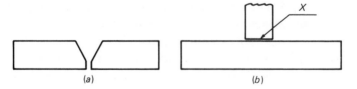

(b)

Question 41

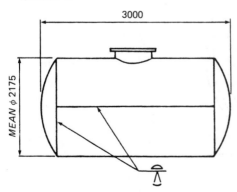

(a)   (b)

Question 42

(a) Calculate the length of plate required to be gas cut and rolled to form the cylindrical section of the tank.

(b) Explain why run-on and run-off plates are used when welding the longitudinal seam.

(c) Explain why welds made using the submerged arc welding process usually have a low level of defects.

(d) If the weld deposit was required to have good impact properties state whether it would be made with either a large number of runs or as few a number of runs as possible.

(e) State *four* variables under the control of the welder which will influence the weld profile.

(f) Name *two* other mechanized welding processes that could be used for welding the joints and state in *each* case *one* reason for using the process in preference to the submerged arc process.

(g) State in *each* case *one* advantage which may be obtained by using

(1) alternating current for submerged arc welding,

(2) direct current for submerged arc welding.

43 In relation to the plasma arc welding process

(a) reproduce the sketch shown in the figure and identify the

(1) negative pole within the circuit,

(2) shielding gas ports.

(b) State whether the parent metal forms part of the welding circuit when a non-transferred arc is used for welding.

(c) State a temperature within the temperature range of the arc.

(d) Sketch a transverse section through the weld made by means of a transferred arc on low-carbon steel 10 mm thick.

44 (a) List *four* variables under the control of the operator which will influence the quality of a resistance spot weld.

Question 43

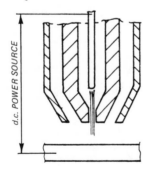

d.c. POWER SOURCE

(*b*) Low-carbon steel sheet has been resistance spot welded. State whether the welding current would need to be increased if this process was to be used for welding aluminium sheet of the same thickness.

(*c*) Describe the operational sequence used for making resistance butt welds in short lengths of 50 mm diameter low-carbon steel pipes of wall thickness 6 mm.

(*d*) (1) A number of steel studs have the form shown in the figure are to be welded to low-carbon steel plate. Describe, with the aid of a sketch, the operational sequence used to make a drawn arc stud welded joint.

(2) The figure shows a ceramic ferrule used in the drawn arc stud welding. Explain the purpose of the serrations.

45 Each term used in column *A* in the table is directly related to *one* of the terms used in the column *B*. Pair *each* of the terms listed in column *A* with its appropriate term in column *B*.

Column *A*	Column *B*
d.c. negative polarity	Force
Nugget	Oxy-fuel cutting
Kerf	Welding standard
API	Resistance spot welding
newton	Hard surfacing

46 A box section column 750 mm square × 6500 mm long is to be fabricated from 40 mm thick low-carbon steel plate.

(*a*) Select and describe a suitable automatic welding process.

(*b*) Sketch the corner joint and show, if necessary, any edge preparation.

Question 44. (*d*) (1)

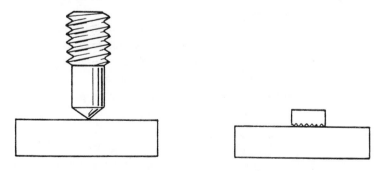

(c) Give *three* advantages of using automatic welding as opposed to manual metal-arc.

47 A spherical vessel, 3.6 m diameter, is fabricated from 75 mm thick low-carbon steel, the welding of the circumferential seams being carried out by the electroslag process.

(a) Sketch the macrostructure of the plate and weld area.

(b) Describe a suitable method of heat treatment to be carried out on completion of the welding.

(c) State *three* advantages to be obtained by the form of heat treatment chosen.

(d) State *two* safety precautions to be taken when pressure testing the vessel.

48 (a) Explain why the effect of electric shock is much less when using a flat characteristic instead of a drooping characteristic power source for welding.

(b) Name *two* gases which may be produced during arc welding operations which are harmful to the health of the welder.

49 From the information given below list in the correct order the welding sequence that should be carried out when friction welding two 15 mm diameter low-carbon steel bars.

(a) Load the machine.

(b) Apply axial force.

(c) Place parts lightly in contact.

(d) Release upset force.

(e) Rotate the chuck and close the gap.

(f) Apply upset force.

(g) Arrest the chuck movement.

(h) Remove specimen.

50 (a) State *two* factors that should be considered before selecting a filler/electrode for a hard surfacing application.

(b) State *three* precautions which may be taken by the welder to control the level of dilution when surfacing using an arc welding process.

51 (a) The figure shows three views of a platform bracket. Using graph isometric paper, make an isometric sketch of the bracket in the direction of arrow *A*. Omit all hidden lines.

(b) Parts of the bracket are to be assembled and welded using the electrical resistance spot welding process.

(1) Outline the principles of electrical resistance spot welding.

(2) State why the stiffener shown at *B* could not be resistance spot welded for the design shown.

(3) Sketch a modification to the stiffener that would enable spot welding to be used.

(4) State *three* advantages of using the resistance spot welding process for welding the bracket in preference to a fusion welded process.

52 State *five* variables, other than electrode diameter, that will influence the production of an acceptable quality weld when using the mechanized flux shielded visible arc process.

53 Name *each* of the welding or thermal cutting processes illustrated in the figure. Identify your answer by stating the appropriate letter and name the process concerned.

54   (*a*) State *one* safety precaution that must be carried out.

(1) when oxy-acetylene cutting inside a steel container,

(2) when manual metal-arc welding galvanized steel,

(3) when gas-shielded metal-arc welding inside a container.

Question 51

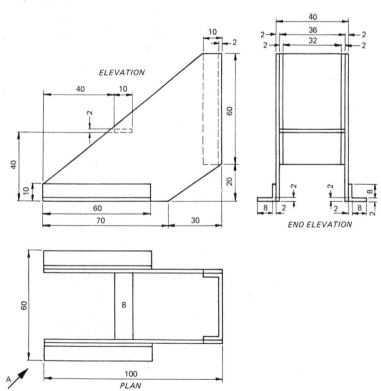

(*b*) State why gas-shielded tungsten arc welding should not be carried out inside a container that has been degreased using carbon tetrachloride until all the degreasing compound and vapours have been removed.

55 The figure shows an aluminium–magnesium alloy safety rail which is to be welded by the gas-shielded metal-arc welding process using 1 mm diameter filler wire.

(*a*) State a suitable welding voltage and amperage setting for making the fillet joint shown at *A*, in the horizontal–vertical position.

Question 53

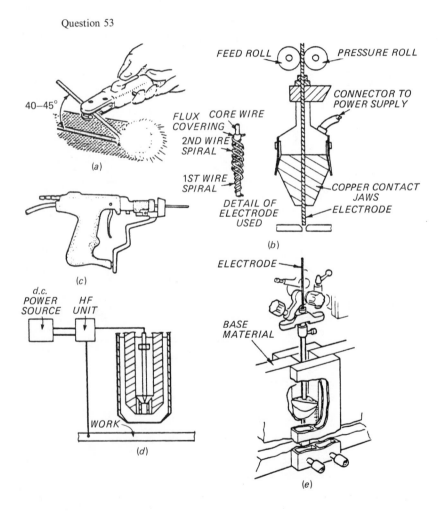

(b) Name the mode of metal transfer which would be used for welding *each* of the following:
    (1) the fillet joint shown at *A* in the horizontal position,
    (2) the fillet joint shown at *B* in the vertical position.
(c) When welding aluminium with this process, electrode wire may be supplied by pull feed from the torch or by push feed from a separate wire feed unit. State *two* advantages and *two* disadvantages of *each* method of wire supply.
(d) List *four* problems which may be encountered when fusion welding aluminium.

Question 55

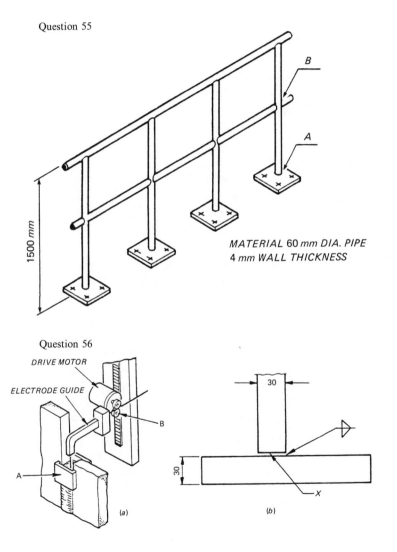

1500 mm

B

A

MATERIAL 60 mm DIA. PIPE
4 mm WALL THICKNESS

Question 56

DRIVE MOTOR

ELECTRODE GUIDE

B

A

(a)

30

30

X

(b)

(e) State *two* reasons why an 80% argon, 20% carbon dioxide gas mixture should *not* be used as the shielding gas for welding aluminium.

(f) Give *two* reasons for using helium as a shielding gas for welding aluminium with the gas-shielded tungsten arc welding process.

56 The figure (a) shows a section of a joint being welded using the electroslag process.

(a) Name the parts identified in the sketch at points *A* and *B*.

(b) State why run-off plates are required to complete the weld.

(c) State whether or not an arc is maintained throughout electroslag welding.

(d) Explain why small additions of flux may require to be added during electroslag welding.

(e) Give *two* reasons why electroslag welds may require to be heat treated after welding.

(f) State *three* basic differences that exist between conventional electroslag welding and consumable guide welding.

(g) Sketch in section a cruciform joint made by electroslag welding.

(h) Figure (b) shows a joint in low-alloy steel set-up for fillet welding using the manual metal-arc welding process. Explain why the edge shown at *X* was oxy-acetylene flame cut, then machined before welding.

57 In relation to welded fabrications:

(a) state what is meant by the term brittle fracture,

(b) state *five* conditions in which an otherwise ductile steel may behave in a brittle manner after welding,

(c) name a test that may be used to assess the notch ductility of a steel,

(d) state *five* factors to be considered when selecting a manual metal-arc electrode to be used for a given application,

(e) BS 639 classifies electrodes for manual metal-arc welding by using index numbers and letters. Thus a typical electrode specification may be: E 43 22 R 160 27. Reconstruct the table given below and interpret the specification by identifying the details listed in column *B* with the appropriate index letter or numbers listed in column *A*.

Electrode specification
E 43 22 R 160 27

Column *A*	Column *B*
E	Type of covering
43 22	Operating characteristics (Type of current and position of welding)
R	Electrode efficiency
160	Mechanical properties
27	Electric arc welding

58 The figure shows a beam to be fabricated from low-alloy steel, using the gas-shielded metal-arc process with flux cored electrode wire.

(*a*) Using the carbon equivalent formula given, calculate the carbon equivalent of the steel beam when its composition is:

Com-position	Carbon	Manganese	Silicon	Niobium	Vanadium	Remainder/ Iron
As a per-centage	0.22	1.2	0.5	0.1	0.2	Sulphur and phos-phorus in acceptable quantities

$$CE = C\% + \frac{Mn\%}{6} \quad \frac{Cr\% + Mu\% + V\%}{5} + \frac{Ni\% + Cu\%}{15}$$

Question 58

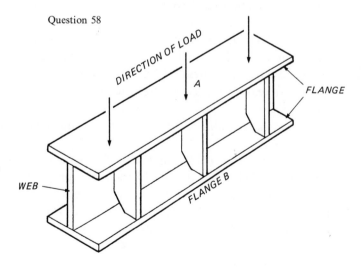

(b) Explain why steel components with a high carbon equivalent should be
    (1) pre-heated before welding is carried out,
    (2) immediately post-heated after welding.
(c) The figure shows that the flanges of the beam have been stiffened using gusset plates. Explain why the corners have been removed.
(d) Name the main stress acting on the underside of flange B if a load is applied to the beam in the direction of arrow A.
(e) State *four* methods that may be used to reduce undercut at the toes of the welds joining the flanges to the web.
(f) Outline, with the aid of a sectional sketch, the effect of using too low a voltage setting for depositing a single run fillet weld by this process.

Fabrication and welding engineering (technical grade)

1 (a) Give the main feature which distinguishes between cylinders for (1) combustible gas, and (2) non-combustible gas.
  (b) Explain the function of a 30 A fuse.
  (c) Why is a cylinder normally rolled to a slight overlap of the butting edges in a plate binding rolls?
  (d) Show, on an appropriate cross-section of rolled steel member, the correct positioning of a tapered washer.
  (e) State *four* factors which would ensure good quality soldering.
  (f) What information may be obtained by an examination of an etched cross-section of a fillet weld?
  (g) A large triangular plate of 30 mm thickness is lying on a workshop floor. Describe how the plate could be marked off to find its centre of area, for lifting purposes.
  (h) Give *five* reasons for using false wired edge on thin sheet metal.
2 (a) Explain the essential differences between ductility and malleability.
  (b) What is meant by (1) hardness and (2) toughness? Give practical examples of the use of these *two* properties in different materials.
3 Compare and contrast the following cutting processes: (1) oxy-fuel gas cutting, (2) guillotine cutting, with respect to:
  (a) cost,
  (b) limitations of cut,

(*c*) type of material to be cut.

4 Discuss the factors which must be considered when deciding whether to form a given metal by hot or cold processes.

5 Describe, with the aid of sketches, *three* methods of stiffening and strengthening metal platework without the attachment of additional members.

6 When producing a riveted watertight butt joint, what are the important features to look for:

(*a*) when setting up the work,

(*b*) after riveting.

7 Explain the meaning of the following terms when used in connection with the control of distortion in welding:

(*a*) preset,

(*b*) restraint,

(*c*) back-step,

(*d*) intermittent,

(*c*) staggered.

8 It is suggested that the following requirements are essential for safe and adequate electrical circuits in a fabrication workshop:

(1) colour coding,

(2) notices,

(3) maintenance,

(4) notification of faults,

(5) earthing,

(6) load protection,

(7) insulation,

(8) protection,

(9) emergency switches.

Write explanatory notes on any *five* of the above.

9 With the aid of sketches of cross-section show *each* of the following weld defects:

(*a*) undercut on a fillet weld,

(*b*) undercut on a butt weld,

(*c*) lack of root penetration in a fillet weld,

(*d*) lack of penetration in an open square-edge butt joint,

(*e*) excessive root penetration in a single bevel butt joint.

10 One hundred welded fabrications each weighing 48 kgf are at a temperature of 25 °C. They are placed in a furnace and heated to 500 °C for the purpose of stress relieving. Calculate the quantity of heat to be supplied so that the fabrications reach the required temperature. Specific heat capacity of the welded fabrication is 0.5 kJ/kg °C. Furnace efficiency is 20%.

Index